Designing the Fitness Program

A Guide for Public Safety Organizations

Designing the Fitness Program

A Guide for Public Safety Organizations

John LeCuyer

Fire Engineering

PennWell

DISCLAIMER

The recommendations, advice, descriptions, and methods in this book are presented solely for educational purposes. The authors and publisher assume no liability whatsoever for any loss or damage that results from the use of any of the material in this book. Use of the material in this book is solely at the risk of the user.

Published by Fire Engineering Books & Videos
A Division of PennWell Corporation
Park 80 West, Plaza 2
Saddle Brook, NJ 07663
United States of America
www.fireengineering.com

PennWell Corporation
Information Products Group
1421 S. Sheridan
Tulsa, OK 74112
1-800-752-9764
sales@pennwell.com
www.pennwell.com
www.pennwell-store.com

Illustrations by Steve Hetzel
Layout and cover design by John Potter

Library of Congress Cataloging-in-Publication Data

LeCuyer, John 1948–
Designing the fitness program: a guide for public safety organizations / John LeCuyer.
p. cm.
Includes bibliographical references and index.
ISBN 0-912212-92-6 (hardcover)
1. Fire fighters—Physical training. 2. Police—Health and hygiene. 3. Physical fitness. 4. Health promotion. I. Title.

TH9128.L43 2000
363.37'068'3—dc21

00-057314

Printed in the United States of America

1 2 3 4 5 05 04 03 02 01

Dedication

To Marie-Jeanne Gervais and Emile L'Ecuyer, who, in their eighth decade of independent living, are testimony to the benefits of hard work and an active lifestyle. Also to my children, Jared and Jenny, who are destined to profit from their example.

Acknowledgments

A special thanks to my wife, Karen, for her encouragement, support, and patience throughout the writing of this book.

My thanks also to friends and colleagues who offered advice and feedback on selected portions of this book: Arthur Dickinson, Ph.D., and Dale Mood, Ph.D., of the University of Colorado; and Richard Gerkin, M.D., physician for the Phoenix Fire Department; Paul Davis, Ph.D., of ARA Human Factors, to name a few, and also to the firefighters of Aurora and its neighboring departments, who, through their commitment to fitness, have supported my efforts over the years.

Table of Contents

APPENDICES

INTRODUCTION

Over the past fourteen years, I have responded to hundreds of questions on the subject of fire and police department fitness programs. Inquiries have come from across the United States and Canada, as well as Europe and Australia. Although the questions often vary in form, their thrust is fundamentally the same: "What is *your* program? What standards do *you* use?" Conversations routinely begin with, "We are thinking of starting a fitness program, *but*," implying that they have no idea as to where to begin.

Although many words have been written on the topic since the early '80s, little of substance has been said on how actually to design a program that will address the needs of public safety officers. The subject of fitness for public safety has always been controversial and, unfortunately, the available literature has contributed more confusion than clarity to the issue. The result is that program fundamentals still remain a mystery. In an effort to rectify that situation, I have set about compiling a basic, how-to approach to the subject; a guide that hopefully provides a systematic, step-by-step method to design, develop, and implement a department fitness program.

This book is targeted for use by whatever official is assigned the task of introducing and managing such a program, usually the health and safety officer or a designated fitness coordinator. Others who will find this book informative are fire department physicians, physical therapists, exercise physiologists, health care providers contracted to public departments, risk managers, attorneys, test consultants, firefighter candidates, and virtually anyone who is involved directly or tangentially with public safety .

This book does not present a specific program; rather, it offers a program model that will apply to any and all fitness programs, including those currently in use. By intention, it is written in a very basic format, without all the ancillary information that tends to clutter and confuse more than it helps. Virtually all fitness books are laden with exercise diagrams, charts, and other information that is redundant if you happen to possess more than two books on the topic. This text will provide the essentials only. For those who desire the extras, I have included references at the end of each section, and even these sources have been carefully selected so as not to send the reader scrambling in too many directions.

PART I ANALYSIS

Chapter 1

Identifying the Problem

In 1998, the American Heart Association released its 1998 Heart and Stroke Statistical Update, giving mortality figures for 1994-95, prevalence statistics from 1995, and economic costs from 1998. The report stated that, during that two-year period, cardiovascular disease (CVD) had claimed the lives of 960,592 people in the United States, meaning an average of 2,600 a day, or one every thirty-three seconds. The AHA estimates that 20 percent of the U.S. population has some form of the disease, which has been the number one killer in United States since 1900, 1918 excepted. By these statistics, sudden cardiac death is the leading cause of mortality in developed nations, accounting for 481,287 U.S. deaths in 1995. Stroke claimed 157,991 lives and is reported to be the leading cause of long-term disability in this country. It is estimated that another 50 million Americans suffer from hypertension, with 39,981 deaths directly attributable to that affliction.

The major risk factors for cardiovascular disease are all a matter of lifestyle, such as smoking, poor diet, and obesity. The latter, a by-product of the good life, is a major contributor to preventable death and is often associated with diets high in fat, combined with physical inactivity. It is responsible for claiming over 300,000 lives annually. According to the AHA, approximately 65 million people are above their recommended weight. Excessive weight triples the risk for coronary heart disease (CHD). It increases the risk of hypertension, and it is a major contributor to the development of diabetes, a condition that can promote fatal heart attack, stroke, or some form of cancer in 85 percent of its victims. Stroke is the third largest cause of death in the United States, ranking only behind heart disease and cancer, and accounting for 150,000 deaths a year.

The relationship between disease and the three deadly conspirators—lifestyle, genetics, and quality of health care—is depicted in Figure 1–1. As you can see, lifestyle is the predominant factor, except in the case of diabetes, which is strongly determined by genetics.

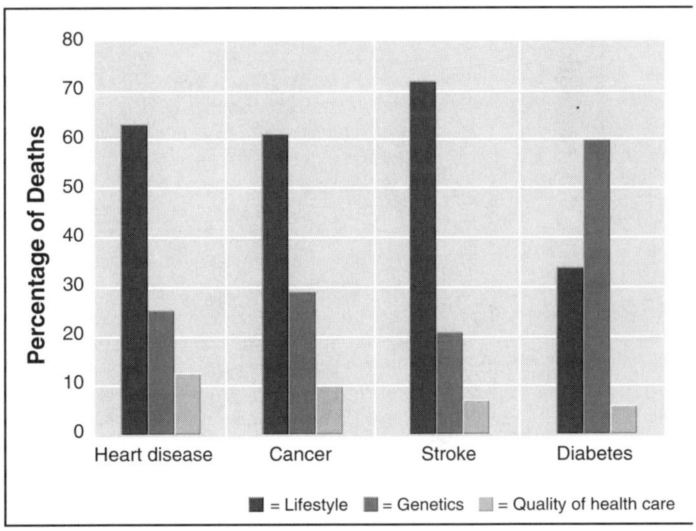

Fig. 1–1 Deadly Conspirators
Source: Adapted from the University of Colorado Wellness Program handout.

Heart disease, diabetes, and hypertension are all hypokinetic diseases (*hypo*, low; *kinetic*, motion), meaning that they are tied to a lack of activity. It is reported that a physically active lifestyle can reduce the risk of heart attack by anywhere from 35 to 55 percent. The relative risk of developing CHD as a result of physical inactivity ranges from 1.5 to 2.4, an increase in risk comparable to other lifestyle factors, such as smoking, hypertension, and hyperlipidemia (high cholesterol) (Staff, 1997). The lack of physical activity places far more people at risk for CHD than any other factor, as shown in Figure 1–2. According to the AHA, approximately 30,000 deaths each year in the U.S. are directly attributable to physical inactivity.

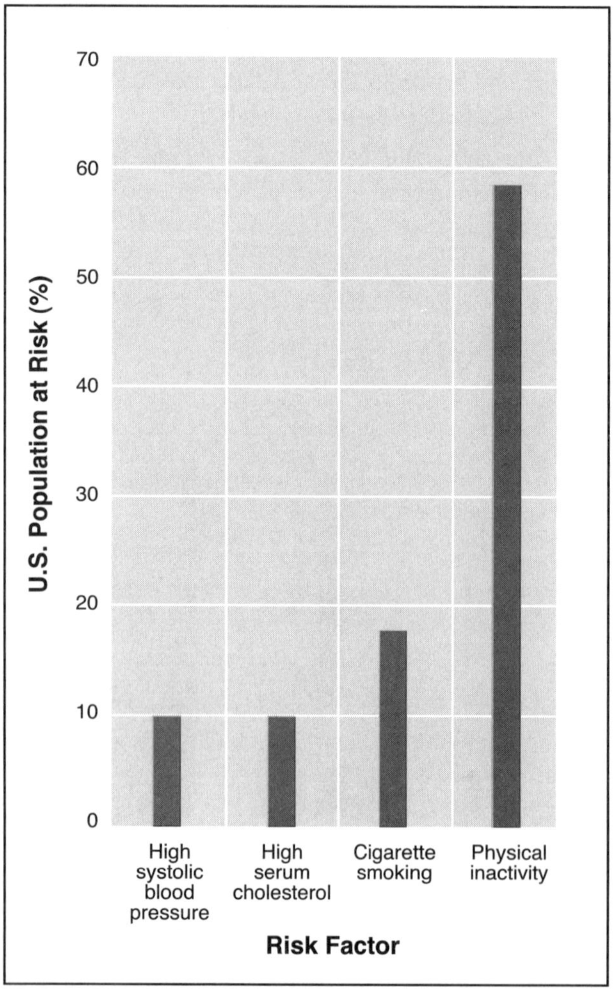

Fig. 1–2 Physical Activity and CHD Risk Factors
Casperson, C.J. (1987) "Physical Inactivity and Coronary Heart Desease," *The Physician and Sports Medicine,* **15** (11): 44. Reproduced with permission of McGraw-Hill, Inc.

THE ECONOMIC PERSPECTIVE

Health care costs continue to climb rapidly, a trend that shows little sign of reversing. In 1998, both the direct and indirect costs of CVD alone exceeded 274 billion. Additionally, personal injury, shown to correlate with lifestyle risk factors, is a significant contributor to escalating health care costs. The relationship between occupational injury, health status, and job performance is well documented in the literature (Adams, *et al.*, 1986; Cady *et al.*, 1979; Cady *et al.*, 1985; Mealy, 1979; Gardner *et al.*, 1974; Cady, 1985; Bahrke, 1982; Davis *et al.*, 1982; Kalkhoff *et al.*, 1983; Collingwood, 1995). The two most frequently reported occupational injuries are cumulative trauma disorders (CTD) and injuries to the back. For public safety officers—police and fire alike—back injuries have historically ranked number one. In 1997, 44.7 percent of the line-of-duty injuries resulting in disability retirements for firefighters were back-related (IAFF, 1998). In the private sector, both back and cumulative trauma disorders prevail. Recent reports claim that upper-extremity injuries cost employers $2.1 billion, while back injuries carry a price of more than a billion in lost production and $250 million in workers' compensation. In 1995, more than 60 percent of the total injuries in private industry were attributable to some type of repeated stress. Cumulative trauma disorders, the result of repetitive motion, cost $563 million that year, despite a drop in reported cases for the first time in fifteen years (OSHA, 1997).

THE CORPORATE RESPONSE

When asked in 1894 as to why he allowed and promoted exercise on company time, John H. Patterson, founder of the National Register

Company, responded, "It pays." At a time when an employee's personal welfare was anything but a priority of the American factory owner, Mr. Patterson began to schedule twice-daily exercise sessions of ten minutes each for his workforce. His employees were also provided with such amenities as coffee, clean wash water, and lockers. The reported results were reduced use of sick leave and increased productivity (Gilmore, 1981). The connection between fitness and profit hasn't gone unnoticed in today's corporate world, particularly given our rising health care costs. Increased sick leave benefits and their use, which result in reduced productivity, are factors often attributable to injury and illness. The corporate response to the problem has been to move toward the development and implementation of fitness/wellness programs. Research attests to the effectiveness of emphasizing fitness, particularly in Canada, Japan, Russia, and in certain European countries where fitness is given a high priority. The Osaka Gas Company of Japan reported in a longitudinal study that a corporate fitness program is effective in improving fitness levels and in preventing the progression of adult diseases (Okada and Iseki, 1990).

In America, hundreds of corporations are addressing their health and injury problems through fitness/wellness programs. A book by Robert Pritchard and Gregory Potter, titled *Fitness Inc.: A Guide to Corporate Health and Wellness Programs,* profiles fifty-three American corporations that have implemented such programs and describes their cost-benefit relationship, as well as the impact on each organization. According to Pritchard and Potter, Fitness Systems, a preeminent provider of fitness program services for corporate America, reports reduced health care costs of from $200 to $600 for fit employees, lower absenteeism, and enhanced productivity. Other providers and organizations, such as Coors Brewery, have reported similar results, with an average return of three to five dollars for every dollar spent. Authors Pritchard and Potter state that the most compelling evidence for the success of such programs is revealed by the insurance industry, which presumably has the most data on health care costs and mortality rates. Proportionally, insurance companies have more fitness/wellness programs than any other industry!

THE FIRE SERVICE RESPONSE

Corporate America's interest in wellness programs began to escalate in the early '80s, around the same time that the International City/County Management Association released a document that reported budgetary savings in cities with physically fit firefighters. Although documentation to support the introduction of fitness programs for public safety personnel has been available for approximately thirty years, the fire service and police have been relatively slow to respond.

One of the earliest studies that examined the feasibility of using a physical fitness program to address the health problems and injuries of firefighters was conducted within the Los Angeles County Fire Department (Cady and Karwasky, 1985). The fourteen-year study, beginning in 1970, reported heart attacks to be 2.6 times more frequent among firefighters of lower work capacity as compared with firefighters who were more physically fit. Additionally, the occurrence of back injuries among the former group was seven times greater. The program report documented a 25-percent decrease in workers' compensation costs per $100 of payroll, a 16-percent increase in work capacity, and a decrease in disabling injuries. The study helped lend support for the implementation of firefighter fitness programs, but it also raised some questions regarding program goals and emphasis. This is because there was no clear increase in muscular strength or change in the number of nondisabling injuries, two of the program's goals.

Another study that evaluated the benefits of fitness programs had almost the opposite results, showing increases in strength and flexibility with no improvement in work capacity (Beam and Davis, 1979). Other studies reported varying results (Puterbaugh *et al.*, 1983; Behnke and Antonacci, 1995; White and Hunt, 1974), although some included data on cost savings to support on-the-job firefighter exercise programs. Some short-term studies have reported improvements with job-specific fitness routines, as measured by pre- and postperformance test times, in as little as ten weeks (Obermeyer *et al.*, 1995). Fire departments using different approaches to health and fitness programs report decreases in the use of sick leave. Some show no reduction in the number of injuries but a decline in the severity of injuries. Others simply report their "satisfaction with beneficial results" (Grieve, 1992). Well-designed programs that include a pilot study prior to implementation report across-the-board improvements (Dubois, 1984). One example that showed improvements in all fitness dimensions, with a concomitant reduction in both occupational injuries and workers' compensation costs, is presented as a case study in Chapter 13.

Although fitness/wellness programs have been around for several decades, displaying varying degrees of effectiveness, they are just beginning to capture serious attention from many police and fire administrators. The mixed comments and results of various studies and programs, as well as their degree of effectiveness, appear directly related to program objectives and, more specifically, to the design of the program.

WHITE COLLAR vs. BLUE COLLAR

Adult diseases such as diabetes mellitus, obesity, hypertension, and heart disease are more prevalent within the sedentary white-collar occupations. The issues of productivity, the containment of health care costs, and the reduction of lost work time due to stress, substance abuse, and illness are primarily the concern of the white-collar worker. Such problems are most effectively addressed through wellness programs. The injury problems of the white-collar group in sedentary occupations, as well as those in light industrial work, tend to be of the repetitive-motion type (CTD), including the simple muscular aches and pains (lower back, shoulders, neck, and hands) common to repetitive, stationary work (OSHA, Feb. 1997). Problems of this type are more easily corrected through ergonomic interventions and job-specific stretching exercises, not unlike Mr. Patterson's approach. Still, employers in the light-work sector of industry typically aren't supportive of adopting exercise programs to address those problems. In reviewing corporate wellness program articles, injury prevention is rarely mentioned as a prime motivator. Moving into blue-collar industries where heavy manual labor is the norm (e.g., construction, lumber, steel), hypokinetic diseases are less prevalent and the emphasis of health programs shifts toward the more pressing concern of preventing injuries.

The typical wellness program is rarely seen in occupations that require heavy manual labor. One prominent Western brewery that provides an excellent fitness/wellness facility for its employees has remarked that few factory workers take advantage of that benefit. Most who used the facility were the spouses of employees or those workers who already displayed a high degree of fitness. The target group, those employees who needed exercise the most, seldom participated. Even so, the cost savings from the program were documented with a reported five-dollar return for every dollar invested. The director of the company's wellness program stated during a college lecture (Univ. Northern Colorado, 1989) that the introduction of an in-house rehabilitation program netted a $2.5 million savings in one year, 1988.

A similar participation problem has been noted in other industries (Halfon *et al.*, 1990), and some companies that employ a sizable contingent of workers who do only light industrial work are aware of the problem. The current trend is to identify methods that will solicit greater involvement from the blue-collar group, such as on-duty exercise time, a common practice with some Russian and Japanese corporations.

During Patterson's era, raw muscle power accounted for one-third of the energy required by industry. Today, mainly due to technological advances, the industrial demand for human effort has dropped to a fraction of a percent. Labor-saving measures mean less physical activity and a resultant decline in basic fitness, the deterioration of health, and an increase in adult diseases and mortality. All of these are the price of a sedentary life (Paffenbarger *et al.*, 1986). Although technology has provided industry and the fire service with efficient, high-quality equipment and tools, the primary work tasks of the fire service have changed little. Firefighters still drag hose into burning buildings, carry out victims, and swing axes. Since today's firefighters, like their predecessors, continue to work in a mode of heavy manual labor, it seems logical that they wouldn't suffer the consequences associated with a sedentary lifestyle. Ironically, they do, and to a greater degree than the general population, including other heavy-labor occupations. Although the physical demands of firefighting have changed little, the off-duty lifestyle has changed dramatically, and the gap between the physical demands of home and work are far more pronounced.

The negative effect of technology, in terms of health and lifestyle, is exemplified by the anecdotal lawn mower. The push-type reel mower was a true calorie burner, but any young couple would soon replace it with a power mower when they could afford one. As the family income grew, the power mower would be upgraded to a self-propelled version, followed in due time by a riding mower. The final act within this American dream would be to hire someone to ride it for you. This progression would relieve one of manual labor, providing free time to pursue other activities, such as exercise, preferably at a health club, paying someone else to watch *you* work, while you were hopefully delaying your own physical decline as a result of inactivity!

Although the gap between home and work may explain a small part of the problem, it doesn't explain the discrepancy between occupations, since all people experience a transition between home and work. In terms of health and fitness, there appears to be a range of problems that vary depending on occupation. These problems raise several questions related to specific work demands, differing between sedentary tasks and heavy manual labor. The problems facing Xerox Corporation aren't likely to be the same as those facing U.S. Steel. A stockbroker's health and fitness priorities certainly differ from those of a bricklayer. The key question to be answered, as far as this text is concerned, is where it is along the fitness/wellness continuum that firefighting belongs. To answer this, we need to look at some of the data, principally statistics related to injury.

INJURY DATA REVIEW

To begin, let's start with the perceived problem, as deduced from injury reports. An opening statement that frequently appears in fire service articles and publications is that firefighting is among the most hazardous of occupations. This statement is often supported and reinforced by reference to published death and injury data. Both the International Association of Fire Fighters (IAFF) and the National Fire Protection Association (NFPA) publish an annual report on injuries sustained the previous year (1998 data is reported in late 1999). The validity of the statement and the methodology used to support it, however, is inconsequential and best left to the reader's opinion. What is important is that the statement is made so frequently, prompting the question as to whether fitness programs should be integrated into the daily activities of firefighters.

In the 1997 November/December issue of *NFPA Journal*, an estimated 87,150 firefighter line-of-duty injuries were reported, the lowest number in the twenty years since the NFPA first began compiling reports (NFPA, 1997). The decline in the number of injuries reported was due in part to a change in the method as to how injuries are categorized. Prior to 1994, exposures (e.g., infectious disease) were included in the total number of injuries, whereas now they have been excluded, giving a false impression that conditions have improved.. The total number of injuries has consistently been around 100,000, even though the total number of fires has been steadily decreasing. The rate of injury per 1,000 fires is the preferred method of analysis, as shown in Table 1–1.

Table 1–1

FIREFIGHTER INJURIES ON THE FIREGROUND AND AT NONFIRE EMERGENCIES, 1987–96

	On the Fireground		At Nonfire Emergencies	
Year	*Injuries*	*Injuries per 1,000 Fires*	*Injuries*	*Injuries per 1,000 Incidents*
1987	57,755	24.8	13,940	1.41
1988	61,790	25.4	12,325	1.13
1989	58,250	27.5	12,580	1.11
1990	57,100	28.3	14,200	1.28
1991	55,830	27.3	15,065	1.20
1992	52,290	26.6	18,140	1.43
1993	52,885	27.1	16,675	1.25
1994	52,875	25.7	11,810	0.84
1995	50,640	25.8	13,500	0.94
1996	45,725	23.1	12,630	0.81

Since 1987, little change has occurred. (Remember that the years 1994–96 exclude exposures.) The IAFF annual report from 1989–96 demonstrates a similar pattern, with a higher rate per 100 firefighters than that reported by the NFPA. The IAFF report (IAFF, 1996) represents a much smaller segment of the total population of firefighters (94,000, or just over one-third of career firefighters, or around 10 percent of all firefighters) and accounts for approximately one-third (31,000) of the total injuries for career and volunteer firefighters, according to the NFPA's total. In a memorandum to the NFPA 1500 committee, October 6, 1993, John Hall of the NFPA stated, "The IAFF totals reflect a career injury rate much higher than other occupations. On the other hand, if they were related to the overall firefighter injury rate as the two death rates are, then they would not be significantly different from other occupations, such as construction and manufacturing." When comparing the rate of injury to other occupations, the IAFF consistently reports the firefighter rate to be around 4 to 4.5 times that of the private sector (4 in the 1996 report). The NFPA report rate, however, is comparable to the private sector, with injury rates ranging from 1.4 percent for smaller departments (population under 2,500) to a high of 17 percent for the largest departments (population 500,000 to 999,999), approximately one-half the rate per hundred reported by the IAFF (32.6 percent). Since reporting methods differ between the two organizations, it is often difficult to interpret the data. The one exception in which the IAFF and NFPA data are consistent is the *trend* in the rate of injury. The injury rate is not declining and has remained relatively stable for the past ten years, even though the total number of incidents has dropped. Firefighters are working fewer fires, but with no change in the injury rate. In 1996, 8.2 percent of all alarms required fire suppression operations, yet they account for 74 percent of the injuries. The death rate per 100 firefighters has also remained relatively constant at about 1 in 10,000 (Table 1–2). In comparison, production or work effort (i.e., employee work hours) in agriculture, construction, and manufacturing have increased, yet the injury rate in those industries is steadily declining (Table 1–3). In fact, the construction industry has been openly applauded by OSHA for successfully reducing its injury rate from 13.1 per 100 workers in 1992 to 10.6 per 100 workers in 1995 (OSHA, April 14, 1997).

Table 1–2

STATISTICS FOR LOCAL FIREFIGHTERS, 1983-96

Year	Career Firefighters	Volunteer Firefighters	Total Firefighters	Firefighter Deaths *Career*	*Volunteer*	*Total*
1983	226,600	884,600	1,111,200	54	51	105
1984	231,600	897,750	1,129,350	43	59	102
1985	238,500	839,450	1,077,950	54	66	120
1986	237,750	808,200	1,045,950	51	55	106
1987	243,200	816,800	1,060,000	48	68	116
1988	252,500	788,250	1,040,750	43	81	124
1989	250,600	770,100	1,020,700	42	65	107
1990	253,000	772,650	1,025,650	25	62	87
1991	261,800	771,800	1,033,600	36	66	102
1992	253,000	805,300	1,058,300	24	44	68
1993	259,650	795,400	1,055,050	21	54	75
1994	265,700	807,900	1,073,600	33	38	71
1995	260,850	838,000	1,098,850	29	57	86
1996	266,300	815,500	1,081,800	27	65	95

Note: These statistics do no include firefighters employed by state and federal agencies, contractors, or temporary employees, nor do they include the employees of private fire departments or fire brigades.

Reprinted with permission from the National Fire Protection Association, Quincy, MA 02269.

Table 1–3

INJURY RATE BY OCCUPATION
Rate per 100 Workers

	1990	1991	1992	1993	1994	1995	1996	1997	1998
Fire	35.3	35.7	35.7	41.4	39.2	37.6	32.6	43.1	32.4
Manufacturing	13.1	13.2	13.2	12.5	12.1	12.2	11.6	10.6	10.3
Construction	14.3	14.2	14.2	13.1	12.2	11.8	10.6	9.9	9.5
Agriculture	10.9	11.6	11.6	11.6	11.2	10.0	9.7	8.7	8.4

The rate per 100 workers appears relatively stable for firefighters over the nine-year period, while other occupations show a pattern of steady decline.

Source: Data taken from the IAFF Death and Injury Surveys, 1990-98.

THE NATURE OF FIREFIGHTER INJURIES

The NFPA has reported that strains, sprains, and muscle pain account for 53.7 percent of all injuries (NFPA, 1997). The leading cause of fireground injuries is overexertion (25.6 percent), followed closely by slips and falls (25.3 percent). Predictably, the type of duty that accounts for the most injuries is activity on the fireground, coming in at 52.5 percent in 1996. By reviewing data from 1986 through 1996, the most prevalent types, causes, and places of injury remain unchanged. In fact, the respective percentages can even be predicted with amazing accuracy. But where does the problem lie? The standard approach is to look at the numbers, compare them to other industries, then arrive at the conclusion that firefighters have a high injury rate, as per the proclamation that theirs is a hazardous occupation. That, I conclude, is the perceived problem, with the data revealing only the symptoms of that problem.

This perception is in part responsible for directing our attention toward the introduction of many of our current safety icons, such as the incident command system, accountability systems, personal alert safety systems, and improved personal protective equipment, to name a few. As a member of the NFPA Occupational Safety and Health Committee (NFPA 1500) since 1992, I submit that I am a staunch supporter of such safety interventions, and I firmly believe that adopting them can impact the incidence of death and injury during firefighting operations. Surprisingly, however, these efforts haven't appeared to alter the data or change the trend in a significant way. If the safety measures previously mentioned were effective, and there is little reason to believe that they aren't, then something other than those measures is preventing the injury rate from dropping. That unrecognized variable, I would contend, is the physical condition of the individual firefighter. When back injuries represent the greatest percentage of overexertion injuries, and 70 to 80 percent of back injuries can be attributed to inadequate muscular strength, then one can only conclude that increasing the strength of the lower back would result in a significant decline in the rate of such injuries. This isn't a cure-all, for even the strongest back will break, but this is the best approach. When the problem is incorrectly identified—attributing the injury to causal factors other than physical condition—the intervention may be totally inappropriate or, worse, ineffective. An excellent example is the knee-jerk response of some employers to issue back-support belts as a preventive measure against back injuries. I will fault manufacturers before employers, for they are largely responsible for the promotion of this device. A back belt will provide some protection if used properly, but with improper and continuous use, it will tend to exacerbate the problem. The problem, weak musculature, isn't addressed, and a belt will simply mask the problem. Similarly, outfitting an unfit firefighter with the best personal protective equipment won't increase his capacity to perform physical work, nor will it protect him from the most frequent type of injury.

The first step in recognizing the core problem is in accepting that the injury profile of the fire service is constant, well defined, and even predictable. As shown in Table 1–4, approximately 80 percent of all firefighter injuries are attributable to 10 percent of the alarms: those that call personnel to the fireground.

The overall rate and type of injury isn't changing. Without changing anything but the date, a study of occupational injuries and illnesses in firefighters conducted in 1982 for the Federal Emergency Management Agency (FEMA) could easily be resubmitted and still be considered accurate (Safety Sciences, 1982). Acknowledging that places us one step closer to solving the problem. When the NFPA, the Bureau of Labor Statistics, and the IAFF collect injury data, one of the most important pieces of information *not* solicited is the physical condition of the injured or deceased employee. The exception is when medical records are made

Table 1–4

Percent of Alarms Accounting for Percent of Injury

	1989	1990	1991	1992	1993	1994	1995	1996	1997	1998
Fire alarms (%)	15.1	12.2	11.9	11.9	11.1	9.0	12.0	8.2	10.0	8.6
Injury (%)	87.9	74.8	80.7	80.2	81.4	77.9	80.5	74.0	68.4	75.4

Source: IAFF Death and Injury Surveys, 1989-1998.

available for line-of-duty deaths. It has been estimated that 80 percent of the firefighters who die while on duty had a related medical history and, according to the NFPA 1582, *Medical Requirements for Fire Fighters,* should not have been assigned to fire suppression (Fahey, 1993).

There is a wealth of information available that has established the association between specific physical deficiencies and injuries, yet firefighter injury surveys fail to identify any such relationship. This is due in part to the failure to use standardized fitness testing and reporting procedures. When and if standardization is achieved, reporting agencies such as the NFPA and IAFF will be capable of incorporating that information in their respective surveys. If standardization cannot be achieved nationwide, it must be pursued at the local level, documented, and correlated to injury data. The second step is to determine the specific problem. We know that the weak link is the physical status of the individual, meaning both his health and his fitness. Overexertion injuries and sudden cardiac events are both symptoms of the problem. We know that a technological quick fix isn't the solution, as exemplified by the use of back belts, nor can we expect a dramatic decline in the injury rate through the safety practices previously mentioned. Although some ergonomic interventions are appropriate and effective, their application in fire and police work is far more limited than it is in the industrial environment (LeCuyer, 1994). If the crux of the problem lies with the human body and the demands placed on it, why don't other blue-collar occupations, with physical demands comparable to the fire service, display the same problems? Although manufacturing, construction, and agriculture are all rated as industries requiring heavy manual labor, the fire service shows twenty times the incidence of back injuries compared with those industries. The answer to the question can be found within the unique nature of the occupation itself. It isn't simply the physical status of the individual, but his physical status relative to the occupational demands.

To respond properly to these issues, we need to look at the occupation in terms of physical demands, comparing it to other occupations in which the requirements for physical labor are similar. The three occupations that I have chosen for comparison are construction, agriculture, and heavy industry. These occupations were selected for two reasons. First, by virtue of their high rate of injury, they are commonly used for comparison by the IAFF. Second, I have personally worked in all three occupations, prior to entering the fire service, and can therefore objectively compare them. The focus of comparison will be on the differences as opposed to the similarities, in terms of physical demand, risk, and the nature or pattern of injury. The similarities between them are quite general. Annually, these industries jockey for the worst injury rate, and they are assigned a comparable strength factor rating based on the Dictionary of Occupational Title Codes (Table 1–5).

Although some data will be included in the analysis, the emphasis will be on four physical factors that I contend best describe the occupations from a physical perspective. These factors, which coincidentally form the basis of exercise programming, are used intentionally to establish an important link between fitness and work. Those who have some familiarity with exercise physiology will readily see the relationship. The four factors are contained in the acronym FITT, namely Frequency, Intensity, Time, and Type.

Table 1–5

DOT OCCUPATIONAL CODES

Occupation	DOT Code	Strength Rating	Job
1. Firefighter	373.364-010	V	Structural
2. Firefighter	452.687-014	H	Forestry
3. Construction	860.381–042	H	Carpenter
4. Construction	861.381–018	H	Bricklayer
5. Industrial	519.686-010	H	General-metal
6. Industrial	502.682-014	M	Cast-machine
7. Industrial	514.362-010	H	Machine-steel
8. Agriculture	410.684-010	H	Dairy farmer
9. Agriculture	421.683-010	H	General

Strength Rating	Occasional	Frequent	Constant
Sedentary (S)	Neg.–10 lbs.	Neg.	n/a
Light (L)	Neg.–20 lbs.	Neg.–10 lbs.	Neg.
Medium (M)	20–50 lbs.	10–25 lbs.	Neg.–10 lbs.
Heavy (H)	50–100 lbs.	25–50 lbs.	10–20 lbs.
Very Heavy (V)	100+ lbs.	50+ lbs.	20+ lbs.

Neg. = negligible.
n/a = not applicable.

Source: (1993) *Selected Characteristics of Occupations Defined in the Revised Dictionary of Occupational Titles.* U.S. Dept. of Labor, U.S. Government Printing Office, Washington, D.C.

OCCUPATIONAL FITT ANALYSIS

Firefighting is described as heavy manual labor. The DOT classification for it is V, representing "very heavy," but is it?

When we look at a job analysis, one item that we rate is frequency, or how often we perform specific tasks. In most occupations, the frequency rating is described as a percentage of work time. One study estimated that public safety officers (police and fire) spend somewhere between thirty and seventy-five minutes a week performing "critical" tasks (Landy, 1992). It must be assumed that the critical tasks are those that require the highest physical exertion, such as structure firefighting and arrest control. This frequency of performance can only be described as minimal, and in organizations outside the realm of public safety, it could only be viewed as insignificant and perhaps not even qualify as a bona-fide occupational requirement. This, of course, isn't an issue in public safety, since the tasks performed to rectify an emergency situation are patently essential, regardless of the frequency with which they're performed. The central point is that firefighters and police officers typically spend only a small portion of their workday actually performing the "heavy" tasks, which naturally implies that the greatest percentage of their time is spent in low-level physical activity. In fact, looking closely at both police and fire activities, aspects of these occupations might well qualify them as being fundamentally sedentary.

There's little that can be done to change that, since personnel cannot fight fires or chase suspects if there aren't any. That's the nature of the job, and that is also part of the problem. Some busy fire companies may see a fire each day, but most do not. Once a month is frequent for many, and even a one-a-day response would be inadequate for the purposes of physical adaptation. If firefighters were battling flames all day long, there would be little need for physical supplementation. In 1996, the NFPA Journal's injury report stated that the rate of injury increases as the number of fires increases. Still, the rate of severe injuries (those requiring hospitalization) is greater among smaller departments that respond to fewer fires. The rate of serious injury requiring hospitalization was 7.3 per 100 firefighters for departments with the fewest fires, more than twice that of larger departments (3.3 per 100 firefighters) that respond to more fires. The risk or chance of injury for the largest departments is twelve times that of the smaller departments, yet the rate of injury isn't proportional, considering that the larger departments attend 300 times as many fires. The conclusion is that the less you do it, the more likely you are to be injured, and with a greater risk of serious injury. Frequency, therefore, is a prime factor in identifying the problem.

To make matters worse, the infrequent calls are often performed at a high level of intensity. There are few other occupations that involve primarily sedentary activities leading to infrequent, unplanned bouts with life-threatening physical challenges. There is nothing more stressful than having to make a transition from rest or sleep to maximal effort at a moment's notice, day or night, rain or shine.

Unless we prepare for that transition through exercise, we will be unprepared to make such physical adjustments without a greater risk of injury. Maintaining a high level of fitness, of course, is no guarantee. Some researchers have documented cardiac abnormalities in a high percentage of healthy firefighters when required to perform at high levels of exertion; however, maintaining one's fitness is the safest bet and realistically the only measure of self-preservation within one's control. In one live-structure fire study, a twenty-seven-year-old firefighter worked for ninety minutes wearing a heart-rate monitor that recorded near-maximal heart rates, with one fifteen-minute period of heart rates reaching 188 bpm (Barnard and Duncan, 1975). A less fit individual could not tolerate this workload without negative consequences.

In terms of duration, there is a significant difference between the three comparison occupations and public safety. The available work time in industry, construction, and agriculture enables one to adapt through repetition, a key advantage that work in the public safety sector doesn't provide. The firefighter's work time is set by the limits of SCBA, minus ten minutes' reserve. Additionally, it has been reported that the extreme effort required is difficult for approximately 70 percent of the *potential* workforce to sustain for more than fifteen to twenty minutes (Rodgers, 1986). The police officer's high-intensity efforts during a foot chase and subsequent arrest are also brief. In a factory setting, I was accustomed to lifting seventy-pound castings to shoulder height at a rate of fifty per hour for eight hours a day. That steady work pattern produced physical adaptations, such as increased strength and endurance. A person will either adapt or break within the first few weeks on this type of job, but once conditioned to the demands, an employee can perform well without undue stress. The problem is getting to that desired level of conditioning.

The type of occupation itself comprises the fourth factor in assessing what sort of exercise programming is required. What other job requires a worker to crawl through a structurally unsound building in the dark while wearing fifty pounds of equipment? In the case of a police officer, what other occupation might entail a fistfight or an exchange of gunfire as part of one's day?

In severe adverse weather, the construction worker and the farmer will call it a day, but even weather is but one component of the overall environmental picture for blue-collar occupations. The construction worker, particularly if he works high steel, is exposed to some unusually risky work situations. The farmer is exposed to hazardous chemicals, such as pesticides and fertilizer. Working around livestock is hazardous, and heavy farm equipment claims the lives of many farmers—579 in 1995 alone (U.S. Dept. of Labor, 1995). The industrial worker is inside and mostly unaffected by weather, but he is often subjected to noise, extremes of temperature, and air pollution. The risks are high for all of these groups, and that fact is supported by the injury data. The greatest difference between them and the public safety sector is the extent of control that the workers have over their work environment and, even more importantly, their familiarity with the workplace.

The work environment of the public safety responder can be anything and everything beyond the front door of the station. It's the unknown human and environmental elements that increase the risks for firefighters and police, but do they really make these occupations more dangerous? The fatality rate (IAFF, 1997) may provide some indication (Figure 1–3).

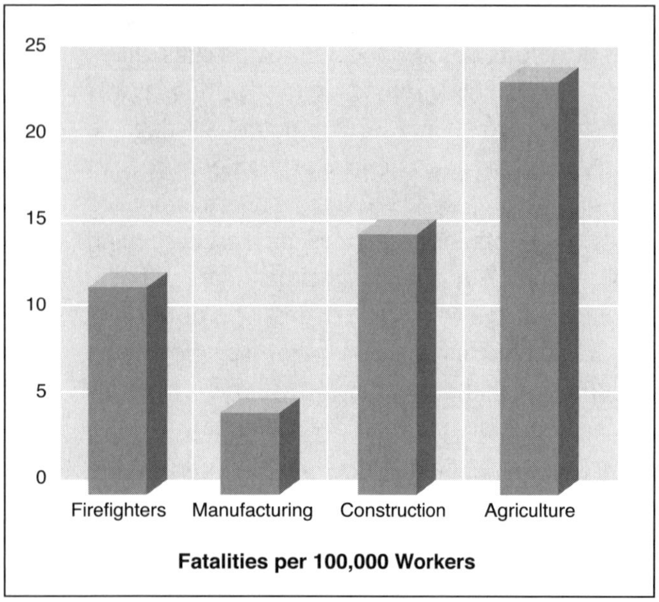

Fig. 1–3 Fatality Rate by Occupation
Source: International Association of Fire Fighters 1996 Death and Injury Survey.

The injury rate for firefighters is significantly greater than it is for the three comparison occupations, which share comparable injury statistics. Even though these occupations don't have the same restrictions and limitations imposed on them as do firefighters and police, workers in manufacturing, construction, and agricultural environments are fundamentally at higher risk. I base this statement on the amount of exposure. Construction and industrial workers are exposed to their unique hazards for eight to ten hours a day, usually five days a week. The farmer has the greatest exposure, working longer days, often of twelve hours or more, and often seven days a week. The length of exposure to occupational hazards is minimal for firefighters compared to the other occupations, yet the injury rate is significantly higher. When comparing the chance of injury among the occupations, based on exposure, firefighters suffer a horrendous rate. In other words, the opportunity for injury or death is greater for occupations outside of public safety, yet their injury rate is far less (Figure 1–4).

One possible explanation is the additional load

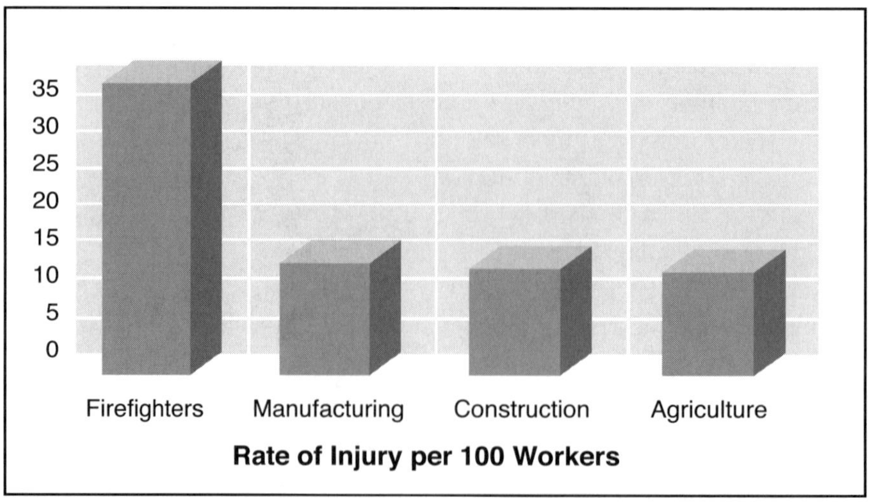

Fig. 1–4 Injury Rate by Occupation
Source: International Association of Fire Fighters 1996 Death and Injury Survey.

imposed on a firefighter by personal protective equipment (PPE). None of the other occupations require wearing fifty pounds of equipment. A firefighter's PPE reduces work capacity by 30 percent, and heat-protective gloves reduce grip strength by 40 to 60 percent. In general, the bulky nature of protective apparel makes the wearer's mobility akin to that of a two-year-old in a snowsuit trying to walk through a drift. Add to the mix a harsh environment, obscured visibility, and psychological stress, and the potential causes for a higher injury rate become more apparent.

How does a firefighter safely cut a hole in a roof over the hottest spot of the fire? How does a police officer engage in pitched battles against criminals bearing all manner of weapons? How does a steel worker safely walk a six-inch-wide beam a hundred stories above the ground? All of these are inherently unsafe acts. Walking high steel is as dangerous as entering a burning building, even assuming that the worker is experienced and knows what he's doing. In terms of strength and power demands, a construction worker pushing an industrial-size wheelbarrow full of mortar expends as much effort as a firefighter does performing a rescue, the most arduous task on the fireground. The difference between these acts isn't as much related to the act itself as it is to how and how often it is performed. Not being able to control the work task is what makes firefighting the more physiologically challenging occupation. To some extent, police work is equally uncontrollable. From the perspective of ergonomic design, what a firefighter does is simply *not recommended.* From the perspective of the FITT factors, firefighting is contrary to all recommended exercise guidelines. No exercise physiologist would advise a person to wake at 2 A.M., don insulated clothing, and work at an intensity level of 100 percent without warming up first, yet that is exactly what firefighters do. The problem, therefore, isn't so much an issue of the danger or hazard, but the public responder's inability to control how and when he must respond to the alarm. The frequency of the experience is key.

TOWARD FORMING A PROGRAM

To return to the central theme of this chapter, identifying the problem, we can produce evidence that firefighters have an occupational problem and that fitness programs have had varying degrees of success with firefighters. Do public responders, then, really need a program? Do the reports and data reveal a lack of personal fitness, or can injuries be more correctly attributed to unsafe practices, environmental exposures, and apathy toward fitness?

According to the Bureau of Labor Statistics, approximately 53 percent of the business sector offers wellness-type programs to employees. That percentage is comparable to the ratio of fire departments that have fitness programs in place, approximately 65 percent, based on a survey reported by Frank J. Landy. Of those, 50 percent require mandatory participation, although only 36 percent of the reporting departments (164 total) included health promotion benefits (Landy, 1992). Since the corporate emphasis has been on wellness as opposed to fitness-based programs, the private sector has documented program effectiveness with an industrywide decline in the occupational injury rate. The fire service, on the other hand, hasn't demonstrated the ability to affect injury rates or health status (based on the incidence of heart attack), even though a comparable proportion of departments have programs. A few departments have reported success, but their number is insufficient to affect the total occupational picture. The reason for the discrepancy between the results reported by the different occupations may be attributed to an inappropriate or ineffective program design, one that is based on the perceived problem rather than the real problem. Of course, the general problem, employee health and fitness, as well as its relationship to health care costs, injuries, and worker productivity, is a common concern of all occupational groups. The more specific problems, however, are unique to each group and must be clearly defined if they are to be addressed. The employee with coronary artery disease will be viewed differently by IBM than by FDNY. Modest improvements in an employee's functional capacity may be considered beneficial in the corporate environment, yet inadequate in the fire service. The design of a fitness program is dictated by the occupational demand, the nature of the problem, the risks associated with the problem, and the consequences of not correcting the problem. I am convinced that the corporate world is more in tune with its specific problems, and therefore has been more effective at developing resolutions. The fire service has failed to identify this important fact or, for that matter, its problem. The corporate response has been to jump in with both feet, committing funds to build facilities and proceeding with an almost blind faith in the benefits, reserving cost/benefit analysis for a future date. Some administrators may argue that a large corporation can afford to take this approach; however, one must remember that companies are in business to make a profit, and investing a large sum for an exercise facility isn't done without some concern for the return on investment. The fire service has been much more conservative, choosing first to investigate the physical aspects (Southman *et al.*, 1992; Barnard and Duncan, 1975; Lemon and Hermiston, 1997; Davis and Dotson, 1987; Duncan *et al.*, 1979; and Jamnik, V., and Gledhill, N., 1989), then attempting to establish a relationship between fitness and the occupational demands (Davis and Dotson, 1982; and Dotson *et al.*, 1983). This methodical approach may be more related to issues of test validation and employee selection procedures than to the perceived benefits of fitness programs and their relationship to occupational diseases and injuries. In any event, fire departments that haven't taken the initiative to introduce a program will in time recognize and respond to the need. There is a difference in the approach taken by the public safety sector and the private sector with regard to fitness/wellness issues. Corporations and industry typically don't design a fitness program with an emphasis on performance. Rather, they tend to focus on wellness concepts and employ ergonomic practices in response to injury problems. The public safety group, in contrast, will go beyond the health/fitness paradigm, with the ultimate goal of matching the work to the worker in terms of physical performance. Although the corporate strategy of implementing safety and ergonomic interventions should also be practiced by public safety agencies, the willingness of the latter group to address the performance aspect is what differentiates the two groups.

Although this discussion has centered on the fire service, the issues are basically the same for police officers. Firefighting has often been described as a form of combat, although having served with the U.S. Marine Corps in combat in Vietnam from 1967–68, I see little comparison, other than the overall physical effort required. (The Department of Labor assigns the same strength rating of V, very heavy, to both the infantryman and the firefighter.) The demands of law enforcement compare more closely to combat. Police engage in hand-to-hand struggles and rely on weaponry as essential job tasks. One-on-one contact with combative suspects accounts for the greatest percentage of police officer injuries. According to U.S. Department of Justice data, the largest percentage (37 percent) of officers killed in the line of duty can be attributed to arrest control situations (U.S. DOJ, 1998). In one review of a local police department's injury profile, data revealed that 45 percent of the occupational injuries were associated with arrest control. Although different, the problem in terms of injury rates for police is overwhelmingly similar to that of firefighters, since both are fundamentally sedentary occupations that require unexpected bursts of maximal effort. Thus, the gap between fitness level and job demands is also the same. The solution for both firefighters and police is to close that gap.

RECOMMENDED READING

The International Association of Fire Fighters annual reports.

The National Fire Protection Association annual report.

The Bureau of Labor Statistics, annual report.

U.S. Department of Justice, *FBI Uniform Crime Reports.*

CHAPTER 2

STANDARDS AND LEGAL ASPECTS

We have identified the need for a fitness program, but the proposal to introduce one cannot be made without considering the legal aspects that will influence its design. The extent of legal influence is directly related to deciding how a department may use and apply the tests and standards of its program. A brief discussion of applicable legislation and standards is necessary prior to developing a purpose statement, which will define our objectives and support the mission of instituting a health and safety program.

NFPA STANDARDS

The National Fire Protection Association was founded in 1896. It is a nonprofit organization dedicated to the protection of life and property from the hazards of fire. The association's primary purpose is to develop and revise standards covering all aspects of fire safety. The standards are developed through a consensus process that allows maximum participation from a wide variety of fields and interests. The NFPA publishes over 286 nationally recognized codes and standards.

In 1987, the NFPA introduced the first edition of NFPA 1500, *Standard on Fire Department Occupational Safety and Health Program.* This was the first consensus standard on this issue for the fire service. The intent was to provide an umbrella document that would serve as a guide for assisting fire departments in the development of a comprehensive program. The second updated and revised edition was introduced in 1992, and the current third edition appeared in 1997. The next edition is scheduled for release in 2002.

There are companion documents that augment specific chapters in the parent document. Three that are relevant to our program are NFPA 1521, *Standard for Fire Department Safety Officer, 1997 Edition;* NFPA 1581, *Standard on Fire Department Infection Control, 1997 Edition;* and NFPA 1582, *Standard on Medical Requirements for Fire Fighters, 1997 Edition,* to be retitled in 2002. All of the standards are appropriate for either mandatory or voluntary compliance and meet or exceed existing standards as they relate to firefighter safety and health.

Although NFPA standards aren't law, they can be adopted as such by the authority having jurisdiction, and they have been cited in case law as an accepted safety and health standard in the field. Noncompliance with NFPA 1500 isn't without risk, and complete avoidance, as some fire executives have chosen, won't be a supportable defense for whatever the reason (Sukay, 1989).

In February 1990, *Fire Chief* magazine published the results of a survey that targeted fire chiefs of both paid and volunteer departments. The purpose was to solicit their response to the new 1500 standard (Baltic, 1990). When questioned about the legal implications of 1500, most emphasized that their decision to follow it wasn't based on the threat of possible litigation. Their express concern was for the safety and health of their members, and they had concluded that the standard provided a reasonable course of action to address those issues. It is in this context that I recommend adoption of 1500 as a program model. The incentive to begin a program shouldn't be based on some legal threat, but rather, in response to our mission and the obligation to protect life and property, including the lives of our own members (Stittleburg, 1988; Craley, 1988; and Diezel, 1988).

OCCUPATIONAL HEALTH AND SAFETY ACT OF 1970

The Occupational Safety and Health Administration was created under the OSHAct (OSHAct; 29 U.S.C. 651 *et seq.*) as directed by Congress in 1970 for the purpose of regulating workplace health and safety. The act covers all employees and employers in the fifty states, the District of Columbia, Puerto Rico, and all other territories under federal jurisdiction. OSHA adopted federal and national consensus standards, and OSHA representatives stated that if a fire department was in compliance with NFPA 1500, 1997 edition, it would be in compliance with OSHA rules in terms of "procedures for interior fire fighting."

A provision of the act exempted state and local agencies from OSHA regulations, thereby excluding state and municipal fire departments from OSHA compliance. The law did, however, allow states desiring OSHA approval to assume responsibility for workplace health and safety. Twenty-five states accepted the responsibility. These "state-OSHA" jurisdictions are required by Congress to set standards that are at least as effective as the federal standards and ensure compliance from both state and local agencies, including fire departments. Of the few OSHA standards applicable to fire departments, the Respiratory Protection Standard, 29 CFR 1910.134, is among the most important. Under this standard, employers are required to certify firefighters as being fit for duty prior to using respiratory protection. Many OSHA standards reference NFPA standards and vice versa, with NFPA integrating OSHA regulations into their documents. Respiratory requirements similar to OSHA 29 CFR 1910.132–140 appear in NFPA 1404, *Standard on Fire Department Self-Contained Breathing Apparatus Program, 1996 Edition,* section 4-2.1, which requires members to be examined and certified as being medically and physically fit in accordance with NFPA 1582, *Standard on Medical Requirements for Fire Fighters.* Also, as per TIA 96–1 NFPA 1404, physical fitness requirements for entry-level personnel shall be developed and validated by the authority having jurisdiction.

Thus, the relationship between OSHA and NFPA standards becomes obvious. Ignoring the legal implications of noncompliance with OSHA can expose departments to significant liability, since standards such as 29 CFR 1910.120 Hazardous Materials, for example, apply to all emergency personnel, volunteer or paid, in every state, with OSHA relying on NFPA as the "industry standard" (Bentivoglio, 1995). All OSHA standards are published in the *Federal Register* when adopted, as are all amendments, corrections, insertions, and deletions. All of the current regulations and standards are published annually in the *Code of Federal Regulations.*

AMERICANS WITH DISABILITIES ACT OF 1990

The Americans With Disabilities Act (ADA) prohibits discrimination against the physically and mentally disabled. The ADA draws on the Rehabilitation Act of 1973, which provides protection for people with disabilities. Under the Rehab Act, non-job-related medical exams or medical decisions aren't allowed unless an employer can demonstrate that a person cannot perform the job. The Rehab Act also mandates that employers consider "reasonable accommodation" of qualified handicapped persons, similar to the ADA. The ADA, which became effective for public-sector departments in 1992, forces departments to reevaluate all testing procedures that affect employment decisions. The ADA prohibits preemployment medical exams prior to a conditional job offer, but it does allow physical fitness or agility tests as long as they don't involve a medical diagnosis. The ADA will allow medical screening and monitoring of incumbents if required to comply with other federal law (*e.g.*, OSHA), to determine disability if an employee requests accommodation, or as a return-to-work exam following injury or illness leave.

The key to compliance with the ADA is in determining whether an individual can perform the essential functions of the job, with or without accommodation (Dezelan, 1992). The act does recognize the employer's concern for workplace health and safety, and it does provide some latitude, particularly for public safety organizations, in determining what constitutes essential functions. Thus, an employer is not required to hire or retain an individual who may pose a "direct threat" to himself or other members. This determination, however, requires a current assessment of the individual's physical status and his ability to perform.

Although the act does not require employers to develop job descriptions, doing so is highly recommended. Employers must be prepared to defend their claims as to what tasks actually are essential and job-related. This implies that a job-task analysis be conducted to identify the essential tasks. Since not all job assignments within a given organization are the same in terms of physical requirements, a single blanket test that requires all members to meet the same standard may not be justified (Jones *et al.*, 1989). Consider the strength-level requirement of the job assignments identified in the Dictionary of Occupational Titles, as depicted in Table 2–1. The table clearly identifies different requirements and therefore implies that a different test or test standard is appropriate.

Table 2–1

DOT OCCUPATIONAL CODES FOR PSO

Occupation	DOT Code	Strength Rating	Job
1. Firefighting	373.364-010	V	Structural
	452.134-010	V	Smoke Jumper
	373.167–010	L	Battalion Chief
	452.687–014	H	Forestry FF
2. Public safety	379.263-014	V	Public safety officer
	375.263-014	M	Police officer

Strength Rating	Occasional	Frequent	Constant
Sedentary (S)	Neg.–10 lbs.	Neg.	n/a
Light (L)	Neg.–20 lbs.	Neg.–10 lbs.	Neg.
Medium (M)	20–50 lbs.	10–25 lbs.	Neg.–10 lbs.
Heavy (H)	50–100 lbs.	25–50 lbs.	10–20 lbs.
Very Heavy (V)	100+ lbs.	50+ lbs.	20+ lbs.

Neg. = negligible.
n/a = not applicable.

Source: (1993) *Selected Characteristics of Occupations Defined in the Revised Dictionary of Occupational Titles.* U.S. Dept. of Labor, U.S. Government Printing Office, Washington, D.C.

There are two types of tests in common use: the fitness test (*e.g.*, aerobic capacity) and the agility or performance test (*e.g.*, drag a hoseline). Both types of tests are acceptable under the ADA, with the same consideration for issues of validity, safety, standards, job-relatedness, and so on. The impact of ADA on test development and choice is directly related to the type of test selected and the purpose or use of that test. A test designed for use as an evaluation tool in a fitness program is very different from one designed to help make employment decisions.

Title I of the Civil Rights Act of 1991

The Civil Rights Act of 1991 (CRA '91) states that "It shall be an unlawful employment practice...to adjust the scores of, use different cutoff scores for, or otherwise alter the results of employment-related tests on the basis of race, color, religion, sex, or national origin." The use of age- and gender-based norms in fitness program assessments has been a common practice in the field of physical education for decades. The same practice has been widely adopted by public safety organizations and supported by many test developers, including the Institute of Aerobic Research (IAR), a former promoter of norm-based fitness standards. The principal difference between the field of physical education and public safety occupations is in the use and application of these standards. Physical educators (as well as coaches, trainers, and exercise physiologists) use normative data as a measurement tool to evaluate and develop fitness programs. Public safety organizations have been using such data to help make employment decisions. Although this practice has been successfully defended in case law history, as in *U.S. v. Wichita Falls, 1988,* it is now in direct conflict with CRA '91 when used to make hiring decisions. Additionally, the ADA lends further support for discontinuing the use of "multiple standards," since the same physical job requirements must be applied to all candidates and current employees. The favored argument in support of age- and gender-based standards is that fitness declines with age, and women typically score lower than men on fitness tests, primarily due to physical size. Although both statements are basically sound, they are not defensible, since the physical demands of the job do not change, regardless of who is performing them (Lusa et al., 1994). If fitness testing and subsequent application of standards are used for purposes other than employment (physical education/training), the act would not apply. Thus, the intended use or purpose of the fitness test is the overriding factor, since there are applications that may or may not be appropriate (Collingwood, 1995). A recent statement from the U.S. Department of Justice, given at a summit meeting held in Oregon, astonishingly supported the use of age- and gender-based standards for selecting employees. The arguments given, in my opinion, are both confused and illogical, and they present a clear misunderstanding of the relationship between fitness and work (Hornick, 1997).

Title VII of the Civil Rights Act of 1964 (CRA '64)

The Civil Rights Act of 1964 prohibits discrimination based on race, color, sex, religion, and national origin, but it does not prohibit physical testing. Title VII of the act specifically addresses employment testing and authorizes the use of any professionally developed test, provided that it is neither intended nor used to discriminate unlawfully. Although the legislation was originally aimed at written tests in the same manner as the guidelines, the requirement applies to all test instruments, whether written or physical, and it has been supported in court decisions (Bentivoglio, 1996; and Rafilson, 1995).

Title VII has had the greatest impact on physical testing in public safety organizations. It is unfortunate that claims of discriminatory practices provoked an assessment of the "way we do business." On the plus side, however, the process did provide the impetus to address fitness issues and test practices that may not otherwise have been on the agenda. The act forced employers to review their tests and procedures, as well as to provide documented support, through validation, for the tests that they select and use. The root cause of the problem has been failure to match carefully the requirements of the job with the ability of the worker, using a valid test instrument derived from a comprehensive job analysis. The need for uniform guidelines to develop test procedures resulted in the development of a set of principles outlined and published in *Uniform Guidelines on Employee Selection Procedures 1978*. These guidelines were issued by the Equal Employment Opportunity Commission (EEOC), the U.S. Department of Justice, and the U.S. Department of Labor to assist employers in the proper use of tests, their development, and validation requirements as they pertain to employment decisions. The guidelines were developed from court decisions. They apply to all tests and other selection procedures used for employment decisions, and they are enforced by the EEOC under Title VII (Title 29 Code of Federal Regulations, Part 1608, Chapter XIV, July 1991).

AGE DISCRIMNATION IN EMPLOYMENT ACT

The Age Discrimination in Employment Act (ADEA, 1967) was designed to protect employees and applicants over the age of forty against arbitrary age discrimination practices. Public safety officers were exempt from the ADEA until a 1983 ruling placed them under the act. If state and local governments wanted to impose a mandatory retirement or minimum hiring age, they were going to have to prove that such criteria were bona fide occupational qualifications.

The significance of ADEA is that legal challenge to the exemption status of public safety officers resulted in a court ruling that proposed using alternative methods, rather than age, of determining physical capability *(EEOC v. County of Los Angeles)*. This brought the issues of physical fitness and testing to the fore. In 1986, both police and fire organizations lobbied and received a temporary exemption from Congress until 1993. In the interim, a Penn State study was commissioned by the EEOC and the Department of Labor to find some way besides age for selecting and retaining public safety officers (Landy, 1992). Although the study confirmed that age isn't a valid indicator of physical ability, it failed to provide any direction as to how organizations might assess public safety officers, since neither tests nor standards were identified.

In 1996, police and fire organizations were again forced to lobby, and their effort produced another exemption, with the provision that the National Institute of Occupational Safety and Health (NIOSH) would conduct a study and develop valid job performance tests within three years. Any evaluation developed is slated to be attached to the law in the year 2000.

A stipulation in the law would require an annual physical assessment of public safety officers who reach retirement age so as to determine their eligibility for continued employment. A similar test requirement would apply to entry-level candidates and all incumbents, since the ADA requires a current assessment of all members. Regardless of the outcome, public safety agencies will need a program and subsequent test. The IAFC/IAFF developed a test to evaluate candidates (CPAT), but it excludes incumbents.

A CASE FOR FITNESS

The *Fire and Police Personnel Reporter* provides a case law digest relating to police and fire disciplinary actions and personnel problems from 1975 to 1997 (Smith, 1997). The case law briefs are categorized by topic and include citations on medical exams, fitness requirements, agility tests, and other health-related subjects as they relate to employment issues. Since fitness is the subject of this book, it is crucial that the association between various case studies and fitness program objectives be examined. Aerobic capacity, one of the fitness components, is often viewed as the single most important component of fitness (Evenson, 1988). Proponents of a fitness test (e.g., push-ups, sit-ups, running) for employment purposes often cite this reasoning. Case law has supported the use of fitness testing and has therefore reinforced the emphasis on aerobic fitness and its relationship to job performance in *U.S. v. Wichita Falls, 1988*. On the other hand, experts who disagree with this opinion will declare that the anaerobic component is both more important and more closely related to job issues, citing case law *Zamlen v. Cleveland* to support their position. The use of both a fitness and an agility test has been challenged and successfully defended in *U.S. v. Wichita Falls (1988a)*. The 1.5-mile run, a common field test of aerobic capacity, has been included in fitness test batteries and as a supplement, combined with agility tests (e.g., dragging a hose, carrying a victim). The 1.5-mile run test, when challenged, has produced different results. It was supported in one case, *U.S. v. Wichita Falls* (fitness test), but in a 1994 case, *Minnesota Department of Human Rights v. City of St. Paul* (fitness and agility combination), it was eliminated in the court settlement. What should be apparent at this point is that there appears to be some conflict between the relevance of one fitness dimension, such as aerobic capacity, and the use of a field test to measure that component, such as the 1.5-mile run. In another case, a run test designed to evaluate aerobic capacity was administered to a police candidate, who was rejected on failure of the test. In *Ferrante v. Niagara County*, the candidate didn't challenge the test itself, but instead objected that he wasn't allowed to take an alternative test that was offered to another candidate. Since different types of aerobic field tests will produce different results, the candidate had a legitimate argument, although this fact wasn't presented in the case. A similar problem on the subject of test alternatives appeared in *Smith v. City of Des Moines (1996)*, in which a fire captain was given a substitute test that did not help his case and was probably inappropriate. The fitness test of aerobic capacity was offered as a substitute for a spirometry test, which is generally classified as a medical test. To further complicate the problem in this case, a rather arbitrary $\dot{V}O_2$ standard of 33.5 $mL \cdot kg^{-1} \cdot min^{-1}$ was selected as the criterion for passing. Although the value was derived from a research study, it was accepted on the basis that 86 percent of the study's firefighters who passed had a $\dot{V}O_2$ of greater than 33.5 and that 60 percent of those who failed had a $\dot{V}O_2$ of less than 33.5. The other interpretation is that 40 percent of the firefighters with values below that standard did pass! Since the law offers some latitude in applying standards to public safety personnel, based on life safety risks, the 33.5 value was accepted in spite of the 40 percent pass rate of those below the 33.5 standard. The case relied on ADEA and ADA, with references to NFPA and other respiratory protection standards. The question that this case raises is one of test validity. If the $\dot{V}O_2$ requirement of 33.5 $mL \cdot kg^{-1} \cdot min^{-1}$, the lowest value recommended by research, is a valid criterion, why did 40 percent with lower scores pass? Part of the answer lies in the nature of the tests. The firefighter task test was primarily anaerobic, and the $\dot{V}O_2$ test was an assessment of aerobic capacity, a distinctly different energy system.

The *Uniform Guidelines on Employee Selection Procedures* requires validation of any test used for employment decisions. These guidelines also require that a test be significantly correlated to job performance, and herein lies the crux of the problem. To validate the 1.5-mile run test, for example, a sample of subjects must perform the field test and a maximal graded exercise test, such as a treadmill. The results are then statistically correlated to arrive at a coefficient of determination. Correlation coefficients for the 1.5-mile run vary with the investigators and the population sampled. Reported coefficients for the 1.5-mile run tests range from a low of $r = 0.27$ to a high of $r = 0.90$ (Heyward, 1998), which indicates that the test may or may not be a valid predictor of aerobic capacity, depending on the population and other test factors (Noakes, 1988). Generally speaking, a coefficient of less than $r = 0.71$ would not be considered significant, since less than 50 percent of the variance would be accounted for in the test. To confound things further, the field test must also be statistically correlated to the job requirements. Establishing this relationship has been difficult, particularly for firefighters, since the test measures aerobic capacity and the essential tasks of firefighting are predominantly anaerobic. Aerobic measures of 242 incumbent firefighters showed a correlation of only 0.25 to a firefighter simulated task-type test (LeCuyer, 1989). The same problem can exist with police officers, where the physically demanding tasks, such as arrest control and sprint chases, are predominantly anaerobic. However, if it is determined that the typical foot chase performed by a police officer is of several blocks or longer, as some investigators report, then the measurement of aerobic capacity may be supported. Since police officers assigned to patrol on foot do run, a run test could be regarded as a task-type test and directly correlated to job performance, independent of aerobic measurements (Davis, 1998).

Thus, if aerobic capacity is expected to be an accurate predictor of job performance, the field test to assess this dimension must be validated as such, then correlated to essential job functions. The conclusion is that the field test is twice removed from the actual job requirement, and the inferential leap from test to job is so great as to render it highly suspect as a method for making employment decisions. The fuzzy nature of this fitness issue is in part responsible for the lack of uniformity seen in numerous court decisions. A summary of the test steps and five potential sources of error are depicted in Table 2–2.

The first error in Table 2–2 would result from failure to match the appropriate test with the job requirements. If the job is fundamentally anaerobic, then the test should measure this dimension. The second potential error would be in selecting a test with poor discriminatory value relative to the demands of the job. The 1.5-mile run has not been shown to correlate significantly with structural firefighting tasks, although it could correlate to some police tasks *(Lanning v. SEPTA, 1998)*. The third is a test design error, allowing two different test modes that would, if given to the same individual, produce different scores, even though they are designed to measure the same fitness component. The fourth stumbling block involves the selection or determination of standards that either aren't population-specific or are age- and gender-based, violating the CRA '91. The last hurdle would be failure to establish the critical link between test and task.

Although the 1.5-mile run standard, typically under twelve minutes, is a reasonable expectation for public safety personnel, it doesn't qualify as a business necessity for firefighters, and it may not qualify for police. Firefighters typically do not run

Table 2–2

POTENTIAL TEST ERRORS

Test Process	Sample Case	Potential Error
Aerobic test	*Zamlen v. City of Cleveland, 1989*	Appropriate test?
Field test—running	*Minn. Dept. of Human Rts. v. City of St. Paul, 1990*	Valid predictor?
Score (aerobic)	*Ferrante v. Niagara County, 1990*	Alternative tests?
Standards	*Smith v. City of Des Moines, IA, 1996, CRA 1991*	Dual standards?
Criterion	*Blake v. City of Los Angeles, 1979, ADA 1990*	Test-job match?

Note: The cases cited are examples of potential challenges that can be expected during the process of selecting and validating a test or standard, and they do not necessarily represent the central issue in each case.

during the performance of firefighting functions, nor is the aerobic component (when expressed in $mL \cdot kg^{-1} \cdot min^{-1}$) as measured by the test significantly correlated to most tasks. If the testing program that you're contemplating can survive the five potential errors in this example, then you may have a valid test instrument. However, one must remember that this is only one component of fitness and that the remaining four that comprise a fitness test battery must undergo the same scrutiny.

In terms of fitness, body composition is best described as the relationship between fat weight (adipose tissue) and lean weight (muscle, bone, etc.), expressed as a percentage relative to overall weight. A 200-pound person with 15 percent body fat would have 30 pounds of fat weight and 170 pounds of lean weight. An individual with an excessive amount of fat weight would be regarded as obese. Different investigators use a variety of terms and percentages to describe levels and types of obesity, and this is one problem. Since case law on the issue of weight is on the rise, the problem is exacerbated. Most fitness test batteries often include an assessment of body composition, and a host of problems have been raised on this one dimension alone. Is obesity a handicap? A California court in *Cassista v. Community Foods Inc.* recognized it as such if it is caused by a physiological disorder, whereas other courts *(Civil Service Commission v. Pennsylvania Human Relations Commission* and *McMillen v. Civil Service Commission)* have declared that obesity is not a handicap and that a city can reject an applicant without proof of job relevance. Is obesity a disability? According to one federal court in *Smaw v. Virginia State Police,* the demotion of a police officer to dispatcher was upheld, stating that the officer was not disabled. The trend appears to support the position that obesity is neither a handicap nor a disability, and that employment decisions need not prove job relatedness. If a fitness test used for employment is valid, it must be job related. If so, why does it appear that weight issues need not show job relatedness? In one Oregon case, *Beck v. Department of State Police,* an overweight police officer's claim for protection under handicap law was rejected, and the court supported the department's "appearance standards," which apparently was sufficient cause. One of the greatest problems with obesity cases is in defining the term. In *Philadelphia Electric Company v. Commonwealth of Pennsylvania,* an applicant was rejected due to "morbid obesity," defined as being 100 pounds above ideal weight. Still, what is considered an ideal weight? Using height and weight charts, which have been obsolete for decades, some professional football players could be classified as being morbidly obese, even though they are highly fit athletes. However, if the classification of these athletes were based on body fat percentage, the results could be entirely different. This brings up the issue of weight standards. The use of weight standards has been supported in a number of cases: *United Paramedics of L.A. v. City of Los Angeles, Armstrong v. City of Dallas,* and *Brletic v. Municipality of Monroeville.* The Brletic case could serve as an example of the football player comparative, since the rejected applicant was a weight lifter. He was probably fit, but not of a proportion to be pigeonholed! If body fat percentage had been used in place of height and weight charts, it is likely that he would have been selected. The use of body fat percentages isn't the solution, however. Determining the proper percentage can be a problem, since 25 percent fat on a 200-pound person isn't the same as 25 percent on a 120-pound person. The larger person would still have adequate lean weight (150 lbs.) to perform, whereas the smaller person (90 lbs.) might not. Thus, the issue of body composition is as dubious as aerobic capacity. Although weight standards have been determined to be a bona fide occupational qualification and justification for discrimination based on "legitimate and rational considerations," "a business necessity," and "a duty to the public," the trend seems to be toward using analytical factual data and away from mere stereotypical generalizations (Neff, 1994). Employing body fat percentage methods in place of height-weight standards and arbitrary definitions of obesity would be a good start, but not the answer. Another problem with this method, commonly used in fitness assessments, is exactly the same as that described with aerobic capacity. The "gold standard" for body composition analysis is hydrostatic weighing. All field test methods of determining body composition, including caliper, body mass index, and bioelectrical impedance methods, are correlated to this one test. The dilemma is the same, one of inferential leap, with a compounding factor of significant variance between test methods. The measurement of a single firefighter, using four different methods of assessing body composition, reported a variance of 12 percent between the lowest and the highest estimate. Other fitness tests suffer the same drawbacks. Sit-ups, a common test of muscular endurance, have been successfully challenged on the grounds of job-relatedness, yet using them has also been upheld in court decisions *Merrimack Co. and State Employees Association*, based on questionable logic that resulted in the termination of an incumbent.

In general, the use of fitness tests for employment decisions can and often will lead to litigation, the results of which can be highly unpredictable. Although fitness tests are an accepted format, their "distance" from the job renders them vulnerable to attack. Establishing that critical link between test and task is difficult using fitness tests. In *Blake v. City of Los Angeles 1979,* the court ruled that some fitness predictors were correlated with criterion measures, yet the relationship was modest and not to the extent that they could be considered a business necessity.

So why use fitness tests at all? For one reason, they do measure underlying capabilities that relate to physical performance, and they are well suited for use as a measure of reasonable assurance that one could perform. In my practice, using a composite fitness test score derived from criterion-based standards, I have been able to predict with good results ($r = 0.71$) how well one will perform on a physical agility test. When using combined measures of upper-body strength (bench press, lat pull-down, preacher curl), I have found a correlation of $r = {}^{-}0.95$ to the CTTest. These are acceptable rates, given the purpose of the fitness program. However, the inability of fitness tests to accurately predict job performance to the extent of determining someone's career is sufficient cause not to rely on them for employment purposes.

Fitness is related to performance, and when fitness tests are used for their intended purpose—to evaluate the components of fitness and develop exercise recommendations—they are very effective. When they are used for reasons other than their intended purpose, such as employment decisions, they will present a formidable challenge, not only in terms of establishing the critical link between test and job, but also in defending that position.

RECOMMENDED READING

Smith, W. (1997) *Fire and Police Reporter: Case Law Digest, 1975–1997.* Chicago, IL: Americans for Effective Law Enforcement Inc.

CHAPTER 3

THE LABOR/MANAGEMENT AGREEMENT

The initial step in establishing a physical fitness program is to arrive at a consensual agreement between labor and management to pursue the project. The decision to take that first step may appear easy, but for many organizations, it is often the most difficult one in the process.

At this stage, I do not recommend assembling a committee to determine whether to introduce a program. The axiom "A camel is a horse designed by a committee" says much about what may result if too many people are involved prematurely. A discussion of standards, for example, could provoke an argument that could delay or even derail the process entirely. Two or three people at the most—the chief, a labor representative, and the department safety officer—are all that are needed at the start.

The greatest stumbling block in any such endeavor is created when individuals or groups focus their attention on personal concerns that needn't be addressed in the preliminary stages. In our context, labor may become unduly focused on what happens if a member doesn't make the grade; management may get caught up in wondering how the program will be funded. The focus during the initial stages should be on acknowledging that a fitness program is necessary and that it is the most logical, most effective way to address the problem. As always, you may need to bring labor and management together by making them realize that they share a common goal: protecting the health and safety of personnel.

Some departments have no member organizations (i.e., unions), whereas others have a variety, some unique to their department, jurisdiction, state, or country. For the purposes of this discussion, I will use the two most prominent fire service organizations: the International Association of Fire Fighters (IAFF) and the International Association of Fire Chiefs (IAFC) as being representative of labor and management, respectively.

In recent years, the controversies surrounding fitness testing, test validation, and the development of standards have prompted both labor and management to try to resolve the issues of dispute. In December 1996, the IAFF and IAFC gathered in Phoenix to begin the development of a wellness/fitness program for the fire service *(International Fire Fighter, 1996)*. This group project, dubbed the IAFF/IAFC Joint Labor Management Fire Service Wellness/Fitness Initiative, consisted of the members of ten fire departments. Ten fire chiefs and ten labor union presidents were selected to participate, based on their personal history of excellent labor-management relations and a mutual commitment to the implementation of fitness programs. One of the key goals of this conference was to develop standards that would be fair, safe, and beneficial to all firefighters *(IAFF Leader, 1997)*. Following the initial meeting, physicians and exercise physiologists were invited to assist with the project and provide expert advice. In August 1997, the results of this work, which included a manual, a brief video on the development process, and a data collection protocol, were presented at the IAFF's Redmond Symposium in Toronto and distributed to the attendees.

The effectiveness of the joint initiative, which in part depends on using the product developed and the extent of a department's participation in it, remains to be seen. Regardless of the outcome, the significance of the event was that both labor and management have realized that physical fitness is an issue that can no longer be ignored. Perhaps more importantly, these two groups have realized that they share responsibility and that a cooperative effort will be required to address the issue properly. One IAFF local representative stated, "...70 percent of what labor wants and management wants is on the same line...and in this case, we both get what we want: healthier firefighters. We don't disagree about that" (Dezelan, 1997).

Presently, the task at hand is to arrive at an agreement to begin the project. The development of a mission or purpose statement is well advised. If you currently have a mission statement written for your health and safety program, review the text to verify that member health and fitness terminology is contained within it. Fitness is a safety issue, and the text should reflect this. The following sample mission statement is an adaptation of a policy statement presented in NFPA 1500, *Standard on Fire Department Occupational Safety and Health Program, 1997 Edition.*

Mission Statement

> *The mission of the health and safety program is to provide for the safety, health, fitness, and general welfare of all members through the implementation of programs and activities designed to reduce occupational exposures; (to reduce) the frequency and severity of injuries, accidents, and illnesses; (to) facilitate the rehabilitation of occupational injuries; and to achieve, maintain, and promote the enhancement of member health and fitness.*

This statement is an example of a policy that would apply to a complete health and safety program, of which fitness is but one part. Although the term mission and purpose are often regarded as synonymous, I recommend adopting a separate program purpose statement in support of the health and safety mission. Since the mission covers numerous programs, each serving a different purpose within the framework of health and safety, a purpose statement will serve as a point of reference for the decisions that you make throughout the development of your wellness/fitness program.

Purpose Statement

> *To achieve and maintain an acceptable level of fitness that is commensurate with assigned duties and that will provide reasonable assurance that the members can perform safely and effectively without undue risk to themselves, their coworkers, and the citizens whom they are sworn to protect.*

Nowhere in this statement is there any reference to employment decisions, nor is such a reference to be found anywhere in the library of fitness literature. That is not the function of a fitness program—at least, not this one. Still, there is no guarantee that a fitness program won't affect employment status. There may be situations in which, as a result of the program, a member is identified as being physically deficient in some respect, possibly leading to transfer, retirement, discipline, or some other action. The idea of cooperation in the development of this program does carry some burden of concession for both labor and management if the goal is to be accomplished. It also emphasizes individual responsibility for maintaining one's level of fitness (O'Connor, 1995). By continually referring back to the purpose statement whenever procedural decisions are to be made, both labor and management can ensure that they will make the right choices for the right reasons. Every aspect of the program, beyond agreement with the two initial steps, is subject to negotiation and secondary to getting the project started.

By now, there should be little doubt as to the need for a program, but there are several questions and concerns frequently raised by either labor or management that should be answered before we proceed further. The most obvious are (1) What are the specific concerns of labor and management with respect to implementing the program? (2) Although the need has been established, why should management provide the program? and (3) How is the program to be accomplished in a cooperative manner?

To answer the first question, the principal labor concern is for the job security of the members. This was addressed in the purpose statement, which clearly excludes reference to employment decisions, thereby removing any direct threat to job status. The principal concern of management is how to pay for a program that labor both endorses and requests. In Chapter 6, several alternatives are provided that will resolve this problem, including a possible no-cost option that will nullify any potential threat of stalemate. The priority joint concern, member fitness, will positively contribute to a lower injury rate, enhanced job performance, and career longevity, thereby offering incentive for mutual cooperation. Other concerns, such as fair treatment, program compliance, and so on are easily answered by following the recommendations and options presented in this text and by staying focused on the purpose.

The response to the second question is best given with four reasons: (1) It is beneficial to the citizens, the department, and all of the members, (2) a fitness program is mandated by the unique nature of the job, (3) a fitness program is a necessity that will not interfere with job productivity, and (4) management has a continual obligation to ensure that its members are fit for duty. The latter reason is perhaps the most important of them, since a public responder who is unfit for his job may pose a life threat to his coworkers and the general public. The moral weight of this alone is sufficient justification to support employer-sponsored fitness programs, although liability is also at issue. There is case law *(City of Northglenn and Colo. Comp. Ins. Auth. v. J. Eltrich and the Indst. Claim Appls. Ofc. of State of Colo.)* that holds employers responsible for off-duty exercise injuries if a mandatory fitness program is instituted and on-duty time to exercise is not allowed. To quote one fire chief from a department that has worked on wellness issues for nineteen years, "For us not to invest in their (firefighters') long-term well-being would be foolhardy."

This draws our attention to a long-standing problem. Historically, departments have been remiss in their responsibility for ensuring a physically capable workforce. The fitness status of members wasn't evaluated, and therefore, in the absence of any standard or requirement, the fitness levels of members were allowed to decline. Labor also shares in the blame for having resisted the introduction of programs and standards, for example; for interpreting programs and standards as being punitive; by demanding that candidates and incumbents meet different standards; by requesting grandfathered standards and accepting that minimum standards be established at the level of the worst-fit member (Grieve, 1992; and Rukivana, 1996). Although some objections appear justified, the conflict has only delayed the introduction of many programs.

In *Work Inc.: A Philosophical Inquiry,* author Edmund Byrne presents two work traditions, one described as Authoritarian and the other as Communitarian (Byrne, 1990). These two approaches are at opposite poles. The Authoritarian approach is inherently elitist, with priority placed on the work, above the needs of the worker. The Communitarian approach gives priority to the people and seeks ways to make the task more appealing. The Authoritarian tradition is typically referred to as the "stick" approach, a method that will always justifiably raise the objections of labor groups. Departments that attempt to gain compliance through coercion lose far more than they think they gain. The Communitarian approach can be viewed as the "carrot" method. Although more appealing to the worker, it isn't without flaws, and it can often fail to achieve the desired goal, particularly in a voluntary program.

Historically, the Authoritarian approach appealed to the work ethic, which in Mr. Byrne's opinion, was a motivational form of forced labor that doesn't apply today. Modern workers value work less for its own sake than for some other reward. That other reward in public safety, I prefer to think, is the satisfaction that one receives when he has preserved, protected, or enhanced the lives of those he serves. A battalion chief once commented to his subordinates, "You don't work for me, you work for the citizens of the community." As public safety officers, we do not have the right to pose a threat to fellow members or citizens. As Herbert Spencer (1820–1903) said, "The preservation of health is a duty."

The role and responsibility of management with regard to the fitness program is to provide the tools and the means to achieve that goal. The responsibility of labor is to ensure that management follows the policies and procedures to which they have agreed. The individual's responsibility is to remain fit. If the individual fails to achieve that goal, his failure should not be attributed to either labor or management, but to himself and his own personal decisions. Let it be said that any member who is medically capable of participating in a sound exercise program can meet the accepted standards, and he will only fail out of personal choice. The only threat to job security is the choice to remain unfit, and this threat is self-imposed.

In answering the last question, how to ensure that labor and management cooperate, recognize that a no-nonsense approach is necessary to accomplish that "shared" goal of both camps. That approach requires total participation (Craig, 1979).

Since the threat of losing one's job has been removed, the choice of a mandatory program becomes nonthreatening. It is generally agreed that mandatory programs are more effective than voluntary ones (O'Conner, 1995; Collingwood, 1995; and Williams and Egvenson, March 1988), and labor has repeatedly expressed support for mandatory fitness programs. If a department is committed to investing in fitness and expects a program to succeed, its management should require total participation. The best results cannot be achieved on a voluntary basis. One survey evaluating voluntary participation from police and fire departments reported that 62 percent of the firefighters participated and only 34 percent of the police volunteered. Participation levels for health promotion programs were even lower, reporting 50 percent for fire and 29 percent for police. In the same report, 65 percent of the fire departments had a fitness program, and only 31 percent were mandatory. Of the police departments surveyed, only 29 percent offered a program, with a meager 3 percent considered mandatory (Landy, 1992). Based on this report, the first order of business isn't only to implement a program, but also to mandate total participation. Although mandatory compliance is recommended and preferred, voluntary participation is better than no participation, and it is acceptable in some situations, such as during the phase-in period. Some departments that have volunteer programs have been able to achieve 95 percent compliance based on peer pressure alone (Grieve, 1992). In my view, even that impressive number isn't acceptable, since the missing 5 percent is typically your target group! When making the decision, ask yourself whether fitness for duty is an option. The answer may clarify the point and help you decide.

RECOMMENDED READING

American Heart Association (1997). *1998 Heart and Stroke Statistical Update.* Dallas, TX: American Heart Association.

Byrne, Edmund F. (1990) *Work Inc.: A Philosophical Inquiry.* Philadelphia, PA: Temple University Press.

International Association of Fire Fighters. *Annual Death and Injury Survey.* Washington, DC: IAFF.

Smith, Wayne (1997). *Fire and Police Personnel Reporter: Case Law Digest, 1975-1997.* Americans for Effective Law Enforcement Inc.

Part I References

Adams, T.D.; Yanowitz, F.G.; Chandler, S.; Specht, P.; Lockwood, R.; and Yeh, M.P. (1986) A study to evaluate and promote total fitness among firefighters. *Journal of Sports Medicine.* 26: 337–345.

AELE Workshop (1997). *Discipline and Labor Problems.* Americans for Effective Law Enforcement Inc., Chicago, IL.

Age Discrimination in Employment Act (ADEA), as amended (1996).

Almand, Lloyd (1984). Fitness at last. *Fire Command.* April, pp. 38–41.

American Heart Association (1997). *1998 Heart and Stroke Statistical Update.* Dallas, TX: American Heart Association.

Americans With Disabilities Act of 1990. 42 U.S.C. § 12101 et seq.

Aoki, K.; Suzuki, Y.; Noji, A.; Yanagibori, R.; and Gunji, A. (1990) Analysis of the relationship between physical fitness level and risk factors of cardiovascular disease in workers. Kaneko, Masahiro, editor. *Fitness for the Aged, Disabled, and Industrial Worker: International Series on Sport Sciences, Vol. 20.* Champaign, IL: Human Kinetics, pp. 239–244.

Armstrong v. City of Dallas, 62 FEP Cases (BNA) 852, 997 F.2d 62 (5th Cir.1993).

Bahrke, M.S. (1982) Voluntary and mandatory fitness program for fire fighters. *Physician in Sports Medicine.* 10: 126–132.

Baltic, S., editor (1990). Coping with NFPA 1500. *Fire Chief.* February, pp. 43–47.

Barnard, R.J.; and Duncan, H. (1975) Heart rate and EKG responses of fire fighting. *Journal of Occupational Medicine.* 17: 247–250.

Barnard, R.J.; Gardner, G.W.; Diaco, N.; and Kattus, A.A. (1972) *Ischemic Response to Sudden Strenuous Exercise.* Paper presented at the annual meeting of the ACSM, Philadelphia.

Beam, J., and Davis, P.O. (1979) Payoffs of a physical fitness program. *The International Fire Chief.* June, pp. 13–17.

Beck v. State Dept. of State Police, Case No. A8603–01816, Multonomah Co. Cir. Ct., Ore (11/16/86) (AELE Ref. No. 5139).

Behnke, K., and Antonacci, R. (1995) The case for fitness screenings. *Fire Chief,* February, pp. 28–29.

Bentivoglio, John (1995). OSHA compliance—fireground commanders cited for violation of safety laws. *Firehouse.* May, pp. 58–61.

Blake v. City of Los Angeles, 1979.

Bentivoglio, John T. (1996) Legal limbo: volunteer fire departments and federal discrimination laws, part II—compliance issues. *Fire Engineering,* April.

Brief Times Reporter 663 (Colo. App. 1995). City of Northglenn and Colo. Comp. Ins. Auth. v. J. Eltrich and the Indst. Claim Appls. Ofc. of State of Colo.

Brletic v. Municipality of Monroeville, 440 A.2d 686 (Pa. Cmwlth. 1980).

Byrne, Edmund F. (1990) *Work Inc.: A Philosophical Inquiry.* Philadelphia, PA: Temple University Press, pp. 52–57.

Cady, Lee D.; Thomas, P.; and Karwasky, R. (1985) Program for increasing health and physical fitness of fire fighters. *Journal of Occupational Medicine.* 27 (2) February, 110–114.

Carmean, G. (1984) Putting fitness to the test. *Fire Command.* February, pp. 15–19.

Caspersen, Carl (1987). Physical inactivity and coronary heart disease. *The Physician and Sports Medicine.* 15 (11): 44.

Cassista v. Community Foods Inc., 2 AD Cases (BNA) 1188 (Cal.1993) (93 FP 169–70).

Civil Rights Act of 1964, 42 U.S.C. 2000e, Title VII, as amended.

Civil Rights Act of 1991 (CRA '91).

Civil Service Comsn. v. Penn. Human Rltns. Comsn., 591 A.2d 281 (Pa.1991).

Collingwood, T. R. (1995) Physical fitness standards: measuring job relatedness. *The Police Chief.* February, pp. 31–34; 46–47.

Craig, G.B. (1979) Mandatory physical conditioning standards. *FBI Law Enforcement Bulletin,* March, pp.13–17.

Craley, M.D. (1988) Legal insights. *Chief Fire Executive.* Aug./Sept./Oct., pp.16–17.

Davis, P.O.; Dotson, C.O.; and Santa Maria, D.L.(1982) Relationship between simulated fire fighting tasks and physical performance measures. *Medicine and Science in Sports and Exercise.* 14 (1): 65–71.

Davis, P.O., and Dotson, C.O. (1987) Physiological aspects of firefighting. *Fire Technology.* 23 (3): 280–291.

Davis, P.O. (1998) Taking the stand for fitness standards. *Fire Chief.* August, pp. 36, 38, 40–41.

Dubois, Paul (1984). Promoting physical fitness in the fire service: a statewide effort. *Fire Journal.* May, pp. 51–55.

Dezelan, L.A. (1997) Fire fighter fitness is now on the front burner. *Fire Chief.* February, pp. 56–58.

Diezel, H. (1988) Safety: a look at NFPA 1500. *Fire Chief.* August, pp. 31–32.

Dotson, C.O.; Santa Maria, D.L.; Davis, P.O.; and Schwartz, R.A. (1978) *The Development of Job-Related Physical Performance Examinations for Fire Fighters.* Washington, DC: U.S. Government Printing Office, Stock No. 003–000–00541–1.

Duncan, H.W.; Gardner, G.W.; and Barnard, R.J. (1979) Physiological responses of men working in fire fighting equipment in heat. *Ergonomics.* 40: 482–489.

EEOC v. County of Los Angeles. 31 FEP Cases 1974.

EEOC (1978) *Uniform Guidelines on Employee Selection Procedures.* Washington, DC: U.S. Government Printing Office.

Eltrich v. City of Northglenn and Colorado Compensation Insurance Authority. 19 Brief Times Reporter 663 (Colo App. 1995).

Fahey, Rita (1993). *The NFPA Journal,* March/April, p. 73.

Ferrante v. Niagara County, 551 N.Y.S. 2d 134 (A.D. 1990).

Frost, S., and Durfor, W. (1986) Confidential fitness reports. *Fire Command.* November, pp. 36–37, 42–44.

Gardner, G.W.; Barnard, R.J.; Disco, N.V.; and Kattus, A.A. (1974) *Near-Maximal ECG Stress Testing and CHD Risk Factor Analysis in Los Angeles City Firemen.* A paper presented at the 21st Annual Meeting ACSM, Knoxville, TN, May.

Gilmore, C.P. (1981) *Exercising for Fitness.* Alexandria, VA: Library of Health, Time Life Books, p. 60.

Gledhill, N., and Jamnik, V.K. (1992) Characterization of the physical demands of fire fighting. *Canadian Journal of Sports Medicine.* 17 (3): 207–213.

Grieve, Curt (1992). *Physical Fitness for the Fire Service: Do We Care About the Subject?* Executive Fire Officer, Outstanding Applied Research Awards National Fire Academy, pp. 37–61.

Hall, John (1993). *Memorandum to Gary Tokle and the NFPA 1500 Committee.* Memorandum on the subject of firefighter deaths and injuries and fireground staffing. October 6.

Halfon, S.T., et al.(1990) Daily physical activity program for industrial employees. Kaneko, Masahiro (editor). *Fitness for the Aged, Disabled, and Industrial Worker.* International Series on Sport Sciences, Vol. 20, pp. 260–265. Champaign, IL: Human Kinetics.

Healy, Thomas (1988). We're getting healthier. *Fire Command.* July, pp. 11–13.

Herr, Monty (1998). Occupational illnesses: where is the outrage? *AIHA Journal,* 59: 83–84.

Heyward, V. (1998) *Advanced Fitness Assessment and Exercise Prescription, Third Edition.* Champaign, IL: Human Kinetics.

Hornick, Chris (1997). Physical ability testing and the DOJ. *WFCA Newsletter,* Vol.7, No. 6, pp.13–14.

IAFF (1997). *1996 Death and Injury Survey.* International Association of Fire Fighters. 1750 New York Ave., N.W. Washington, DC 20006.

IAFF (1998). *1997 Death and Injury Survey.* International Association of Fire Fighters. 1750 New York Ave. N.W., Washington, DC 20006.

IAFF Leader (1997). January-February. International Association of Fire Fighters: Washington, DC, p.2.

International Fire Fighter (1996) November-December, Vol. 79, No. 5.

Jones, G.R.; Spitler, D.L.; Swan, P.D.; and Hawkins, J.D. (1989) Health and fitness assessment criteria for public service agencies. *Journal of Physical Education Recreation and Dance.* October, pp. 12–16.

Kalkhoff, R.K.; Hartz, A.; Rupley, D.; Kissebah, A.H.; and Kelber, S. (1983) Relationship of body fat distribution to blood pressure, carbohydrate tolerance, and plasma lipids in healthy obese women. *J. Lab. Clin. Med.* 102: 621–627.

Landy, Frank (1992). Alternatives to Chronological Age in Determining Standards of Suitability for Public Safety Jobs, Volume I. Technical Report. Pennsylvania State University, pp. 6–53.

Lanning v. SEPTA U.S. District, Lexus 9388 (E.D.P.A.), 1998.

LeCuyer, John (1989). Unpublished data. Aurora, CO: Aurora Fire Department.

LeCuyer, John (1994) Where tools and people meet. *Fire Chief.* February, pp. 34–41.

Lemon, P.W.R., and Hermiston, R.T. (1977) The human energy cost of fire fighting. *Journal of Occupational Medicine.* 19 (8): 558–562.

Lusa, Sirpa; Louhevaara, V.; and Kinnunen, K. (1994) Are the job demands on physical work capacity equal for young and aging firefighters? *Journal of Occupational Medicine.* 36 (1): 70–74.

McMillen v. Civil Service Cmsn., 8 Cal. Rptr. 2d 548,1992.

Mealy, M. (1979) New fitness for police and fire fighters. *Physician in Sports Medicine.* 7: 96–100.

Merrimack Co. and State Employees Assn., 102 LA (BNA) 1096 (McCausland, 1994) (95 FP 5).

Minn. Dept. of Human Rts. v. City of St. Paul, No. 8–1700–3224–2; 28 (1380) G.E.R.R. 1143 (M.O.A.H., 1990) and 32 (1556) G.E.R.R. 362 (M.O.A.H., 1994).

Mostardi, R.A.; Porterfield, J.A.; King, S.; Wiedman, K.; and Sherman, B. (1986) Preemployment screening and cardiovascular intervention program. *Journal of Orthopaedic and Sports Physical Therapy,* pp. 42–45.

Mullen v. Princess Anne Vol. Fire Co., 853 F.2d 1130 (4th Cir.), 1988.

Neff, Barbara (1994). Expanding liability of employment discrimination: obesity as a handicap. *For the Defense.* April, pp. 14–18.

NFPA (1997). U.S. fire fighter injuries—1996. *NFPA Journal.* November/December 1997, pp. 67–71.

NFPA 1500 (1997) *Fire Department Occupational Safety and Health Program, 1997 Edition.* Quincy, MA: National Fire Protection Association.

NFPA 1521 (1997) *Fire Department Safety Officer, 1997 Edition.* Quincy, MA: National Fire Protection Association.

NFPA 1582 (1997). *Medical Requirements for Fire Fighters, 1997 Edition.* Quincy, MA: National Fire Protection Association.

Noakes, T.D. (1988) Implications of exercise testing for prediction of athletic performance: a contemporary perspective. *Medicine and Science in Sport and Exercise.* 20 (4): 319–330.

Norris, D.A. (1993) The where and how of firefighter injuries: a long-term study. *Fire Chief.* February, pp. 30–32.

Obermeyer, D.; Harshaw, C.; and Ellis, A. (1995). Steps, pec decks, and fire combat. *Fire Chief.* February, pp. 26–27.

O'Conner, J. (1996). Top-five fitness questions. *Fire News,* April/May, pp. 22–26.

Okada. K., and Iseki, T. (1990) The effects of a ten-year corporate fitness program on employee health. Kaneko, Masahiro (editor). *Fitness for the Aged, Disabled, and Industrial Worker.* International Series on Sport Sciences, Vol. 20, pp. 249–253. Champaign, IL: Human Kinetics.

Occupational Hazards (April 1997). How much do back claims cost? Vol. 59, No. 4, p. 65.

Occupational Safety and Health Act of 1970.

OSHA Compliance Advisor (Feb 10, 1997). *News Roundup.* No. 317.

OSHA Compliance Advisor (April 14, 1997). *Washington Watch.* No. 321.

Paffenbarger, R.S.; Hyde, R.T.; Wing, A.L.; and Hsieh, C.C. (1986) Physical activity, all-cause mortality, and longevity of college alumni. *New England Journal of Medicine.* 314: 605–613.

Parker v. Washington, DC (1998), 850 F. 2d 708 (D.C. Cir. 1998).

Philadelphia Electric Co. v. Commonwealth of Pennsylvania, 68 Pa. Commw. 212, 448 A.2d 701 (1982).

Puterbaugh, James S., and Lawyer, Carl H. (1983) Cardiovascular effects of an exercise program: a controlled study among firemen. *Journal of Occupational Medicine.* 25 (8): 581–586.

Rafilson, Fred (1995). Legislative impact on fire service physical fitness testing. *Fire Engineering.* April, pp. 83–89.

Rodgers, Suzanne, H. (1986) *Ergonomic Design for People at Work, Volume II.* New York: Van Nostrand Reinhold Co., p. 485.

Rukavina, J. (1996) Lex de incendiis: an imperfect tool for gradual change. *Fire Chief.* December, pp.17–18.

Rukavina, J. (1997) Lex de incendiis: the fire service has one more chance. *Fire Chief.* March, pp. 22–24.

Safety Sciences, San Diego, CA. (1982). *A Study of Occupational Injuries and Illnesses Among Fire Fighters: Based on Analysis of Injury/Illness Data From 43 Fire Departments.* Conducted for the Federal Emergency Management Administration, U.S. Fire Administration. Vol. I, May.

Siegrist, J., and Klein D. (1990) Occupational stress and cardiovascular reactivity in blue-collar workers. *Work and Stress,* 4: 294.

Smaw v. Va. State Police, 862 F. Supp. 1469 (D. Va. 1994).

Southman, M.S.; Saupe, K.; Jasenof, D.; and Blaney, J. (1992) Heart rate responses of firefighters to actual emergencies. *Journal of Occupational Medicine,* 34 (8): 797–800.

Smith v. City of Des Moines, Iowa (1996). 99 F.3d 1466, 1472.

Smith, Wayne (1997). *Fire and Police Personnel Reporter: Case Law Digest 1975–1997.* Americans for Effective Law Enforcement Inc.

Staff (1997). *City of Aurora, Colorado, Employee Health Fund.* Winter, Vol. XV, No. 2. Edmonds, WA: Scott Publishing, p. 2.

Stittleburg, P.C. (1988) Coping with NFPA 1500. *Fire Chief.* December, pp. 29–30.

Sukay, Lawrence (1989). Only the beginning. *Chief Fire Executive.* Jan./Feb./Mar., pp 39–43.

United Paramedics of L.A. v. City of Los Angeles, No. 89–1182–R (C.D. Cal. 1989).

U.S. Department of Justice (1998). Law enforcement officers killed and assaulted, 1996. *FBI Uniform Crime Reports.* Washington, DC: U.S. Dept. of Justice, p. 30.

U.S. Department of Labor (1993). *Selected Characteristics of Occupations Defined in the Revised Dictionary of Occupational Titles.* U.S. Dept. of Labor Employment and Training Administration. Washington, DC: U.S. Government Printing Office.

U.S. Department of Labor (1995). *Ready Facts* (816) 426–3152. Bureau of Labor Statistics.

U.S. v. Wichita Falls, 704 F. Supp.709, 711 (N.D. Texas 1988a).

U.S. v. Wichita Falls, Texas, 47 FEP Cases 1629 (N.D. Texas 1988b).

White, James R., and Hunt, Howard F. (1974) University teaches fitness to fire fighters. *International Fire Chief,* April.

Williams, T., and Evenson, S. (March, 1988) Physically fit for duty, part I: mandatory physical fitness programs. *Fire Chief,* pp. 43–46.

Williams, T., and Evenson, S. (April, 1988) Physically fit for duty, part II: sliding-scale fitness standards. *Fire Chief,* pp. 58–61.

Williams, T., and Evenson, S. (May, 1988) Physically fit for duty, part III: enforcing fitness standards. *Fire Chief,* pp. 55–59.

Zamlen v. City of Cleveland, 906 F. 2d 209 (6th Cir. 1990).

PART II DESIGN

Chapter 4

Fitness Defined

The first step in designing a fitness program is to define what we mean by fitness. That the word has a different meaning for different people is evident in public safety occupations, where entry tests are referred to as fitness, agility, ability, criterion, and performance tests, using these terms interchangeably and with no particular concern for the test mode or type. The test developers and "experts" are partially responsible for the confusion. There is no consensus as to the definition of fitness, and there appear to be as many variations as there are texts on the subject.

A definition offered in *Mosby's Medical and Nursing Dictionary* (Mosby, 1994) seems to incorporate a few of the phrases most frequently heard.

> *"The ability to carry out daily tasks with alertness and vigor, without undue fatigue, and with enough reserve to meet emergencies or to enjoy leisure-time pursuits."*

By this definition, the unfit are those who become fatigued and recover slowly. But could the average couch potato presumably become more fit by reducing his activity level, thereby increasing his resistance to fatigue? (Morehouse and Miller, 1963.) Something is missing. The American College of Sports Medicine has offered two versions in its *Resource Manual* (Gettman, 1993).

> *"A set of attributes that people have or achieve that relates to the ability to perform physical activity."*

> *"A state characterized by (a) an ability to perform daily activities with vigor, and (b) the demonstration of traits and capacities that are associated with low risk of premature development of hypokinetic diseases."*

The definition in the *StairMaster Fitness Handbook* (StairMaster, 1995) includes the health aspect of the second ACSM version, plus fitness components.

> *"The physiological contribution to wellness through exercise and nutrition behaviors that maintain high aerobic capacity, balanced body composition, and adequate strength and flexibility to minimize risk of chronic health problems and to enhance the enjoyment of life."*

The American Alliance for Health, Physical Education, Recreation, and Dance (AAHPERD) defines fitness as having four components, namely (1) cardiorespiratory endurance, (2) muscular strength and muscular endurance, (3) flexibility, and (4) body composition. This is the most popular approach—to define fitness by its parts. Although some experts question the inclusion of flexibility, and some prefer five components, separating muscular endurance from muscular strength, most agree on three general categories, namely (1) energy fitness, (2) muscular fitness, and (3) body composition.

Although the component model is used throughout this text, it is in opposition to my personal philosophy, which is to define a construct such as fitness by its parts. The Cartesian model, which declares that a whole is the sum of its parts, fails to account for a variety of unknown factors that often disprove the conclusion. This approach—to take it apart, analyze it, quantify it, categorize it, then reassemble it to arrive at a conclusion—works well for machinery but poorly when applied to physiology. So why use the component model at all? The answer is that using it is necessary to achieve our intended purpose. Evaluation is an essential part of the program, and since there are no general, all-encompassing tests of fitness available, each component must be evaluated independently. This unit-by-unit analysis will prove useful when we discuss fitness in relation to activity and exercise programming. In Chapter 12, I will present a scoring method that puts the pieces back together and produces a more accurate composite picture. Additionally, this particular approach is widely accepted by the legal profession, where fitness issues are frequently debated.

The word activity appears or is implied in all of the definitions listed. Perhaps the missing link in our definition is related to that term. The average person may not be able to render a precise definition of fitness, but all of us know what it looks like. We know by appearance and performance. To use pieces of the previous definitions, fitness implies a relationship between the individual's capacities and activities. Are you fit? One must know the context before being able to answer the question. If a warehouse job requires the lifting of 100-pound boxes, and if an applicant with the aerobic capacity of an elite marathon runner ($\dot{V}O_2$ >70) lacks the strength to lift the load, then he would have a work capacity of zero and therefore be classified as unfit. The limiting factor in this case is strength, thereby overruling high aerobic capacity as irrelevant. Similarly, a strength-trained Olympic weight lifter could also be classified as unfit when charged with the task of having to compete in a 10,000-meter race. Fitness is relative to the task; in fact, fitness is defined by the task.

The present trend is to classify the fitness components cardiorespiratory endurance, muscular strength, muscular endurance, flexibility, and body composition as health-related fitness. The fitness components not included in our definition are as shown in Figure 4–1. Power, speed, agility, coordination, balance, and kinesthetic awareness are cate-

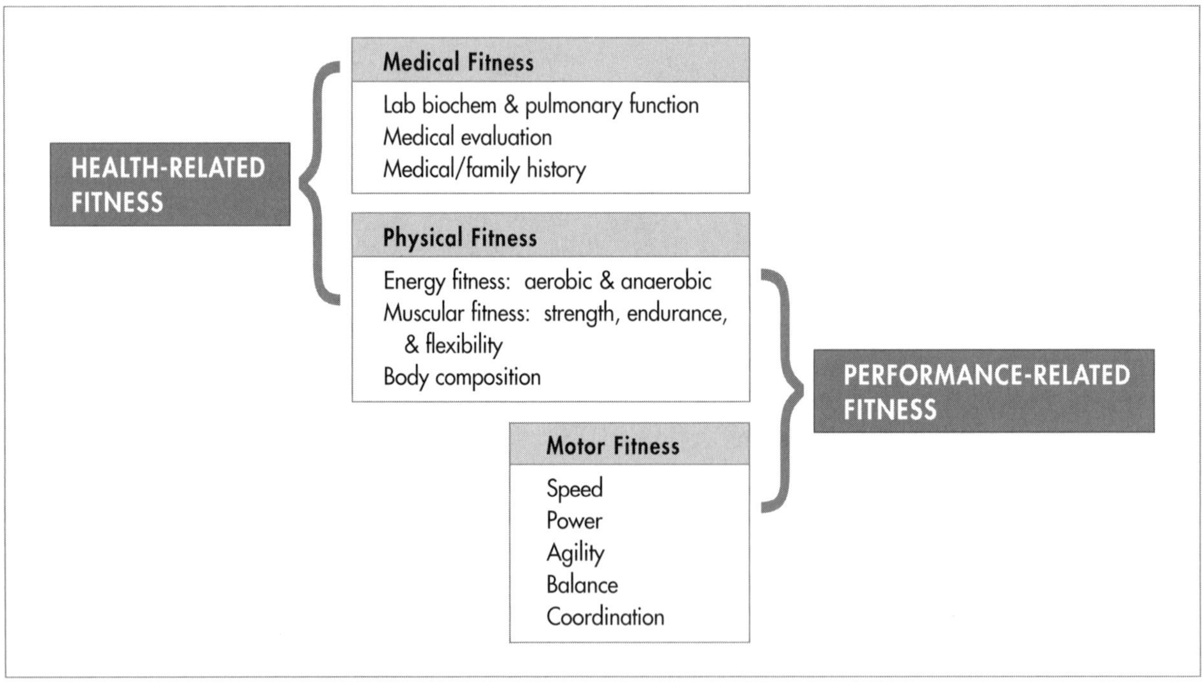

Fig. 4–1 Physical Fitness Diagram

gorized as performance-related fitness, or motor fitness. The motor-related components are necessary for successful job performance and are not regarded by most experts as being modifiable. The health-related components can be modified and are highly responsive to exercise; therefore, these should be the target of our program. For the purposes of this text, we will use the three categories of physical fitness given in our operational definition: energy fitness, muscular fitness, and body composition.

ENERGY FITNESS

Energy fitness refers to the body's ability to generate the energy needed for muscle activity through a system of energy transfer. There are three systems of energy transfer. Two are anaerobic and one is aerobic. The body's energy source comes from the breakdown of nutrients (carbohydrates, fat, and protein) through an enzymatic process to form adenosine triphosphate (ATP). The potential energy within ATP is used for all cellular processes that require energy. The contribution of energy systems and the fuel used vary according to the intensity and duration of the effort, as well as the physical condition of the muscle and the muscle fiber type.

At rest, we predominantly use fat as a fuel source. When we move from a resting position to a walk, slowly increasing our pace to a jog, and then running with an all-out, high-intensity sprint effort, the energy substrate switches from fat to a fat-carbohydrate mix to pure glycogen (stored carbohydrate), as shown in Figure 4–2.

There are three energy systems at work. The ATP-PC (phosphate) system supplies immediate energy for a few seconds, for example, while moving from rest to a quick-run start. The glycolytic system supplies energy for up to four minutes. The aerobic system supports effort lasting beyond five minutes. The first two systems are anaerobic, meaning without oxygen; the third is aerobic, meaning with oxygen. We are never completely in one system or another—the systems are in flux, but at the extremes, one system dominates. The point at which we make the transition from aerobic to anaer-

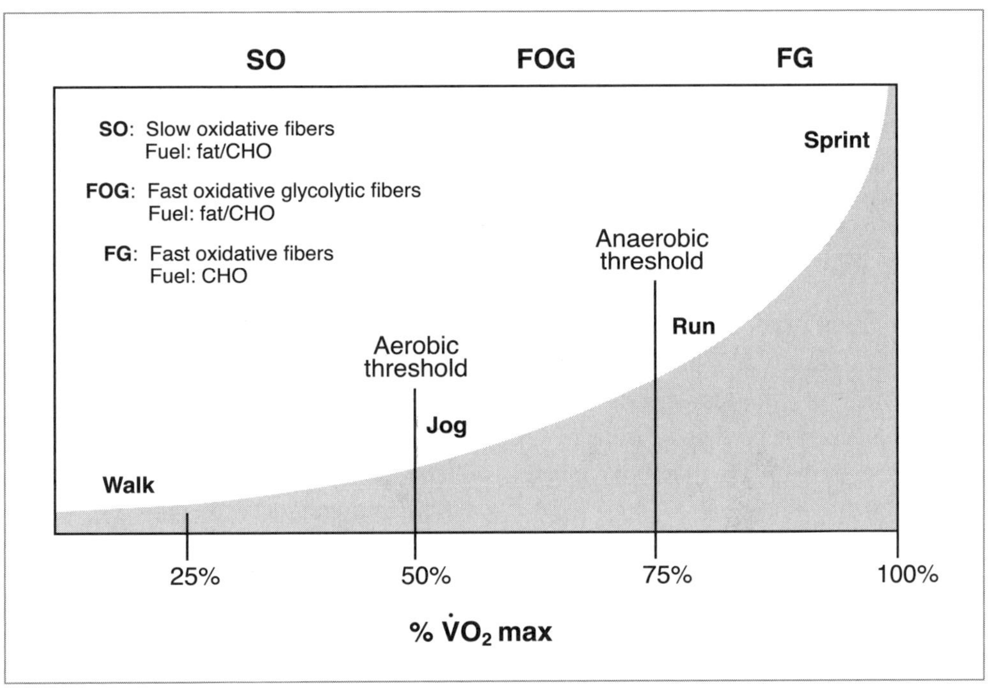

Fig. 4–2 Relationship Between Energy Substrate Muscle Fiber and Intensity
Source: Adapted from Sharkey, B.J. (1997) *Fitness and Work Capacity, Second Edition.* USDA Forest Service, Missoula, MT, p. 6. Reprinted by permission.

obic support is called the anaerobic threshold. The preferred term for this is the lactate threshold. Blood lactate or lactic acid is a by-product of anaerobic metabolism. As lactic acid accumulates in the blood in response to intense effort, it initiates a response that forces one to slow down or stop. When the intensity of the work is reduced, the pathway is switched from one that is anaerobic to one that is aerobic, introducing oxygen, which is necessary to dissipate the lactic acid and continue the activity.

The transition from a low-intensity activity to higher intensity levels results in an energy system transfer, a change in fuel source, and the recruitment of different muscle fibers. There are three basic fiber types: (1) slow twitch, an oxidative type (SO), (2) a fast twitch, both oxidative and glycolytic (FOG), and (3) a fast twitch that is primarily glycolytic (FG). The different fibers are recruited in response to the energy substrate and the energy system employed. At low-intensity levels (walk), the aerobic system dominates, fat is the principle fuel, and the SO fibers are activated. At moderate-intensity levels (jog, run), the aerobic system remains dominant, carbohydrate and fat are the fuel sources, and the FOG fibers are recruited. At the high-intensity level (run), we may cross the lactate threshold using both aerobic (run) and anaerobic (sprint) systems, using carbohydrate (stored glycogen) and fat, while recruiting FG fibers.

AEROBIC METABOLISM

Only a small quantity of ATP is stored within the cell, and the breakdown of nutrients is necessary to provide energy for the resynthesis of ATP. Although oxygen does not participate directly in the synthesis of ATP, it enters the metabolic mill at the end of an oxidative process to produce thirty-eight ATP. The ability of the aerobic system to produce a greater quantity of high-energy compounds (thirty-eight ATP versus two ATP anaerobic) makes it a more efficient system than the anaerobic process, allowing sustained activity for an extended period of time. The aerobic system can use all three fuel sources for the resynthesis of ATP. Although carbohydrate is used initially (carbohydrate must be present to use fat), the preferred fuel of the aerobic system is fat.

Although the aerobic system is the most efficient method of producing energy, the energy system activated is dependent on the intensity of the activity and dictated by the specific demands of the task. We do not consciously select a system. If the task demands a brief all-out effort, such as forcefully swinging an ax, charging up a stairway, or forcing a door, we cannot choose to perform these tasks aerobically, since they are primarily anaerobic events. The importance of the aerobic system in such activities is the ability of the body to recover, getting rid of lactic acid, so that we can repeat or continue the activity. In long-duration tasks, such as during overhaul and salvage operations, a firefighter will rely predominantly on the aerobic system for energy. During these low-key operations, a firefighter, given the opportunity to do so, can and will self-regulate his pace. Unfortunately, this isn't always possible. The critical nature of a task may set the pace for him. There is a common misconception among some firefighters that, by working at a slower rate, they can work longer (work aerobically). This is true if the situation affords the option. Some tasks must be completed at higher levels of effort, and the desire to conserve energy by working at a slower pace may be impractical and less efficient (Morehouse and Miller, 1963). Still, given the limited supply of available energy from anaerobic sources, oxygen via the aerobic system is essential for continued activity. The ability to work aerobically at higher levels of intensity (having a high lactate threshold) and avoid reliance on the anaerobic system is a key benefit of aerobic training, in addition to improved recovery time and enhanced thermal regulation, which is of extreme importance to a firefighter.

The ability to work aerobically is of course determined by the task and the individual's capacity, as well as the individual's physical size and strength. In one physical performance test that I administered, a large firefighter (270 lbs.) performed a simulated victim rescue, dragging a 175-pound dummy with one hand, walking as if he were pulling a grocery cart. In other words, this specific task for him was easy, giving the appearance of an aerobic walk even though the anaerobic system was contributing a fair percentage of the time. A smaller firefighter (150 lbs.) grasped the same mannequin with both arms, moved slowly, and paused every fifteen to twenty seconds. For this person, the task was 98 percent anaerobic, since he had to pause to recover aerobically between the efforts. The smaller firefighter had a higher aerobic capacity than the larger man, but the intensity of the effort forced him to work anaerobically. This would be typical for most of us, but it serves to point out the limitation in thinking that aerobic capacity is the answer. Without question, it is very important; however, its importance hinges on a number of variables, such as body size, strength, workload, and pace. In a separate test, another large firefighter (225 lbs.) practiced the individual events and performed each one with relative ease. Following the practice session, he proceeded to perform the complete test battery in continuous fashion, as required. He faltered after each of the first three events, then eventually resigned in fatigue. In this case, the firefighter had the size, strength, and anaerobic capacity, but he didn't have the aerobic conditioning necessary to recover and repeatedly perform anaerobic activities. He lacked the aerobic capacity to effectively remove blood lactate between events (aerobic recovery), and his lactate level continued to rise, forcing him to quit.

Firefighters as a group have been categorized as average in aerobic capacity when compared with the general population, although test results from some departments prove otherwise (Davis, 1995). I contend that the lower values reported by some investigators are not indicative of poor conditioning, but rather, are more reflective of job demands. People adapt physiologically in response to work demands. If a job demands high aerobic involvement, workers will respond by demonstrating high aerobic capacities. As in most occupations requiring heavy manual labor, the anaerobic system predominates in most tasks, and the development of this energy system will predominate. During day-to-day operations, the aerobic system isn't taxed sufficiently enough for firefighters to respond with an increase in aerobic capacity, nor demonstrate high levels when tested. Although desirable, the job simply doesn't require a high level of aerobic capacity. (Wildland firefighters do require a higher aerobic capacity than their structure counterparts.)

The aerobic system is the most efficient, yielding thirty-eight ATP versus two ATP anaerobic. It produces less lactic acid, and it removes the lactate produced by anaerobic activity. If one's aerobic capacity is high, one can work at higher levels of intensity, delaying the involvement of the anaerobic system. Superior aerobic capacity also provides for quicker recovery between anaerobic tasks, thereby extending the work effort. Of significant importance to firefighters is the relationship of aerobic capacity (higher is better) to thermal regulation, which enhances the dissipation of internal metabolic heat production. Although somewhat less encumbered, a police officer in a dark uniform with a protective vest or motorcycle attire on a hot day can experience the same problem. Unlike a firefighter, the police officer can't remove these protective garments until the end of the work shift. A program designed to develop the aerobic system commensurate with job duties is essential.

ANAEROBIC METABOLISM

The ATP-PC (phosphate) system provides immediate energy from muscle stores. It's a nonaerobic process that uses phosphates to form ATP. This energy is available for about ten seconds. For activity to continue beyond and up to about four minutes, the body relies on glycolysis to produce ATP from glucose (stored glycogen). This process is also nonaerobic and produces two ATP. The anaerobic system is essential in the performance of high-intensity, short-duration activities.

In occupations requiring heavy work, such as firefighting, the ability to perform successfully is often determined by one's anaerobic capacity. This is essential in the performance of most firefighting and police tasks, particularly those critical tasks that require high-intensity effort. The ability to perform anaerobic work is directly related to muscle size (which determines ATP storage capacity), fiber type (fast twitch), and the condition of the muscle (trained anaerobic). The ability to sustain arduous work depends on the individual's aerobic/anaerobic capacity relative to the metabolic demands of the tasks. The anaerobic/aerobic continuum in Figure 4–3 depicts the relative contribution of both energy systems in public safety activities.

The anaerobic system predominates in high-intensity, short-duration tasks. A focus on improving this system using specific anaerobic and strength-training methods will enhance job-related performance for all public safety officers (Gledhill and Jamnik, 1992; and Marcinik et al., 1989).

MUSCULAR FITNESS

Muscular fitness includes both the health-related components and the motor-related components. The health-related components include strength, endurance, and flexibility. The motor-related components include power, speed, agility, coordination, balance, and kinesthetic awareness.

The motor-related components, for practical purposes, are considered nonmodifiable, although power, for instance, depends on strength, which can be improved, as well as muscle speed, which is genetic and cannot be improved. Most occupations do not require much more than average ability for the motor-related components; therefore, these are presented in brief. The health-related components are the most responsive to training. They have the greatest impact on performance; thus, they should be the focus of our fitness program.

HEALTH-RELATED COMPONENTS

Strength can be defined as the degree to which a muscle or muscle group can apply maximal force against a resistance. Strength is the single best predictor of job performance in occupations that require heavy manual labor, firefighting in particular. If only one fitness component were to be selected for training, it should be strength (Noakes, T.D., 1988; LeCuyer, J., 1988; and O'Conner, 1996). Strength provides the foundation on which power and endurance are built (Hickson et al., 1988).

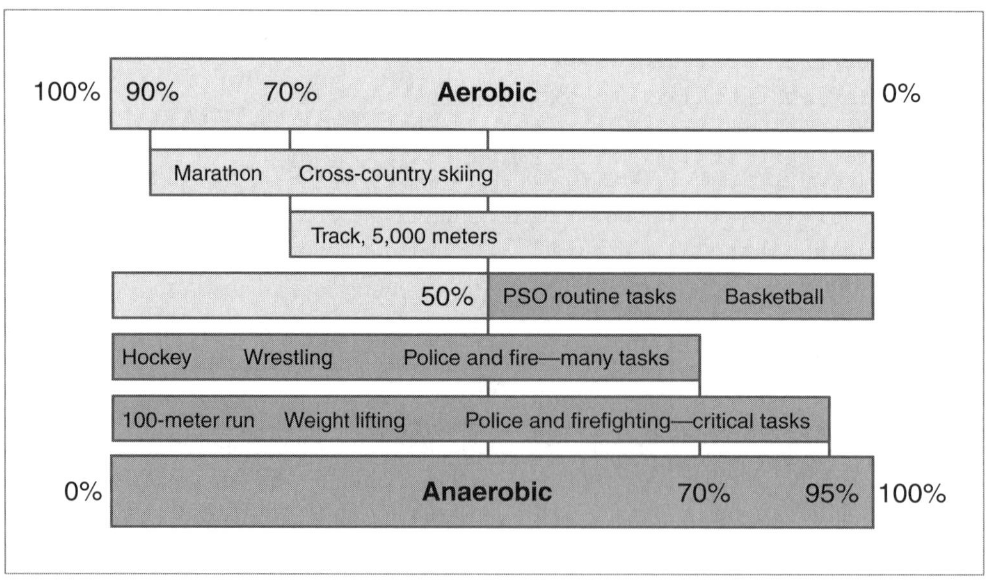

Fig. 4–3 Aerobic/Anaerobic Continuum

As a group, firefighters are above average in strength when compared to the general population. The Dictionary of Occupational Titles classifies the work of firefighting as "very heavy," defined as the ability to lift 100 pounds or more as a minimum strength requirement. As stated earlier, firefighting requires a significant involvement of the anaerobic system, and an individual's strength is directly related to his anaerobic capacity. A firefighter with high muscular strength can sustain anaerobic work longer than others with lower strength levels. Strong firefighters can work more effectively, efficiently, safely, easily, and at a lower stress level, reducing their chances of overexertion injuries. Given that 80 percent of firefighter injuries are associated with the muscular system, improving muscular fitness, particularly strength, should be the first priority.

Muscular endurance can be defined as the ability to sustain repeated submaximal contractions over a period of time. Although muscular endurance differs from muscular strength, it is directly related to strength in that strength provides the basis for endurance development (deVries, 1974).

The importance of muscular endurance is in the ability to work for an extended period of time. This applies most to those long-term scenarios, in which extensive overhaul or a continued defensive fire attack is necessary. In such situations, proper rehabilitation procedures should be implemented. Firefighters should not be required to work for extended periods without rest and recuperation. Because of rehab protocols, the endurance factor indigenous to firefighting is different from that of occupations that require continuous submaximal efforts during the course of a normal workday. Therefore, I contend that if our efforts are focused on developing strength, the endurance issue will be a nonfactor. In other words, muscular endurance training is secondary to strength training. For example, if two firefighters are required to perform a repetitive task, such as stowing 50-pound hose rolls on a rack, the stronger firefighter will persist longer without fatigue than the weaker one. Firefighter A is capable of lifting 200 pounds in this overhead motion, and Firefighter B is capable of lifting 100 pounds in the same operation. Since the load is 50 pounds, Firefighter A is working at 25 percent of his maximal capacity, and Firefighter B is working at 50 percent. In this situation, Firefighter A can be expected to work longer and demonstrate greater muscular endurance due to the lower relative workload. The relationship to strength becomes obvious. If both firefighters were to work at 50 percent of the maximum (Firefighter A lifting 100 pounds), the relationship to strength would become insignificant. Theoretically, both would demonstrate similar endurance capabilities; however, Firefighter A would be capable of performing twice as much work.

In the first scenario, with both firefighters handling the same weight, the correlation between strength and endurance is high. This is reflective of the real world. All firefighters and police officers wear the same personal protective equipment and therefore carry the same weight. The practical application of endurance relative to public safety can easily be addressed through strength development, which will produce adequate muscular endurance relative to occupational demands.

Flexibility may be defined as the range of motion (ROM) available in a joint or group of joints. There are three types of flexibility: (1) static, or ROM without speed, meaning slow, (2) ballistic, or a fast, rebounding motion, and (3) dynamic, or movement as it relates to performance or activity throughout the speed range. There is no such thing as a flexible person. One may exhibit extreme ROM in the low-back muscle group, yet have limited ROM in the shoulder girdle. Specific to each joint or group, flexibility is associated with improved muscular efficiency and reduced risk of injury. At both extremes, the range from "tight" to "loose," the potential for injury increases (Alter, 1996).

Flexibility is not only specific to a joint or group of joints, however, for it is also specific to the activity. Gymnastics typically require great flexibility combined with a high level of strength. The hypermobile gymnast must be strong to protect any lax joint from injury. The runner must focus on increasing the ROM of the hamstring group, which never fully extends during running. In doing so, the runner reduces the flexibility of that group, which could lead to injury. Specific flexibility patterns are often developed in response to various activities. Public safety occupations require good flexibility, particularly in the lower back, and the high incidence of back injuries among firefighters and police officers is often attributed to flexibility limitations and strength. Although the verdict is still out, I have found significant limitations in the low-back/hamstring group to be characteristic among members who suffer serious back injuries. I have tested hundreds of firefighters and have determined that the classic flexibility profile is a loose back and tight hamstrings; i.e., firefighters exhibit good ROM in the back but somewhat less than desirable ROM in the hamstring group. To reduce back injuries, the firefighter must strengthen the loose group (lower back) and stretch the tight group (hamstring). I believe that the loose-back, tight-hamstring combination is reflective of an adaptation to occupational demands, and the imbalance leaves one vulnerable to injury. Since the most frequent injury experienced by public safety officials is associated with the lower back, both strength training and flexibility exercises are recommended. Exercises such as abdominal crunches and lying-back extensions can help strengthen the midsection and improve flexibility in this region, as shown in Figure 4–4.

Since police and firefighters are infrequently required to perform at high levels of exertion, a predominantly sedentary work environment contributes to a higher likelihood of overexertion injury when finally called to respond. The majority of calls that require high exertion occur in the late evening and early morning hours, when flexibility levels are reduced. When a late-night alarm is received, there is little or no time to prepare oneself physically prior to responding. Police officers have some advantage in that they aren't asleep when called. The firefighter, however, often responds from a state of sleep with no time to stretch before responding. Although some activity can be incorporated into the response, such as torso twists, knee-ups, and back stretches while seated in the apparatus, the best approach is to practice and maintain a high level of flexibility throughout the shift. The hip-flexor stretch is one of the most important exercises that will help reduce the susceptibility of the lower back to injury. The stretch is particularly important for police officers seated in a vehicle all day long. The hip-flexor shortens while sitting, reducing the range of motion, as depicted in Figure 4–5. A basic stretch routine is provided in Appendix A.

Fig. 4–4 Crunch-Style Sit-Up and Back Raise

Fig. 4–5 Hip-Flexor Stretch

MOTOR-RELATED COMPONENTS

Power is strength with speed, and it is defined as force multiplied by distance, then divided by time. An increase in either strength or speed will increase power and produce more work in less time. The need for power is limited to sports and occupations that require manual labor. Its application for police is in forcing doors and manhandling suspects during arrest. Its application in firefighting is usually within the context of using lightweight tools. The ventilation of a roof using an ax requires a forceful swing to penetrate the material. An ax that is swung without authority will require a greater amount of time to complete the task, if the arms wielding it don't first succumb to fatigue. Charging up a stairway while carrying a load, forcing a door, penetrating a ceiling with a pike pole, or merely starting a power saw are all common tasks that require a power effort. Strength is the basis of power, and any improvement in the power dimension can be derived from a properly designed strength-training program.

Speed is defined relative to both body movement and reaction time. Body movement refers to how quickly one can move his body, as when running a 40-yard dash, or move a body part, as when throwing a baseball. Reaction time is the interval between the stimulus and the response; between hearing the starting gun and launching out of the blocks. In a track-and-field running event, both body movement speed and reaction time are inseparable. In considering the personal protective equipment worn by firefighters, it is difficult to relate to the concept of speed. Although reaction time may save your life in a given situation, it isn't a foremost component of fireground tasks. In the case of police officers, reaction time and speed are very important, and they are often discussed in training videos. The most obvious use of speed for a police officer is during a foot chase. Reaction time is crucial when drawing a weapon or reacting to a suspect's movements when physically engaged.

Agility is a popular term in the realm of physical testing. It refers to the ability to move and change directions rapidly. Agility is related to skill, and to a degree, it is dependent on other dimensions of fitness. Although the firefighting environment doesn't lend itself to agile maneuvers, firefighters have been known to perform great feats in the face of impending doom. Functioning in a hostile environment while wearing clumsy clothing does require some agility, however. Agility is very important for police officers, and consideration should be given to it in any police fitness test battery.

Coordination is highly related to skill, and it involves the integration of several movements to produce a smooth action. Ground-ladder drills require a coordinated effort. When executed properly, firefighters can make this clumsy task appear effortless. Similarly, handcuffing a suspect requires some coordination, particularly if the suspect resists.

Balance is described as both static and dynamic. Static balance refers to the ability to hold a stationary position, as when standing on one foot. Dynamic balance is the ability to maintain balance while in motion. The semicircular canals of the inner ear assist with the coordination of visual and muscular sensations so that one can maintain balance (deVries, 1974). The ability to maintain one's balance during firefighting operations is extremely important when we consider that slips and falls account for more firefighter injuries than any other cause. Falls are also frequently reported among police officers who sustain injuries while giving chase on foot and making forcible arrests.

Kinesthetic awareness refers to spatial awareness of the body in space. It is sometimes referred to as the sixth sense, or muscle sense. This awareness of the orientation of the body, as well as the limbs, is dependent on muscle receptors, providing a feel for movement. Since tests of this dimension often involve blindfolding, the best correlate in firefighting is a search-and-rescue drill while wearing a blacked-out mask, commonly used in the training of recruits. Some candidates simply don't have a developed awareness of their orientation in space. This occupation offers fertile ground for kinesthetic research that may benefit firefighters.

Muscular fitness is paramount. As stated in our definition of fitness, all is relative to the status of this system.

BODY COMPOSITON

For research purposes, a scientist will typically analyze body composition using a four-component model of water, bone, protein tissue, and fat. For the purposes of a fitness program, we will use a two-component model of fat mass (FM) and fat-free mass (FFM) (Williams, 1988).

The FM component consists of essential fat and stored fat. The Behnke two-compartment model included essential fat in the FFM compartment and referred to this compartment as lean body weight (LBW) (Wilmore and Costill, 1994). Since it is virtually impossible to differentiate between essential and nonessential fat, the Brozek and Siri model has gained favor (Williams, 1983; Siri, 1961). This model includes the essential fat in the FM compartment. The FFM consists primarily of muscle but also includes the organs and bone. The body is about 60 percent water. The FFM compartment contains approximately 70 percent of the water, whereas the FM compartment contains about 10 percent water. A comparison of the percentage of total body weight attributed to muscle, bone, and fat for the average adult male and female appears in Table 4–1.

TABLE 4–1

BODY COMPOSITION ANALYSIS

	Adult Male	Adult Female
Muscle	43%	36%
Bone	15%	12%
Total fat	15%	26%
Essential fat	3%	15%
Stored fat	12%	11%
Other tissue	27%	26%
Total %	100%	100%

Source: Williams, Melvin H. (1983), *Nutrition for Fitness and Sport, Second Edition*. Dubuque, IA: Wm. C. Brown Publishers, reprinted with permission.

The most notable difference between male and female is the percentage of FM. The female has a much larger percentage of essential fat that is classified as gender-specific, and therefore will have a higher total amount of fat proportionally as compared to the male. The additional essential fat is necessary for a female to maintain proper health; therefore, ideal body fat recommendations are higher for women. The differences in FFM between the male and female are primarily due to muscle mass and physical size.

Body composition, then, is expressed as a percentage of total body mass that is FM. A 180-pound individual at 15 percent fat would have 27 pounds of FM and 153 pounds of FFM. The individual's body composition, expressed as a percentage of fat, is perhaps the single most influential factor in determining one's overall fitness level, for it affects all three areas: the health-related components, performance-related components, and one's medical health status.

The most obvious effect is the negative influence of excess body fat on performance (Davis and Starck, 1980). Many studies have reported a decline in performance associated with higher body fat estimates. Although body composition isn't a performance measure in and of itself, it affects performance on all measures of fitness: strength, aerobic capacity, flexibility, and all motor skills. Excess body fat also serves as an insulator, thereby inhibiting the body's ability to cool itself, which lowers the threshold for heat injury.

A 220-pound firefighter at 25 percent body fat is carrying an extra 55 pounds of dead weight. That extra weight is equal to the total weight of his PPE, which has been determined to reduce work capacity by almost 30 percent. If that same 220-pound firefighter were to reduce his body fat to a more acceptable level of 15 percent, he would remove 22 pounds of dead weight—the equivalent of a lightweight SCBA. Without question, any firefighter can relate to the difference between working with and without SCBA.

Although the effect on job performance is important, perhaps the most important detriment of excess body fat is its association with the development of CHD, diabetes, and hypertension. No other fitness dimension affects one's health status more than body composition, which is strongly related to all CHD risk factors (K. Aoki et al., 1990).

FITNESS, HEALTH, AND PERFORMANCE

The emphasis of most fitness programs has been directed toward the health end of the spectrum, with the use of the term health-related fitness becoming more prevalent. We are concerned about the health status of our members, and we should make every effort to supplement the basic fitness program with health promotion and wellness activities, but we should also not overlook the nature of our occupation and inadvertently ignore the performance aspect. Many physical-task firefighting tests have been challenged on the basis that they do not assess the reportedly more important dimension of aerobic capacity (Williams and Evenson, March, 1988). These challenges, although confused and inappropriate for the test format, exemplify the fitness versus performance conundrum. Historically, aerobic exercise was not prescribed for conditioning purposes; rather, it was intended to address the health problem of coronary heart disease. The incidence of CHD in the fire service, although classified as an occupational disease, is comparable in terms of rate to police officers and that of the general population. Heart disease may predispose one to sudden incapacitation, but the physiological demands of firefighting, that infrequent but sudden maximal effort, is the true culprit. Fitness training, which includes the aerobic component, must be responsive to occupational demands in addition to basic health concerns. It is difficult to separate fitness and health, or fitness from performance, but disuniting them isn't necessary. If we intend to address the death and injury problem of public safety officers, then we must respond appropriately and focus on the fitness-performance relationship.

While discussing the fitness components, we stated that aerobic capacity is related to the ability of muscle to utilize oxygen; that strength is related to the contractibility of muscle; that endurance is the capacity for repeated submaximal muscular contractions; and that flexibility is related to the elasticity of muscle. The common denominator is muscle, the source of movement and metabolism. Therefore, I shall offer one last possible definition of fitness. Physical fitness can be defined as "the state or condition of the muscular system relative to the demands of the activity." The goal of our fitness program is to develop and improve that muscular system in response to those demands.

RECOMMENDED READING

Sharkey, B.J. (1997), *Fitness and Health, Fourth Edition.* Champaign, IL: Human Kinetics.

CHAPTER 5

BASIC DESIGN PLAN

The health and safety program director should be assigned the responsibility for designing and developing the fitness program and integrating it into the department's health and safety program. He may choose to form a task group to assist in the planning phase, assigning tasks based on goals and objectives derived from the design matrix. The program director may solicit ideas and opinions from the safety committee, the task group, and other department members, but he should avoid getting bogged down with petty details and instead maintain overall control of the process.

The program director, manager, and fitness coordinator may be one person or three, depending on available staff. If you are the designated program director, you will need to decide on a program model, select the one presented, or develop your own, based on the structure of your department. This will help you organize and develop a strategic plan. The strategic plan follows the outline of this text: analysis, design, development, implementation, and evaluation.

PROGRAM MODEL

Both OSHA and the NFPA have acknowledged that there are many ways to design a health and safety program. NFPA 1500, *Standard on Fire Department Occupational Safety and Health Program,* has survived more than a decade of scrutiny, including three revisions, soliciting input from hundreds of professionals around the world. It is the NFPA consensus process that provides the best argument for selecting and using the standard as your guide. The 1500 model is the most logical choice, since it allows for flexibility in design and application. Although the NFPA standard model targets fire departments, the basic model is appropriate for police departments as well. The diagram in Figure 5–1 presents one model of design that I believe is an effective way to structure the program and separate specific areas. The right branch of the diagram depicts the health component and its relationship to the total program. The subject of this book includes everything listed within the lineage emanating under the heading of "Health." Departments may choose to modify the diagram to suit their individual needs based on available resources and staff. A larger department may separate the individual programs further, expanding the model, whereas a smaller department may find that a condensed version is more suited to its needs.

It is neither desirable nor recommended that you separate health from safety. These two areas are interdependent, and actions taken on the safety side affect health, and vice versa. Although they are connected, however, it is advisable to assign branch responsibilities to separate individuals, based on their roles and expertise. The chief officer who has strong fireground skills and a firm grasp of the incident management system may flounder when confronted with medical issues presented in the health program. Similarly, an officer with a health-related

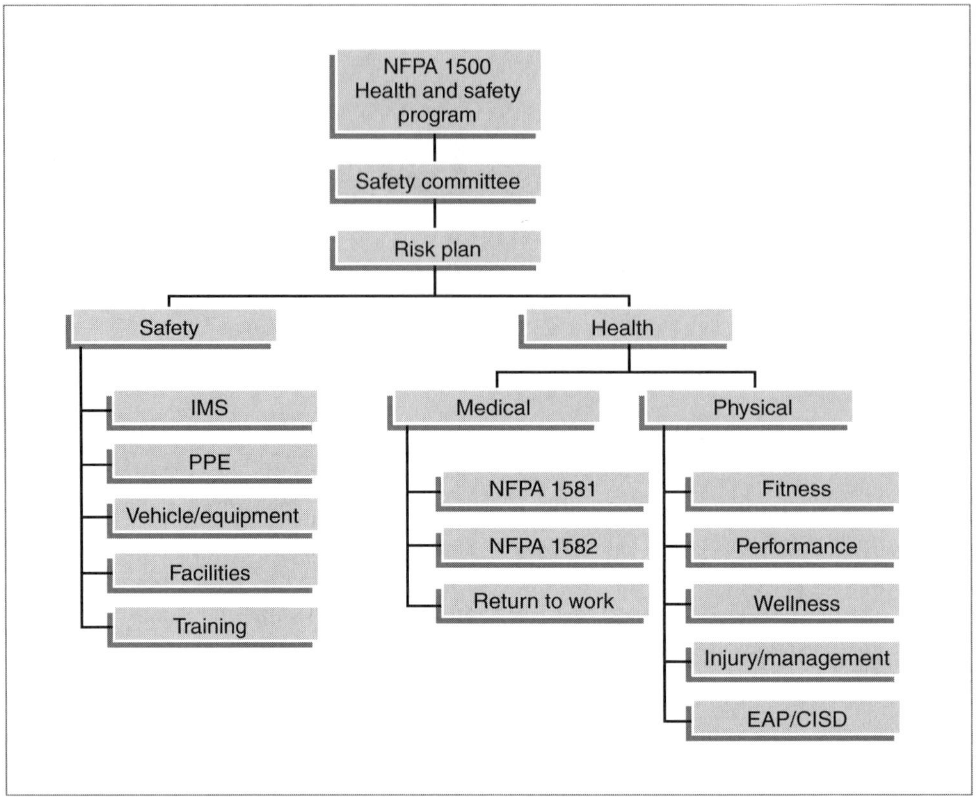

Fig. 5–1 An NFPA Program Model

background might be uncomfortable with the safety role. These are two separate roles that demand careful thought when assigning personnel. The NFPA committee that produced the safety officer standard, NFPA 1521, *Fire Department Safety Officer, 1997 Edition,* recognized the need for this division and therefore designated a health and safety officer (HSO) and an incident safety officer (ISO) with separate roles and responsibilities.

The model presented here deviates from 1521 in that the ISO is assigned responsibilities beyond those indigenous to operations, such as PPE, respiratory protection, training functions, tools, equipment, and facilities. The ISO role is best assumed by the department training program director. The HSO is then assigned the balance, as described in NFPA 1500. This would include medical exams, physical fitness, physical performance, employee assistance programs (EAP), wellness, critical incident stress debriefing (CISD), injury/illness rehabilitation, and infection control. The HSO would work under the supervision of the health and safety program director with the department physician and in liaison with all medical service providers, such as physical therapists, occupational therapists, physicians, nurses, psychologists, and CISD teams. Both health and safety positions would interact with department administrators, city attorneys, risk managers, and human resources. In a smaller department, both roles may be assumed by one individual.

THE SAFETY COMMITTEE

The HSO and the ISO both serve on the department health and safety committee. Some program consultants recommend the organization of a health and fitness committee when developing a fitness program. I see little need for yet another committee, but you could choose to form a task group for startup purposes, then dissolve the group once the projects have been completed. There should be no more than two or three people involved with the basic program design at this stage. Once the program has been implemented, the department's safety committee should become involved, and the fitness program should be integrated with the department safety program. The key issues related to health, fitness, and safety can now be handled within this single-committee format. For instance, the respiratory protection program manager would need to coordinate with the manager of the medical program to certify members for respirator use. Another example would be an injury investigation by the health program manager that requires input from other safety groups to redesign specifications for equipment. Specific fitness program details and day-to-day operations of the fitness program can be handled through routine informal meetings among the fitness staff, reserving the key issues for the general health and safety committee. The general committee must remain informed about fitness program objectives and their impact on the total health and safety program, but it need not be involved in program details.

THE PLANNING PROCESS

The program director must develop a plan of action to provide a framework for constructing goals and objectives, as well as determining strategies, needs, and resources. The planning process will clarify program procedure and requirements.

The design plan includes the following steps.

1. Identifying the problem by means of an occupational analysis.
2. Proposing a solution: a fitness program.
3. Arriving at a consensual agreement and mission statement.
4. Drawing an organizational model.
5. Developing goals and objectives.
6. Using the design matrix.
7. Determining staff and resource needs.
8. Identifying alternatives, options, and obstacles.
9. Planning and developing an estimate of budget requirements.
10. Designing and developing physical fitness tests.
11. Conducting a pilot study prior to implementation.
12. Designing an evaluation strategy, one that is both formative and summative.

The first three items in the design plan were addressed in Part I. The problem has been identified, and a mission statement has been adopted. From the information contained in Chapter 4, you should now have a working definition of fitness. You should also be able to position the fitness program within the context of the health and safety program, per NFPA 1500; thus, you understand what it is that you're preparing to design. The next step is the development of goals and objectives. Program goals are written as general descriptions regarding how to proceed with the design of your program. The design matrix in Chapter 6 will help you develop specific objectives in support of the goals.

Two goal samples are as follows.

Goal: *To identify a preexercise screen.*

Objective: *Investigate the screen options available. Evaluate each, then make a selection.*

Goal: *To identify fitness program standards.*

Objective: *The fitness coordinator will determine standard options and their applications, then select the appropriate alternative based on the objectives of the program.*

The development of a design matrix will help to establish goals and assist in the decision-making process for each element of the fitness program. Items five through eight of the planning process are handled through the use of the matrix. You will not have to develop a matrix. It has been designed for you and is presented in the next chapter.

RECOMMENDED READING

NFA course, *Training Program Management (1997)*. Emmitsburg, MD: National Fire Academy.

Chapter 6

The Design Matrix

As true today as in the past, administrators are continually challenged to provide increasing services at lower cost. Their problems and their response to those problems vary from one organization to the next and are usually determined by the community needs. For instance, the trend toward providing emergency medical services in addition to fire suppression may require firefighters to perform dual functions; for example, being qualified as both a firefighter and a paramedic. In some cases, dual roles include police and fire combinations. Such personnel are typically referred to as public safety officers.

The increasing cost of buying and maintaining fire apparatus forces administrators to look at vehicles that can perform multiple functions, such as combination apparatus—a pumper and an aerial device mated together. Other career departments with limited staffing may sign mutual-aid agreements to provide better coverage, or perhaps enlist volunteers to augment their resources. In any given situation, the administrator is forced to develop a solution and find a way to meet the demand. Staffing for career departments involves the greatest investment and consumes the largest percentage of the budget (more than 90 percent for many), yet little attention is paid to improving the ability of the workforce to operate both effectively and efficiently in a safe manner.

Productivity is directly related to physical ability, yet when funds are limited, administrators are reluctant to commit to new programs that they don't fully understand or perhaps appreciate. If well maintained, a twenty-year-old apparatus will get the job done as well as a new piece of equipment, yet this attention to maintenance is seldom applied to fire personnel, our most valuable resource. Equipment and tools do not suppress fires or perform rescues—human beings do that, and those human beings must be fit. In a time of staff reductions, limited funds, and increasing demands for services, there is no room for mediocre performance. The best solution is to ensure that your members are capable of performing at their optimum level, and that is achievable through the implementation of a department fitness program.

The greatest challenge for the public safety administrator is in accomplishing the goal within the limits of the available resources. That is the principal challenge for the program director, and the solution is to provide options. The design matrix in Table 6–1 is provided to help in the planning and decision-making process. The specific categories in the matrix will be discussed individually as we proceed and in greater detail in the next chapter.

The concept of level of participation (LOP) refers to the degree of involvement that you select with respect to each element of the program. The LOP is intended to provide options to administrators and program managers that will enable them to introduce an effective program without exceeding their available resources. The program elements are divided into ten categories. Six categories are cost-related aspects, and the remaining four are no-cost personal choices. There are three levels for each of the ten different categories. The six cost-item categories include equipment and facilities;

TABLE 6–1

DESIGN MATRIX: LEVELS OF PARTICIPATION

Items of Cost	Level 1	Level 2	Level 3
Medical screen	Minimum level.	Preferred level.	Ideal level.
	PAR-Q, no pretest exam. Test high-risk members at 60% of Vo_2max.	PAR-Q or HealthScreen. Examine high-risk members. Basic exam + GXT.	Complete exams: all.
	Level 1A	*Level 2A*	*Level 3A*
	PAR-Q and refer all positives for medical exam.	PAR-Q or HealthScreen. Examine high-risk members and all males over 40, all females over 50. Basic exam + GXT.	Complete exams: all + GXT.
Staff			
Director/manager	Collateral duty.	Full time.	Full-time dir/mgr positions.
Fitness coordinator	CFC certified: collateral duty.	CFC + degree (multiple CFCs).	CFC/FTE: advanced degree.
Physician	By referral.	Contract basis: Set fee.	Full-time physician.
Facility	Station (in-house).	In-house or central facility.	In-house/by contract/ multiple facility.
Equipment	None—calisthenics.	Basic free weights only or plus aerobic equipment.	Full complement of free weights, aerobic equipment, and weight machines.

No-Cost Items	Level 1	Level 2	Level 3
Tests	Calisthenics	Tests employ independent equipment and don't match exercise training mode.	Test specific to training mode and exercise equipment.
Standards	Available norms.	Develop department norms.	Criterion standards.
Exercise prescription	ACSM handout.	Predeveloped cards.	Tailored to the individual.
Compliance	Voluntary.	Voluntary with phase-in period.	Mandatory.

fire department physician; fitness coordinator; medical exams; wellness; and work-injury management. Each category is ranked from the least expensive option (Level 1) through moderate costs (Level 2) to the most expensive (Level 3). The potential no-cost items that involve personal preference decisions are exercise recommendation, fitness standards, program compliance, and fitness tests. Level 1 is considered adequate, Level 2 preferred, and Level 3 ideal. Level 1 in all ten categories represents the minimum requirement to establish a fitness program, and this is attainable by all departments.

The program developer has a host of options and can fund at whatever level for any category. The choices are based on individual preference and projected costs. It is possible to introduce a program at virtually no cost; however, I should emphasize that the willingness to fund a program is directly proportional to one's commitment. If your decision is to support a program, then your fiscal action should reflect your belief in it. Fitness programs do pay!

The ten categories under LOP will be presented and discussed, some in more detail than others, depending on their complexity. The four no-cost items, as well as wellness and work-injury management, will be presented in brief for identification purposes only, since they will be described in more detail later within each specific chapter. For the purposes of design, you need only review the options categorized in the LOP and determine whether a program is even feasible. In the Part III, Development, we will investigate our resources, apply some numbers to each category, then make a selection.

The categories under LOP were derived from complaints voiced by fire departments as to why they cannot introduce or maintain a fitness program. The U.S. Fire Administration published the complaints (or "reasons," as they are described in the text) in Physical Fitness Coordinator's Manual for Fire Departments, FA-95/September 1990. These identified problems appear in Table 6–2. In a survey conducted by the Institute of Human Performance, seventy-six fire departments were asked to identify the factors that most frequently contribute to the failure to develop and implement a department-based physical fitness program. The percentages given in Table 6–2 denote the frequency of response given to each in this survey.

The number one problem, as expected, was funding, followed by a lack of facilities or equipment. The second item is directly related to the first, in that it is fundamentally a budget issue. The remaining five items are educational problems. Essentially, two categories, one financial and the other educational, account for virtually all of the reasons for not having a program. While responding to these comments, I will identify and incorporate each component of the complete program within one of the seven problems listed. This survey is general, but it does encompass every program component. Since the intent of this chapter is to address the causes of resistance separately and hopefully reduce each *reason* to an *excuse*, then provide alternatives through the use of the LOP concept, I will begin with the chief complaint, funding, and proceed down the list.

TABLE 6–2

MAJOR FACTORS INTERFERING WITH PROGRAM IMPLEMENTATION

Problem	*% of Departments Citing Problem*
Lack of funding	66%
Lack of equipment or facilities	53%
No standards or adequate tests	50%
No appreciation for the benefits of the program	45%
No support from employee organization	38%
Legal liabilities	36%
No support from department officials	22%

Reprinted with permission of the National Fire Protection Association, Quincy, MA 02269.

FUNDING

It is no surprise that funding is the most frequently cited reason for not having a fitness program. The greatest contributor to the funding dilemma is the cost of the medical examination. One international survey conducted by the Ottawa Fire Department, Ontario, Canada, revealed an average cost of approximately $195 per exam, with quotes ranging from a low of $70 to a high of $600 (Safety Division, 1997). The specific services incorporated into the examination were not reported, thereby making it difficult to draw any cost comparison. The $195 average appears to be somewhat low, based on an informal phone survey of several midsize U.S. departments, in which the average was $400 and included a graded exercise test (GXT). Since exams that include a GXT can often double the cost, the $195 for a base exam appears to be a modest but fair estimate. If we use the $400 estimate for each exam, multiplied by the number of members employed, the total cost can approach six figures for a midsize department. For both large and small departments, this represents a substantial investment, particularly if it is conducted on an annual basis.

Although a complete medical examination for all firefighters is highly desirable, it may be impractical and, more importantly, unnecessary. In a report prepared by R.J. Shepard of the University of Toronto, *The Costs and Benefits of Exercise: An Industrial Perspective,* the author states, "...over-stringent requirements for preliminary medical clearance can rapidly make a fitness program uneconomical...(and that) exercise scientists do a disservice to potential exercise participants if they follow a common U.S. practice of insisting upon a costly and rather ineffective medical clearance prior to participating in an exercise program." (Shephard, 1990.)

Virtually every text and article that describes an exercise program will include a disclaimer: Consult your physician prior to beginning an exercise program. The advice, although sound, is driven more by legal authorities than by the medical community. For example, an alleged negligence case in California involving a forty-year-old male who sought physician approval to begin an exercise program, and who subsequently suffered a fatal heart attack while exercising, raised the issue of performing a GXT as part of the clearance procedure. Although the physician performed a resting electrocardiogram and blood tests, no stress test was administered. The victim's family brought a claim of negligence against the physician for not conducting the test, and an arbitration panel ruled in favor of the estate. The plaintiff's attorney declared that this was a good case for requiring a stress test for all men over forty prior to starting an exercise program (Herbert and Herbert, 1995). These precautionary measures driven by the legal system have gone to the extreme, according to Dr. Stephen Yarnall. A patient of Dr. Yarnall's requested, on advice from a health professional, "medical clearance" to have her toenails trimmed (Yarnall, 1997). Professional organizations such as the American College of Sports Medicine (ACSM) and the American Heart Association (AHA) provide guidelines and standards delineating a "standard of care." (ACSM, 1991.) Legal decisions are often based on these standards and, in turn, standards are affected by the outcome of such proceedings. An early recommendation from the ACSM, 1986 edition, was age forty-five or older for a GXT, and that has been reduced to age forty in a later *ACSM Guidelines, Second Edition,* 1993. Many physicians object to such decisions. In a sense, the attorneys are setting the standard of care, and most physicians today agree that people with no signs or symptoms of disease do not need an annual medical examination. The need (and for that matter, the value) of a GXT is currently suspect, given the chance of false positives and false negatives—failure to accurately diagnose problems that create unnecessary anxiety (Sharkey, 1990a).

In my own experience, I was once an avid distance runner, and I remember a case in which a physician runner requested his own physician to give him a stress test, "just to be on the safe side," prior to entering the New York City Marathon. Yes, he suffered a fatal heart attack on race day. Fortunately, this isn't a frequent occurrence, but it also isn't uncommon, and it clearly demonstrates that a GXT or a medical clearance is no guarantee of anything.

The legal-medical concern surrounding medical exams is brought to the fore with the Americans with Disabilities Act. The ADA strictly prohibits the use of preemployment screens and medical exams, yet it allows employers to administer physical agility tests without screening. This is in direct conflict with the accepted and recommended standard of care. The ADA suggests that an employer can handle the situation by asking the applicant to provide a statement from his physician that says he can participate. The opinion of one legal workshop is that this approach is to be avoided (AELE, 1997). The second alternative is to make a job offer, then conduct a medical exam. If departments are strapped financially to the extent that they cannot provide exams for incumbents, what are they to do with five hundred applicants? If you screen applicants, you violate the ADA; if you do not, you are acting contrary to the accepted standard of care—clearly a catch-22 situation. Consider the title of one article by Keifetz and Watkins that appeared in *Medical Trends,* 30 (14):1,8 (1989), "Routine Stress Tests Now a Defensive Tactic." Is this an appropriate response based on sound medical principles or simply a CYA tactic?

Returning to the dilemma of providing medical exams for incumbents prior to introducing our exercise program, we must again visit the ADA. An employer may conduct a voluntary medical exam or inquiry as part of the health and fitness program. It must be voluntary! (ADA, 1990.) There are exceptions, such as compliance with other state, local, or federal laws (DOT or OSHA), or if an employee poses a "direct threat" to the health and safety of other coworkers or himself. So, from both the legal and medical perspective, we are caught in a dilemma, one that we presumably can't afford regardless of the choice we make. Are there any alternatives? I believe there are several options, and I propose the alternatives based on the reality of the situation. The options should not be construed as either legal or medical opinion, for I am neither an attorney nor a physician. Rather, consider them to be the personal opinion of one who has safely administered thousands of fitness assessments for over fifteen years. Before I present some alternatives, I must first discuss an interesting aspect of the situation with respect to what we are attempting to accomplish with a fitness program and its relationship to the occupation in general.

Consider first the idea of conducting a prescreening or a medical exam before administering a fitness test to public safety officers, who, by profession, are expected to respond at a moment's notice and perform hazardous duties at high levels of exertion. We certainly do not conduct regular screening prior to performing our normal day-to-day operations, yet for some reason, we find it necessary before we exercise or play. Exercise is simply a uniquely structured form of work. If we raise a fifty-pound hose roll in an overhead motion to place it on a hose rack, we call it work. If we raise a fifty-pound dumbbell in the same overhead motion, we call it exercise. The difference is that one (exercise) requires a medical exam prior to being performed, and the other does not. In reality, there is fundamentally no difference between the two. I realize that there are some who will disagree with this comparison, but I contend that the differences are minimal. Basically, exercise is structured and uses so-called scientific principles, whereas work is neither as specific nor planned. However, the basic exercise principles are a component of work, and in fact, some have been derived from work observations, such as intervals. Workers who engage in heavy manual labor may not know what the overload principle is, but they know how to apply it!

If a firefighter or a police officer by his own admission believes that he is fully capable of working without prior consultation with a physician, then he is fully capable of participating in an exercise program. If he were to believe that participating in an exercise session would pose a direct threat to his personal health, then he should immediately remove himself from active duty until he has received the proper medical attention and clearance to continue working. If he cannot exercise, he certainly can't work. A properly designed exercise program is perfectly safe—far safer than fire or police work. Consider the comparison shown in Table 6–3.

Firefighting is conducted in the most hazardous of environments, whereas exercise is performed in a safe, controlled environment. The intensity and duration of exercise is self-regulated as determined by the participant. Both the intensity and duration of firefighting tasks often exceed the physical capability of the firefighter. The potential for sudden incapacitation, such as myocardial infarction, is significantly greater in firefighting than during exercise. The result of such an event during firefighting operations poses a direct threat to the life safety of others, in addition to the member experiencing the trauma. Quite simply, the health risks are greater during emergency responses than during exercise, and the potential for additional complications is significant. Remember that the purpose of our program is to condition ourselves to reduce these risks.

In the performance of public safety work, there is an assumption of higher risks that are generally not acceptable in other occupations. This concept has been presented and supported in case law, and the risks are considered acceptable in the performance of one's duties. The 1996 NFPA Journal's annual injury issue reported that almost half of the heart attacks and strokes occur in nonfire or other duty situations, such as training exercises (Karter and LeBlanc, 1997). My personal exposure to firefighters experiencing health problems on the job has been predominantly during training exercises. Typically, we do not screen before training sessions, which is why we screen and fit-test annually, to reduce the chances of such adverse events occurring. I am in no way attempting to justify a decision not to screen, based on the preceding discussion, nor do I propose that no premedical evaluation be considered. What I am proposing is that some reasonable, logical approach be considered,

TABLE 6–3

A COMPARISON OF WORK AND EXERCISE DEMANDS

	Fire/Police Work	*Exercise*
Environment	Unknown Hazardous Extreme conditions Clothing restrictions All hours	Known Safe Regulated/controlled conditions Appropriate attire Scheduled, convenient hours
Intensity	Dictated by the job demands 100% heart rate or more	Self-regulated per ability 60–85%, per goals
Health Risk	No prework screen MI* risk is 1 in 10,000 No warm-up	Medical clearance required MI* risk is 1 in 15,000–20,000 Warm-up

*MI = Miocardial infarction (heart attack).

with the understanding that basic risk management practices recognize that not all hazards can be removed, but they should be minimized as much as possible within our limitations. If we choose not to participate in exercise programs, we can remove all threat associated with such activities, and we can also eliminate all public safety officer deaths by not responding to the call or the alarm. On the subject of risks, inventor Grace Hopper once remarked, "A ship in port is safe, but that's not what ships are built for." The benefits of having a program far outweigh the risks associated with having a program, and the greatest benefit is the ability to lower the death and injury statistics.

We begin by identifying our options. Before any department can make the claim that "We can't afford medical exams," we need to know approximately how much we can't afford! Then we can choose our approach, Levels 1 through 3, based on available funds and staff expertise. The first step is to adopt NFPA 1582, *Standard on Medical Requirements for Fire Fighters, 1997 Edition.* Some chief officers have used this document to gain support for additional funding for a fitness program (Staff, 1990). The 1582 document will provide the necessary guidelines and requirements for medical evaluations and examinations. Although targeted for firefighters, some organizations use the same document for police, and this, in my opinion, is acceptable. It is a nationally recognized standard in the field and likely to be given considerable weight by the courts; for example, in *Miller v. Sioux Gateway Fire Department,* 497 N.W. 2nd 838 (1993). It must be noted that complete reliance on the standard may not be sufficient to withstand a legal challenge, but it is the best choice of reference. The second step is to define in terms of content both the medical exam and the medical evaluation. These terms are often confused and sometimes used interchangeably. The distinction is in the requirements, both in content and frequency of use. The appendix of NFPA 1582, pages 17 and 18, provides a physician guide and recommendations. The periodic medical evaluation is designed to evaluate the person's continued ability to perform his duties and to detect any other significant changes in condition of his or her health. The medical exam is more comprehensive and includes the medical evaluation.

Appendix B-3.2 Periodic Medical Evaluations

> *Every year, each fire fighter will be medically evaluated by the fire department physician. This medical evaluation includes an update on the fire fighter's medical history, including any significant changes, a brief review of symptoms, and a report on any significant job-related chemical or infectious disease exposures experienced during the past year. Height, weight, visual acuity, and blood pressure are measured and recorded. The extent of the medical evaluation and additional testing will depend on the fire fighter's medical condition.*

Appendix B-3.3.2 Medical Examination

> *The medical includes an examination of the organ systems, laboratory testing [CBC, biochemical test batter (SMA), and urinalysis], X-rays—baseline only, pulmonary function testing, audiometry, and resting electrocardiography. No firm guidelines for stress testing in asymptomatic individuals are identified. Refer to ACSM guidelines.*
>
> *The annual medical examination is recommended on the following schedule:*
>
> *(a) Ages 29 and under—every 3 years*
>
> *(b) Ages 30-39—every 2 years*
>
> *(c) Ages 40 and above—every year*

The medical evaluation as defined above can easily be conducted in-house and at virtually no cost. The medical history portion that requires a review of significant changes in health due to illness, injury, or exposures can be assumed to have been addressed at the time of their occurrence. Any such changes would indicate medical intervention and result in a return-to-work determination following the incident. Thus, the need for an additional review is eliminated, since the remaining measurements can easily be performed by the health and safety officer or the fitness coordinator, whichever is responsible for conducting the fitness assessment. The visual acuity measure is best left to the employee. Few people who are in need of corrective lenses are apt to delay an appointment for an eye exam. The prescreening and aerobic fitness test, although technically not medical, can effectively serve as a substitute for the annual medical evaluation.

A copy of NFPA 1582 should be provided to the attending physician. Using the above information for medical exams, submit the necessary information to bid, or to a physician of choice, preferably the designated department physician, if you have one. The next step is simply to calculate the cost of the exam based on the number of members. In the next chapter, we will put numbers to our options, but first, let's investigate the options.

Remember that Level 1 is considered to be the minimum requirement. You probably won't have zero cost, but you may come very close to that. At Level 1, first investigate the feasibility of using the employee's health benefits for funding. Many health care providers do cover the costs of annual medical exams, in part or in whole. Remember that a medical exam is for the benefit of the employee, and he should welcome the opportunity to protect himself. To the employee, an added benefit of using his insurance is that it will remove the chance of a breach of confidentiality, which is possible with department-funded examinations. Such a breach should never occur, regardless of who provides the service. The employee health plan also allows one to use his own private physician, which most people prefer. If your insurance provider doesn't cover such exams, consider investigating health plans that do. This method will help minimize cost.

If insurance assistance is unavailable, then the next best option is the selective referral; i.e., referring only those who, through the use of a medical evaluation or preexercise screen, are considered to be high risk. The health screen refers to standard health risk appraisal forms and fitness screens, such as the RISKO, HealthScreen, AHA Checklist, and the Physical Activity Readiness Questionnaire, or PAR-Q (Appendix B). The health risk appraisal RISKO is designed to determine the overall risk for CHD and isn't considered an exercise screen. It is often used in conjunction with PAR-Q. The health risk appraisal is designed to inform individuals of their health risk status and to encourage lifestyle modifications. The PAR-Q and HealthScreen, a computerized screen, are designed to screen for health risk prior to exercise. The PAR-Q, developed by the Canadian Society for Exercise Physiology, was found to have a sensitivity of 100 percent (percent with medical problems answering "yes") and specificity of about 80 percent (percent without medical problems answering "no") for detecting medical contraindications for exercise (Shephard, 1991). The American Heart Association provides a free guide to physical activity, *Exercise and Your Heart,* that includes a checklist to determine whether you need to consult a physician prior to exercising. The checklist was derived from several sources, including the PAR-Q form. A comparison is shown in Table 6–4. Both of these questionnaires are similar and easy to complete.

The PAR-Q also has a companion form, the PARmed-X, which is a physical activity checklist designed to be used by a physician with patients who have had a positive response on the PAR-Q. It can also be used as a medical clearance or exercise referral form.

LEVEL 1: PAR-Q AND RISKO

At the very basic Level 1, the health heart appraisal RISKO is used alone to identify high-risk individuals (a RISKO score of >32 on the old form or >8 on the new form). This group is tested using the walk test at 60 percent $\dot{V}O_2$. All other members use the standard test procedure. No exams are

TABLE 6–4

COMPARISON OF THE AHA CHECKLIST AND THE PAR-Q

AHA	PAR-Q
1. Your doctor said that you have a heart condition and recommended only medically supervised physical activity.	*1.* Has your doctor ever said that you have a heart condition and that you should only do physical activity recommended by a doctor?
2. During or right after you exercise, you frequently have pains or pressure in the left or mid-chest area, left neck, shoulder, or arm.	*2.* Do you feel pain in your chest when you do physical activity?
3. You have developed chest pain in the past month.	*3.* In the past month, have you had chest pain when you were doing physical activity?
4. You tend to lose consciousness or fall over due to dizziness.	*4.* Do you lose your balance because of dizziness or do you lose consciousness?
5. Your doctor recommended that you take medicine for your blood pressure or a heart condition.	*5.* Is your doctor currently prescribing drugs (for example, water pills) for your blood pressure or pressure or heart condition?
6. Your doctor said that you have bone or joint problems that could be made worse by the proposed physical activity.	*6.* Do you have a bone or joint problem that could be made worse by a change in your physical activity?
7. You have a medical condition or other physical reason not mentioned here that might need special attention in an exercise program (for example, insulin-dependent diabetes).	*7.* Do you know of another other reason why you should not do physical activity?
8. You are middle-aged or older, have not been physically active, and plan a relatively vigorous exercise program.	
9. You feel extremely breathless after mild exertion.	

Note: The forms recommend that you contact a physician if you check any item on the AHA checklist or answer yes to any item on the PAR-Q. With the exception of the two additional items on the AHA checklist, both forms are very similar.

Source: Excerpts from the American Heart Association's *Exercise and Your Heart* booklet (1993) and the Canadian Society for Exercise Physiology's *Physiology Activity Readiness Questionnaire—PAR-Q,* reprinted with permission.

required, unless a member experiences shortness of breath, chest pain, or dizziness. Although RISKO isn't a preexercise screen, I have found it to be very effective in identifying high-risk, low-end performers. One step up to Level 1A is to administer the PAR-Q or the AHA Checklist and RISKO, then refer anyone who is considered a positive for physician clearance for a complete medical exam. If you choose this approach, the at-risk people will be the only members who require an exam, and the annual cost of such exams will be reduced significantly. If you choose to include the RISKO form, the remaining people who answer no on PAR-Q but who are considered high-risk using the RISKO would be tested at 60 percent $\dot{V}O_2$. The ACSM guidelines on medical exams state that "asymptomatic men under 40 and women under 50 with fewer than two CAD risk factors need not have a medical exam before beginning a program at intensities above 60 percent $\dot{V}O_2$; asymptomatic men and women regardless of age or CAD risk factors can exercise at intensities of 40 to 60 percent $\dot{V}O_2$ without an exam. For vigorous exercise—above 60 percent $\dot{V}O_2$—men over age 40 and women over age 50 are recommended (*sic*) a medical exam." (ACSM, 1993.) (Table 6–5.)

In Chapter 8, several options for determining aerobic capacity at intensity levels within the 40 to 60 percent $\dot{V}O_2$ range are provided, and even a nonphysical written test is offered for the most timid. The ACSM guidelines are recommendations, and the PAR-Q option is a widely accepted practice that is used quite frequently in the field of exercise science. This is a very cost-effective and practical approach.

A rough estimate using the medical evaluation in conjunction with a RISKO and PAR-Q may show that a high estimate of 25 percent of your members

TABLE 6–5

ACSM GUIDELINES FOR EXERCISE TESTING AND PARTICIPATION

	Apparently Healthy		Increased Risk*		Known
	Younger**	Older	No symptoms	Symptoms	Disease***
A. Medical exam and clinical exercise test recommended prior to:					
Moderate exercise (40-60% $\dot{V}O_2$max)	No†	No	No	Yes††	Yes
Vigorous exercise (>60% $\dot{V}O_2$max)	No	Yes	Yes	Yes	Yes
B. Physician supervision recommended during exercise test†††					
Submaximal testing	No	No	No	Yes	Yes
Maximal testing	No	Yes	Yes	Yes	Yes

* Persons with two or more risk factors or one or more signs or symptoms of cardiac, pulmonary, or metabolic diseases.
** Younger: >40 for men, >50 for women.
*** Persons with known cardiac, pulmonary, or metabolic disease.
† The "no" response means that an item isn't necessary. The no response doesn't mean that the item shouldn't be done.
†† The "yes" response means that an item is recommended.
††† "Supervision" suggests that a physician be in close proximity and readily available should there be an emergent need.

Reprinted by permission of the American College of Sports Medicine (1995), *Guidelines for Exercise Testing and Prescription* (Baltimore: Williams and Wilkins).

require complete exams. If you exceed this percentage, your department is in drastic need of a medical/physical program, and you should proceed directly to the bank for full funding. On a more positive side, I can say from experience that a result of less than 10 percent is more realistic. Once you know the cost of the basic exam, you need only apply that cost to the resultant percentage of your department membership to arrive at an estimate. If it is realistic, you can expect to reduce your total cost by 75 percent if you use the high estimate of 25 percent, or by 90 percent using the lower figure, before you even begin calculating.

LEVEL 2: PAR-Q AND ACSM GUIDELINES

A Level 2 approach would employ either the computerized HealthScreen, which interprets ACSM guidelines for you, or the basic PAR-Q or AHA Checklist screen as mentioned for Level 1A. The same procedures would apply, with the addition of a graded exercise test (GXT) to the complete medical exam.

Level 2A is the same as above, but it includes the exam for all men over 40 and women over 50 per the ACSM guide. The HealthScreen will automatically identify this group for you. You still don't have to provide medical exams for all, but you may have to provide GXTs when indicated and refer more members for complete exams.

LEVEL 3: COMPLETE MEDICAL EXAMS

Level 3 is the highest level of participation. It represents the ideal in that it provides a complete medical exam and, at Level 3A, includes a yearly stress test component for all members. You may reduce your total costs while providing the same service by employing a full-time physician. Don't overlook this option, particularly if yours is a large department. There are many other benefits to having a full-time physician, beyond reducing exam costs.

EQUIPMENT AND FACILITIES

The second most frequent response given as to why fitness programs aren't implemented is related to facilities and equipment. Both of these issues are interdependent, in that a decision with respect to one will affect the other. For example, if you choose to have one facility at a central location, you'll be able to save on costs while providing a greater compliment of equipment. On the other hand, if you choose to confine the exercise area to each firehouse or police substation, you'll need to furnish each, thereby increasing either your equipment investment or by reducing the basic inventory. Therefore, the first decision should be in choosing the facility.

The cost of providing a health and fitness facility entails a significant investment for the private sector. Although many corporations do provide facilities, some choose health club memberships as an alternative. Unfortunately, these memberships are typically reserved for management, leaving the lower echelons of the workforce out. A health club membership is a possibility for a fire or police department, but over the long haul, it'll be more expensive.

At Level 1, each fire or police station would have an in-house facility. This is both the preferred option and the easiest solution, although it may require more money to outfit each station rather than one facility, depending on the choice of equipment. The biggest drawback is in finding room to place the equipment. Usually it's relegated to the apparatus room, which isn't the most desirable location, although it will serve. Finding space in part depends on how much equipment you have and how much you desire you have to find room for it.

A centrally located structure would be at a Level 2 commitment. This could be a training center, firehouse, police station, public health club, or rented building. Having a central location will enable you to stock that facility with more equipment of better quality and at lower cost. The disadvantage is that firefighters or police officers will be leaving their stations to exercise, presenting problems related to response time and convenience, more so for firefighters than police. The disadvantages of having a single facility can be removed either by requiring members to work out prior to their shift change or perhaps by providing a basic complement of equipment at the station for use during those times when the central facility is unavailable to on-duty personnel. A central facility is an excellent option for police and volunteer firefighters.

At Level 3, you would have multiple facilities, such as an uptown and a downtown facility, that would be shared by all of the members. This follows the same concept as the one described above, but it offers greater convenience and reduces travel time. It also doubles the initial investment in equipment. These facilities would be fully equipped and available to both on- and off-duty members. Additionally, each station could have a basic complement of equipment as well. Another advantage of having centralized facilities is that family members could have access to them.

Once, while sitting in a conference room consisting of a table and several chairs, a chief officer asked me, "What is the absolute minimum amount of equipment that you would need to begin a fitness program?" My answer was, "Nothing more than what is currently in this room." A fitness program consists mainly of people exercising, and the only prerequisite is the desire to do so. Equipment is nonessential. Thus, any excuse for not initiating a fitness program based on a lack of equipment is unwarranted. The idea of exercising without equipment may thrust the younger members into a catatonic state, but in truth, this is an effective, no-cost alternative that also resolves the facilities issue. Remember that the military has always been able to train and condition soldiers with neither expensive facilities nor equipment. As stated earlier in this text, an infantry soldier has the same DOT strength code rating as a firefighter (V = very heavy), and the mode of training for infantry is basic calisthenics. In my own experience, the highest physical condition that I ever achieved came when I was serving with the U.S. Marine Corps and participating in a task-specific calisthenic training program. A complete total body workout can be accomplished using calisthenics. A former Marine has designed a boot camp routine called Body Pump that has become popular at some gyms (Shackelford, 1998) and, in a similar fashion, a New York City firefighter named Eric Torres has gained media attention with a routine that uses standard fire equipment.

The chief complaint with calisthenic or bodyweight resistance programs is in gaining strength, which requires an increase in the resistance. This can be resolved by altering positions, performing very slow repetitions (e.g., a sixty-second push-up), adding weight to the body, or by working with a partner, who can apply increasing resistance. The partner-assisted version has been practiced in the National Football League by the Phoenix Cardinals. Many professional athletes perform calisthenic-type workouts and have little problem developing adequate strength or endurance, so this option should neither be overlooked nor underrated.

The fitness trail is another option. It was designed by the U.S. Forest Service and is found in many parks (Sharkey, 1990b). It incorporates a series of calisthenic stations with modest props and signs that include instruction and recommendations for each event. Some departments, particularly police, may have access to one of these trails, which can be used as a fair-weather option. Not only do they provide good exercise, they're also quite fun. The aerobic component is easily satisfied through walking, stair-stepping (nonequipment version), or running routines. Other options include a host of exercise videos, some of which incorporate the use of everyday objects to add resistance.

All of this is to say that a complete exercise program is possible without any financial commitment or equipment, and that a well-designed routine is more than enough to condition any police officer or firefighter to meet the physical requirements of his job.

At Level 2, we can add some basic equipment to the program. When making these purchases, I highly recommend starting with weight-training equipment. The basic weight-training complement is less expensive than most aerobic conditioning equipment, and other aerobic exercises can easily be substituted until more funds are made available. Still, the priority of purchase is best determined by the members who will be using the equipment. If aerobic equipment is preferred, then strength training can be accomplished through calisthenics.

The basic complement of equipment is as follows.

- *Essential weight-training equipment*
  1. Olympic bench-press bench
  2. Olympic weight set of 300 lbs.
  3. Weight tree
  4. Power rack
  5. Roman chair (back raises)
- *Optional weight-training equipment*
  1. Set of dumbbells, 10 to 50 lbs.
  2. Incline bench
  3. Curl bar
  4. Lat-pulldown, plate-load type
  5. Leg press (45-degree hip sled), plate-load type
- *Choice aerobic equipment*
  1. Exercise bike
  2. Stair machine
  3. Rowing machine
  4. Ski machine
  5. Pair of running shoes

With the exception of the two plate-load devices, the weight equipment listed is of the free-weight type, as opposed to weight machines. The basic Olympic setup is all that is necessary to perform a complete total-body workout. The optional items are nice to have, but they're not necessary and more a matter of convenience. Free weights provide the least expensive strength-training system, and they're also the most versatile and most effective. The aerobic equipment is a matter of preference and often selected based on cost. The exercise bike and stair machine are the most popular choices. The most important aspect is to buy *quality.* Exercise equipment that will be in a twenty-four-hour facility will receive some abuse. Don't forget warranties or repair services when you consider equipment options.

Any inventory that is greater than this basic setup would qualify as being at a Level 3 commitment. The addition of weight machines or multiple aerobic devices would constitute a full complement. When adding equipment to your inventory, be very selective. Invariably, if cost is no concern, then space will be the limiting factor. Although resistance machines are nice, they are expensive space eaters and have limited use. Since aerobic devices can only be used by one person at a time, I would recommend purchasing additional items in this category before investing in expensive variable-resistance machines.

TESTS AND STANDARDS

Tests and standards rank third in the list of reasons blocking the implementation of a department fitness program, cited by half of the departments interviewed. In response to this "reason," I will state that both tests and standards are available and that Parts III and V of this text address these issues in greater detail. In terms of levels of participation, I offer three approaches to standards, test selection, and exercise recommendation (EXRX).

At Level 1, adopt and use available normative data based on general-population studies and other select groups. Some normative data for police and firefighters is provided in Appendix C. Tests are selected for each fitness component, and the mode of testing reflects the type of program being used. For example, use calisthenic tests for a calisthenic program. The exercise recommendation at Level 1 is to distribute a one-page ACSM guideline, a modified version of which appears in this text. The original recommendations, published as *position stands,* are available from the college.

At Level 2, you should develop your own normative data based on the testing of your membership. Tests again should reflect fitness components, and equipment can be purchased for testing purposes only. The exercise recommendation is to use a predeveloped handout corresponding to each fitness level and category. An EXRX card is provided in Appendix D for this purpose. It corresponds to the standards table recommended in Chapter 11.

At Level 3, adopt criterion-referenced standards. This text and its appendices offer a variety of tests and criterion standards. The concept is presented in detail in Chapter 11. The tests are specific to the exercise mode. Fitness component and criterion standards are provided in Appendix C. The exercise recommendation is for a qualified staff member to provide personalized and individually developed programs.

No Appreciation for Program Benefits

The lack of appreciation for a fitness program can only be attributed to a lack of information. Hopefully, this book will provide the incentive for many departments to develop such a program. Under this heading, I will present two supplemental programs in the LOP format: a work-injury management program and a wellness program. Both of these appear in later chapters and should be developed once the basic fitness program is in operation. For now, it is unnecessary to consider allocating resources for these programs, particularly if funding is limited. At the outset, you should accept the Level 1 recommendation in each category.

One word of caution, however. Attempting to develop all of the programs in the realm of health and fitness can result in system overload and mass confusion. Each program requires a systematic approach and good planning. The total health and fitness program should be developed in a manner that will allow you to focus on one program at a time. Start with the fitness program.

In terms of wellness, Level 1 involves the dissemination of health-related information via newsletters, brochures, and other information packets, available through a variety of organizations, such as the American Heart Association. The critical incident stress debriefing team should be assembled and introduced at this basic level, if not already active.

At Level 2, the CISD and an employee assistance program (EAP) would address behavioral issues related to stress, substance abuse, smoking, and other such problems. Consulting services with regard to nutrition and weight loss can be provided, in addition to educational literature. Chapters 14 and 15 offer more specific information.

Level 3 participation would include the CISD and EAP programs, plus consulting services available for all aspects of the wellness program. Full- or part-time professionals may also be employed and serve on the wellness program staff.

Work-injury management includes accident and injury investigation; ergonomics; work-hardening programs; monitoring of the workplace environment; transitional work programs; safety inspections; and injury rehabilitation. The basic minimal level, accident and injury investigation, should currently be in practice, for such investigations should include recommendations to correct any identified problems.

At Level 2, professional physical and occupational therapists are consulted to provide accident and injury interventions, as well as rehabilitation programs. Both work-hardening and transitional programs can be developed at this level. The fitness coordinator would work with health providers to develop and implement rehab-type programs (LeCuyer, 1993).

At Level 3, an in-house rehabilitation and conditioning clinic may be incorporated into the health and fitness facility. In one Colorado corporation, such an in-house rehab facility returned approximately six dollars for every dollar invested. Larger departments should seriously investigate this option, particularly if they have substantial workers' compensation costs. Contract services with therapists, ergonomic experts, and safety engineers may also be arranged.

NO SUPPORT FROM LABOR OR MANAGEMENT

Chapter 2 hopefully resolved this issue with a joint labor/management agreement. If such an agreement cannot be made, shelve this text and wait three years, then revisit the book again when the light comes on.

There are two other issues that I will include under this category that require mutual support and agreement, namely program staff and program compliance. Obviously, a department cannot introduce and administer a fitness program without assigning personnel and their specific duties and responsibilities. Management's commitment to the program will be most expressed in its staffing decisions. The first position is that of department physician. NFPA 1500 states in 8-6.1 that "The fire department shall have an officially designated physician..." and in 8-6.2 that "The fire department physician shall provide medical guidance in the management of the occupational safety and health program." The department physician can provide services at three levels, consistent with the other program categories: (1) on a referral basis, (2) on a contract basis, or (3) in a full-time capacity.

The second staff position is that of program director, manager, or both. NFPA 1521, *Fire Department Safety Officer,* identifies a health and safety officer (HSO) to manage the program. Again the three levels apply: (1) a uniformed member is assigned collateral duty, (2) a uniformed member is assigned on a full-time basis, or (3) the department employs a full-time health professional, who may or may not be assisted by a uniformed member. If a full-time physician is employed, the department may elect to assign him the role of program director and eliminate the need for additional staff. Other staff positions are filled by department members assigned as fitness coordinators. NFPA 1500, chapter 8-3.3, assigns fitness program administrative duties to the health and fitness coordinator (HFC). The HFC or CFC (certified fitness coordinator) is responsible for the day-to-day activities, such as exercise programming, testing, and training. The number of CFCs depends on the size and needs of the department. In terms of staffing, there are many options, ranging from the basic requirement to the ideal. The final decision rests on the size of the department, the services provided, staff qualifications, and organizational structure.

Even the best program will be valueless if no one participates. The target group, consisting of those at the lower end of the fitness spectrum, accounts for the largest percentage of injuries. It is they who are most vulnerable to injury. Closing the gap between fitness and job demands must include this group. Typically, they do not volunteer; therefore, labor's contribution to the program effort is to support mandatory participation. With the threat of punitive action removed, members should be receptive to compliance requirements.

By general consensus, a mandatory requirement is preferable to voluntary participation. Voluntary programs simply aren't as effective at gaining participation from the group that is most at risk. Those who will volunteer tend to be the most fit individuals in the organization, and they're the ones who would exercise whether required to or not. The concept of voluntary participation implies an option. Sedentary individuals have already exercised their option to be inactive. If fitness is considered optional in your department, then your program won't be very successful.

There are situations, however, in which a voluntary approach is the most appropriate. The first is in a pilot study when developing the program, and the second is when you implement the program by means of a phase-in period. A phase-in period typically lasts for one year, during which members may either choose or decline to participate. This approach will often entice some of the holdouts to join in of their own free will, once they discover that there is little to fear and much to gain. In the pilot and phase-in situations, you will be involved with people who are supportive of the program, which is beneficial during the introductory period. The volunteers can provide valuable feedback while helping to sell the program to the remaining members. If those who are in greatest need don't participate, you can't achieve the goal of the program. The data that results from selective participation will be skewed, reflecting the condition of the most physically fit while ignoring those who statistically account for the largest percentage of on-duty injuries.

My first pilot program attracted sixty-seven volunteers from one department, too many for a pilot study, considering that it represented 30 percent of the total membership. Still, the most interesting aspect was that this group accounted for only 10 percent of the department's injuries. Although some members from the low-fit category will volunteer, they represent an exception that will offer no reason to support the voluntary method. I recommend that you use volunteers for the pilot study, then proceed with either a voluntary or mandatory implementation, depending on the response of the members. If you don't get any volunteers, or if the opposite occurs and 90 percent of the members volunteer, proceed with a mandatory program after the pilot. If you pick up about 30 to 50 percent, continue in a voluntary mode, with a specific date set for total participation.

Some program developers recommend using incentives to encourage participation. This may be an effective approach in the private sector, but it is of limited value in public safety, where fitness is such an important aspect of the job and risks accrue to those who don't participate. I am not a proponent of providing an incentive for something that is required. To me, an incentive is a form of bribery not too far removed from rewarding children with ice cream if they behave themselves. I don't believe that people who by their appearance demonstrate that they prefer a sedentary lifestyle will respond to the offer of a free T-shirt. There may be exceptions, given individual values and motivations, but I just don't see the carrot working.

In my early days on the running circuit, I would see people sign up and pay an entry fee, only to collect their T-shirt and go home without participating. Some race organizers apparently noticed this peculiar tendency and began to withhold the T-shirt until the race was over. Two styles of T-shirt were issued for the Denver marathon in 1979. Both had a logo, but only one had the word *finisher* stenciled on it. You had to cross the finish line for that shirt.

If your department consists of several hundred firefighters, money will be better spent on exercise equipment that will benefit all the members. The price of two hundred T-shirts can buy a quality stair machine. If there is enough money in the budget for incentives, and if the members want them, then by all means, go for it! Still, you should ask the members for a recommendation. Don't buy incentive trinkets that no one wants.

Awards are similar to incentives but somewhat different. An award comes after the fact as opposed to up-front. Awards acknowledge rather than entice. Giving an award is a way to recognize individual effort, and the effectiveness of an award is highly personal. An award may be a material gift, such as a T-shirt or a pin (which the members will lose and seldom wear), or it may have little or no cost whatsoever, such as a certificate or mention in the department newsletter. The most useful and effective method of recognition is to incorporate the fitness score in the member's annual performance appraisal. This serves several purposes. First, it sends the message that fitness is important and that management is watching. Second, it says that the department is committed to the program. Third, it serves as a basis for personal development and setting goals. The assessment of one's level of fitness belongs in the performance appraisal. It is both a good indicator and a valid predictor of job performance. This is a key relationship that test developers focus on when conducting a job analysis or developing and validating tests. A study conducted by Dr. Art Dickinson of the University of Colorado identified a significant correlation (r = 0.97) between personal performance ratings and their performance in a test battery of simulated tasks (see Appendix E). As expected, the best performers were those who were more highly fit; the two variables are inseparable.

LEGAL LIABILITIES

As stated in Chapter 2, the risk of not having a program far outweighs the risks of introducing one. As a result of litigation and new legislation, the fire administrator must develop some method of determining fit-for-duty status. The physical demands of the occupation must be identified and incorporated into the job description, and a test of some kind must be used to determine the ability of a candidate or an incumbent to perform essential job functions. Since the fitness program, as presented in this test, does not affect a member's job status, the threat of legal challenge is nonexistent unless you choose not to have a program.

Test development can become a legal battleground, since the sole purpose of a physical performance assessment (PPA) is to make job decisions. The testing of applicants isn't part of the fitness program, so don't confuse the two. As stated earlier with wellness and work-injury management, this program should be introduced when the basic fitness program has been completed. The fitness program will provide a means of addressing any physical problems identified in the performance program.

At Level 1 in developing a PPA, a department can develop an entry-level test in-house and play the odds of not being sued. If the test does not have disparate impact, you will not need to validate the test, nor is it likely that the department will be challenged legally. It is also unlikely that you will be able to select high-quality candidates.

At Level 2, a department should contract with a professional developer to design and validate an entry-level test. You can use the same test to evaluate incumbents and make fit-for-duty decisions.

At Level 3, you can contract with a professional test developer to design and validate an entry-level test. Use the concept of company performance standards to evaluate incumbents in lieu of the entry-level test format. Chapter 16 will address these issues in greater detail.

RECOMMENDED READING

ACSM (1991), *American College of Sports Medicine: Guidelines for Exercise Testing and Prescription, Fourth Edition.* Philadelphia, PA: Lea and Febiger.

Part II References

ACSM (1991). *American College of Sports Medicine: Guidelines for Exercise Testing and Prescription, Fourth Edition*. Philadelphia, PA: Lea and Febiger.

ACSM (1993). *Resource Manual for Guidelines for Exercise Testing and Prescription, Second Edition*. Gordon, N.F., and Mitchell, B.S., editors (p. 220). Health Appraisal in the Nonmedical Setting. American College of Sports Medicine. Philadelphia, PA: Lea and Febiger.

ADA (1990). Americans With Disabilities Act of 1990, 42 U.S.C. § 12101, et seq.

AELE Workshop (1997). *Discipline and Labor Problems*. Americans for Effective Law Enforcement Inc. Chicago, IL.

Alter, Michael (1996). *The Science of Flexibility*. Champaign, IL: Human Kinetics, p. 210.

Anderson, D.R., and Staufacher, M.J. (1996) The impact of worksite-based health-risk appraisal on health-related outcomes: a review of the literature. *American Journal of Health Promotion*. July/August, 10 (6), pp. 499–508.

CSEP (1997). *The Canadian Physical Activity, Fitness, and Lifestyle Appraisal*. Ottawa, Canada: Canadian Society of Exercise Physiology.

Davis, Paul O., and Starck, Albert, R. (1980) Excess body fat, not age, viewed as the culprit in fitness decline. *Fire Engineering*. June, pp. 33–37.

Davis, Paul O. (1995) Fitness issues: physical training shouldn't be remedial. *Fire Chief*. February, pp. 15–16.

deVries, Herbert A. (1974a) *Physiology of Exercise*. Dubuque, IA: Wm. C. Brown Company Publishers, p. 81.

deVries, Herbert A. (1974b) *Physiology of Exercise*. Dubuque, IA: Wm. C. Brown Company Publishers, pp. 387–388.

Dickinson, Arthur (1996). *Combining Kinesiology, Aging Science, and Research Design to Produce Valid and Defensible Job Analyses: Human Factors in Organizational Design and Management*. Proceedings from the Fifth International Symposium held in Breckenridge, CO, July 31–August 3.

Fahey, R.F. (1993) Fire fighter fatalities resulting from heart attacks. *NFPA Journal*, March/April, p. 73.

Gettman, Larry R. (1993) *Resource Manual for Guidelines for Exercise Testing and Prescription, Second Edition*. Gordon, N.F., and Mitchell, B.S., editors (p. 229). Chapter 19: Fitness Testing. American College of Sports Medicine. Philadelphia, PA: Lea and Febiger.

Gledhill, N., and Jamnik, V.K. (1992) Characterization of the physical demands of fire fighting. *Canadian Journal of Sports Science*, 17 (3), pp. 207–213.

Herbert, D.L., and Herbert, W.G. (1995) Legal considerations for graded exercise testing and exercise prescription. *The Fitness Handbook, Second Edition*. Peterson, J.A., and Bryant, C.X., editors (p. 55). St. Louis, MO: Wellness Bookshelf.

Hickson, R.C.; Dvorak, B.A.; Gorostiaga, E.M.; Kurowski, T.T.; and Foster, C. (1988) Potential for strength and endurance training to amplify endurance performance. *Journal of Applied Physiology, 65 (*5) 2285–2290.

K. Aoki; Y. Suzuki; A. Noji; R. Yanagibori; and A. Gunji (1990). Analysis of the relationship between physical fitness level and risk factors of cardiovascular disease in workers. Kaneko, Masahiro, editors (pp. 239–248). *Fitness for the Aged, Disabled, and Industrial Worker, Vol. 20*. Champaign IL: Human Kinetics Publishers Inc.

Karter, Michael J., and LeBlanc, Paul R. (1997) U.S. firefighter injuries 1996. *NFPA Journal,* Nov./Dec. 1997, p. 70.

Keifetz and Watkins (1989).Routine stress tests now a defensive tactic. *Medical Trends,* 30 (14): 1, 8.

LeCuyer, John (1988*).* Unpublished report. Aurora, CO: Aurora Fire Department.

LeCuyer, John (1993). A bridge between industry and rehabilitation. *Isernhagen Work Injury Report.* Fall, pp. 1 and 4.

Marcinik, E.J.; Potts, J.; Schlabach, G.; Will, S.; Dawson, P.; and Hurley, B.F. (1989) Strength training improves endurance performance and lactate threshold without altering VO2max. *Medicine and Science in Sports and Exercise*: *Supplement to 21*. No. 2, p. 523 (Section B–8: Poster Presentation, Respiration, Lactate, and Ventilatory Threshold, Abstract no. 138).

Morehouse, L., and Miller, A. (1963) *Physiology of Exercise, Fourth Edition.* Saint Louis: The C.V. Mosby Co.

Mosby's Medical, Nursing, and Allied Health Dictionary, Fourth Edition (1994). Saint Louis: The C.V. Mosby Co.

NFA course *Training Program Management (1997).* Emmitsburg, MD: National Fire Academy.

NFPA 1582 (1997) *Medical Requirements for Fire Fighters, 1997 Edition.* Quincy, MA: National Fire Protection Association. Appendix B, p. 1582–17–18.

Noakes, T. D. (1988) Implications of exercise testing for the prediction of athletic performance. *Medicine and Science in Sports and Exercise*. 20 (4) 319–330.

O'Conner, Jack (1996). Fire fit: top-five questions. *Firefighter News.* April/May, pp. 22–26.

Peterson, J.A., and Bryant, C.X. (editors)(1995), *The Fitness Handbook, Second Edition.* St. Louis: Wellness Bookshelf, p. 309.

Safety Division (1997). *Emergency Services Medical/Physical Testing*. Ottawa Fire Department, March 5. Ottawa, Ontario.

Shackelford, Larry (1998). Basic training. *Muscle and Fitness Magazine.* May, pp. 79–84.

Sharkey, B.J. (1990a) *Physiology of Fitness, Third Edition*. Champaign, IL: Human Kinetics, p. 172.

Sharkey, B.J. (1990b) *Physiology of Fitness, Third Edition*. Champaign, IL: Human Kinetics, p. 399.

Sharkey, Brian J. (1997) *Fitness and Health, Fourth Edition*. Champaign, IL: Human Kinetics.

Shephard, R.J. (1990) The costs and benefits of exercise: an industrial perspective. Kaneko Masahiro, editor (pp. 189–201). *Fitness for the Aged, Disabled, and Industrial Worker, Vol. 20*. Champaign IL: Human Kinetics.

Shephard, R.J., and Weller, T.S. (1991) The Canadian home fitness test. *Sports Medicine*, 11: 358.

Siri, W.E. (1961) Body composition from fluid spaces and density. *Techniques for Measuring Body Composition*. Brozek, J., and Henschel, A. (editors). Washington, DC: National Academy of Sciences.

Staff (1988). Litigation and court rules: are GXTs required for screening of all men over 40? *The Exercise Standards and Malpractice Reporter*. 2 (2): 30

Staff (1990). Coping with NFPA 1500. *Fire Chief*. Feb., pp. 43–47.

Williams, Melvin (1988). *Nutrition for Fitness and Sport, Second Edition*. Dubuque, IA: Wm. C. Brown Publishers, p. 202.

Williams, T., and Evenson (1988). Physically fit for duty? by whose standards? part I: mandatory physical fitness programs. *Fire Chief*. March, pp. 43–48.

Wilmore, J.H., and Costill, D.L. (1994) *Physiology of Sport and Exercise*. Champaign, IL: Human Kinetics, p. 383.

Yarnall, S. (1997) *Health and Safety*. Vol. XVII, No. 12, December. Seattle, WA: The Hope Institute, p. 5.

PART III DEVELOPMENT

CHAPTER 7

STAFF AND RESOURCES

As stated in the preceding chapter, the budget is the link between the resources needed and the services provided. The budget estimate will be used to justify the requests based on the level of the program. The design matrix presented in Chapter 6 will be used as a worksheet for the financial plan. It includes all of the necessary information as to the resources needed; options and alternatives; and prioritization, based on the levels of participation. The matrix includes ten categories, only three of which are cost-related. Four of the items—standards, tests, exercise recommendation, and compliance—are preference or policy decisions that will be addressed later, independently of the cost items. Three items—wellness, work-injury management, and physical performance—are independent programs that should be developed after the fitness program has been introduced. They have been included in Part VI as a supplement because of their close relationship with the basic fitness program.

The expense plan, therefore, will focus on the remaining three items: staff, medical exams, and equipment/facilities. There are three levels for each of these categories, and a cost estimate should be determined for each. Budgeting then becomes a matter of plugging in the numbers. Worksheets are provided for each element. Once the sheets have been completed, the information is transferred to a budget estimate sheet for a one-page comparison of costs at all three levels.

It may be helpful at this time to enlist assistance from the safety committee or potential staff members. The program director can then assign the three cost-category estimates to coworkers. The information presented in the LOP should be provided to your assistants to use as a guide for investigating service costs and alternatives.

SELECTING THE STAFF

All members of the health and safety program staff should have written job descriptions. Each job description should spell out the assignment and title; the qualifications required; and the duties and responsibilities of the position, as well as the express authority to perform them. A sample description for a health and safety officer (program manager) is provided in Appendix F.

You should prepare interviews for each potential position; namely, that of physician, manager, and fitness coordinator. I say "potential" because, at this time, we haven't yet gathered a cost estimate for the positions, nor have we even determined that all of the positions are necessary. Depending on the organization, some of the assignments may overlap.

The director, who is responsible for managing the health and safety program, was identified in Part I and assigned the project as part of the initial agreement. Unless otherwise directed, it is assumed that the director (department safety officer) remains the responsible party and will therefore prepare and conduct the interviews with assistance from the health and safety committee, once the staffing needs have been determined.

The first step in the development process is staff selection. The procedure for selecting the staff involves determining personnel needs, based primarily on available resources and the size of the department. In the smallest department, the requirement may be one individual only, in which instance the staff issue need only identify the program manager and physician. In most cases, however, there will be at least one certified fitness coordinator (CFC) assigned, in addition to the program director.

You should first conduct an internal inventory of qualified personnel. Although there are no specific educational requirements necessary, it is best to seek members who presently have the expertise, education, or qualifications. There is little reason to hire from outside if there are potential candidates on staff. You may find a more qualified individual from the private sector (some at a lower salary than incumbents), but a prospect will have a distinct disadvantage in his lack of occupational experience, which is very important. A combination of outside professionals and department staff is an option, however, depending on resources and desire.

The program director should be a commissioned officer, whereas the CFC can be of any rank desired. The use of rank as a criterion is only a recommendation, however. Depending on your organizational structure, rank may be beneficial, but it should never precede other significant qualifications, such as adequate education and at least five years' job experience.

Physician

The department should designate a physician to provide guidance to the program. The physician should have specific expertise and experience related to departmental operations. A background in occupational medicine is preferred. The American Board of Preventive Medicine certifies physicians as specialists in occupational medicine. Any departmental experience, if lacking, can be provided once the physician is on board.

The physician should serve as the primary medical contact and work directly or in liaison with the health and safety program director. The relationship may be established on a referral basis (Level 1), by contractual arrangement (Level 2), or as a full-time position (Level 3), depending on the needs and resources of the department.

It is recommended that the physician become familiar with department operations through direct participation in safety committee meetings, training exercises, and even ride-alongs. Occupational information should be available to him, including job analyses, job descriptions, and SOPs, as well as the basic medical or physical standards.

Depending on the level of services and estimated costs, it may be economically feasible for departments with as few as three hundred members to employ a full-time physician. Larger departments may anticipate even greater benefits from having a physician on staff. Beyond annual medical exams, other responsibilities that may be assigned to the physician include entry-level exams, injury rehabilitation, and return-to-work determinations, as well as wellness programs and infectious-disease control. A department that estimates the cost of providing annual exams to be above $150,000 should seriously investigate the full-time option.

Program Director

The program director is the designated uniformed officer appointed by the chief to manage the health and safety program. The director may be the department safety officer or some other similar entity, depending on the departmental structure. The director may serve as the program manager, or he may supervise a manager/fitness coordinator. Uniformed staff assignments will vary depending on the organization and the level of the program. In some organizations, the department physician may assume the director role, whereas in others, the safety officer may be assigned a dual-role capacity.

A nonuniformed, graduate-level candidate (master's or doctorate degree) with expertise in exercise physiology, kinesiology, or some other health-related field may be employed as program manager, *in addition to* a designated uniformed member, to function as an assistant. Occupational expertise is a must! You should also consider temporarily using local university interns to assist a uniformed member with program activities. I have found internship to be a rewarding and mutually beneficial relationship. The options for staffing are many, but fundamentally, someone must be designated manager.

HEALTH AND FITNESS COORDINATOR (HFC)

The following qualifications for the fitness coordinator are as identified in the appendix of NFPA 1500, *Fire Department Occupational Safety and Health Program, 1997 Edition.*

> *A-8-7.2 The minimum qualifications for the health and fitness coordinator should include emergency medical technician (EMT) state certification; health and fitness coordinator certification, which includes topics such as physical fitness training, exercise physiology, kinesiology, nutrition, weight control counseling, critical incident stress management, and substance abuse training and education; and NFPA 1041, "Standard for Fire Service Instructor Professional Qualifications, Instructor II."*

If qualified personnel are scarce, nonexistent, or simply not interested in participating, then you will have to examine the alternatives. Members who express an interest should be interviewed. Those who are selected should then be provided the necessary training. Training courses for a CFC are available nationwide, and some internationally, with course prerequisites ranging from no background whatsoever to a four-year college degree. The level of certification of a CFC will dictate what duties he is able to perform. For example, the Canadian Society for Exercise Physiology offers certification and accreditation for a professional fitness and lifestyle consultant (PFLC) and a certified fitness consultant (CFC). A PFLC candidate needs a university degree to enroll, whereas a CFC candidate does not. The ACSM also offers varying levels of certification, from fitness leader to program director, with different program entry requirements. Qualifications will determine training requirements and the type of certification needed. The duties and responsibilities of the CFC are then based on that certification level. (See Appendix G for a listing.)

The job tasks of a CFC (or HFC) will vary with the size of the department, the number of coordinators, qualifications, the length of a tour, and individual expertise. Some of the duties that should be considered are as follows.

1. Health risk appraisal.
2. Test administration.
3. Exercise recommendation.
4. Exercise programming and training.
5. Instruction.
6. The purchasing, maintenance, and repair of exercise equipment.
7. Injury investigation.
8. Medical evaluations.
9. Supervision of work-hardening programs.
10. Supervise transitional work programs.
11. Injury rehabilitation.
12. Ergonomic evaluations.
13. Counseling and referrals related to weight control and nutrition.
14. Research projects on fitness issues.
15. Administration of performance tests.

The recommended ratio of CFCs to department members is ideally 1:50, although fewer can be employed effectively if the job tasks are limited. For example, if the CFC is only required to conduct tests, then a large staff isn't required, since a test session may require only one or two administrators. This is true regardless of the size of the department, since the number that can be safely tested in one session is limited, about fifteen to twenty at best. However, if the CFC is expected to develop fitness profiles, write exercise recommendations, and monitor training sessions, as well as administer tests, then the 1:50 ratio becomes a more reasonable expectation.

Options

The preceding has been a brief description of the possible staff positions. Determining how many to fill is the next step. The first position is easy, since designating a department physician is mandatory. The absolute minimum requirement beyond the physician is a program manager. A CFC could serve as the manager and also conduct all program activities in a small organization (<100); however, larger departments will need additional CFCs to function without compromising services. Basically, the larger the organization, the more "layers" required. Some possible configurations appear in Figure 7–1.

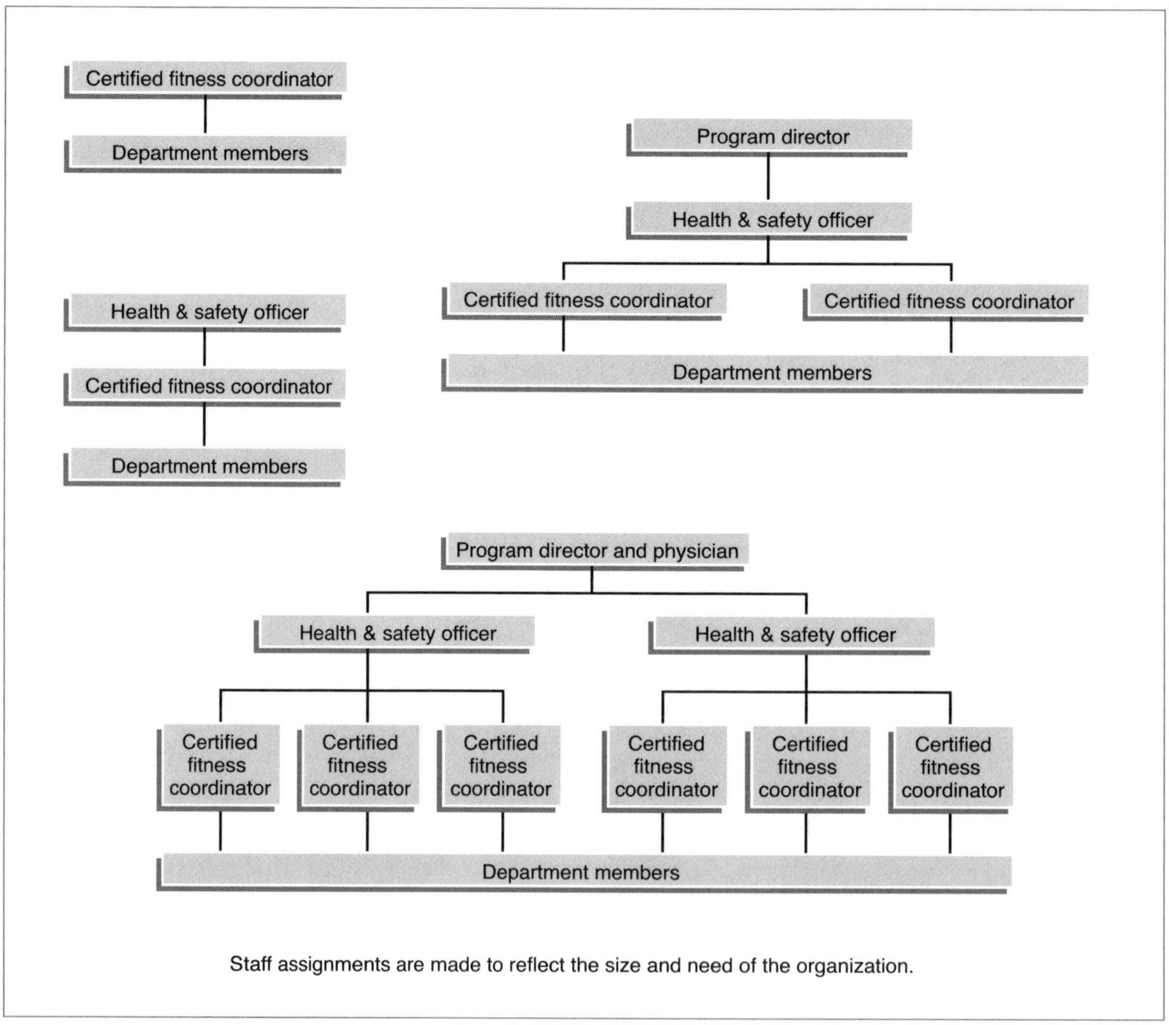

Fig. 7–1 Organization Chart Staff Options

A Staff Worksheet is provided in Table 7–1. Using this worksheet, determine cost estimates for staffing all positions at the three levels of participation (LOP). Don't forget to inventory your department for in-house experts.

TABLE 7–1

SAMPLE STAFF WORK SHEET (LOP MATRIX)

A small department may require only one CFC to manage and conduct the program. If the CFC performs collateral duty, the total staffing cost may be $500 (for certification) or less if he is prequalified (already holds a degree).

Staff Requirements	***Cost/Staff***	***Total Cost***	***Notes***
Physician on a referral basis. No staffing cost at the basic level. Referral fee is included on the medical work sheet.	N/A	Level 1 = $	Since the physician is not employed, the cost at Level 1 is the same figure for a Level 1 medical exam option.
Physician on a contract basis.	Fixed annual fee.	Level 2 = $	Contract basis entails a fixed annual fee of $10,000 to $20,000 (estimated).
Full-time physician.	Estimated $150,000 salary.	Level 3 = $	Full-time physician start at estimated $150,000.
Program director (collateral duty)	20% of salary.	Level 1 = $	Physician may assume role of program director if FTE.
Full-time manager, uniformed.	FTE, salary.	Level 2 = $	Salary based on rank.
FTE civilian, professional	FTE, salary.	Level 3 + $	Salary varies with experience and education. May be more or less than a uniformed member at Level 2. Civilian at about $36,000 plus benefits.
Certified fitness coordinator—department member certified for collateral duty.	Estimate $0-$500.	Level 1= $500	A small department may have one CFC performing all duties. If the CFC is prequalified (has a degree), no training cost required.
Fitness coordinator—degree. Uniformed or civilian.	Estimate $500 and percent of salary.	Level 2 = $	In-house survey may identify qualified member who can perform collateral duty.
Collateral or part-time duty. Additional CFCs on duty.	Estimate $500 each for training costs for part timers.		The number of CFCs required is dictated by the size of the department, typically on a 1:50 ratio.
FTE, fitness coordinator. Graduate-level education, civilian. Additional CFCs.	FTE, salary. Estimate $500 each for training costs of additional CFCs.	Level 3 = $	The full-time CFC may assume the function of program director/manager alone or with assistance from noncompensated on-duty coordinators.

MEDICAL EXAM

The medical exam is perhaps the most expensive item on the matrix inventory, since the costs may be incurred on an annual basis. The actual cost will depend on the level of service and the demographics of your department. Since the objective is to provide several cost options, it is necessary to collect the basic data, then plug in the appropriate numbers for the three levels identified in the matrix under the medical category.

Remember, there are also several options within the levels. For example, a basic exam may be given with or without a stress test and other alternatives based on age. Gather the essential information listed below, then work your options. If you prefer exact numbers in your work sheet versus the percentage recommended (10 to 25 percent), you can distribute the PAR-Q for an exact count of those members who require a referral for a medical exam prior to introducing the program. This survey will be used for estimating costs only, and you will have to administer the PAR-Q again on test day. A blank Medical Exam Work Sheet is provided in Appendix H, and a sample work sheet from an actual department, using 1998 cost estimates, appears in Table 7–2. The work sheet may appear somewhat confusing at first, but it is really rather straightforward and easy to complete. The variables are as follows.

Exams

1. Basic exam with resting EKG $__________
2. Basic exam with graded exercise test $__________

Membership

1. Total number of department members __________
2. Total number of members needing referral (10 percent of department), or use actual number based on a preprogram survey __________
3. Total number of male members forty years of age and older, and total number of females age fifty and older (add together) __________

The above figures are all you'll need to complete the Medical Exam Work Sheet. Gather the data, then plug it into the work sheet. On completion, you will have a range of costs and six exam options from which to choose. Figure 7–2 shows the relationship between exam costs and the level of risk or liability for each exam option. Simply apply some risk management principles and choose your comfort level.

TABLE 7–2

MEDICAL EXAM WORK SHEET, SAMPLE DATA

Sample department profile

N = 260 members — Exam A cost = $253.00
Age ≥ 40 male and ≥ 50 female = 74 — Exam B cost = $380.00
10% = 26 members

LOP Options	Cost and No. of Members	Total Cost of Options	Notes
Level 1, Health Risk Appraisal. Member deemed high risk with RISKO score of >31 at 60% $\dot{V}o_2$max. All others use standard test procedure. All asymptomatic members are tested.	N = 260	$0	Caution: If at any time prior to, during, or after the test a member experiences dizziness, shortness of breath, or chest pain, terminate the test and refer him to a physician.

LOP Options	Cost and No. of Members	Total Cost of Options	Notes
Level 1A, PAR-Q and RISKO. Administer PAR-Q, and refer all positives for exam. Estimate 10% of members, if unknown. Basic exam (resting EKG). Test RISK scores >31 at 60% $\dot{V}o_2$max.	M1 = 26 A = $253.00	M1 × A = $6,578.00	Administer PAR-Q to all members prior to program start to determine the exact number needing referral in place of the 10% estimate recommended.
Level 2, PAR-Q/RISK or HealthScreen. Administer PAR-Q and HS. Refer all positives for exam. Estimate 10% of members. Basic exam plus graded exercise test.	Basic exam + GXT M1 = 26 B + $380.00	Level 2 cost = $6,578 M1 × B = $9,880.00	Or, use the ACSM guide for specific requirements. The HealthScreen program in Appendix B interprets ACSM for you.
Level 2A, PAR-Q or HS, and refer 40 (50). Administer PAR-Q. Refer all all positives for exam. Estimate 10% of members. Refer all men over 40 and all women over 50. Determine number of members. Basic exam, resting EKG. Basic exam plus graded exercise test.	Basic exam + GXT and >40 (50). M1 = 26 M2 = 74 B = $380.00	Level 2A cost = $34,698 M1 × A = $6,578.00 M2 × B = $28,120.00 Total M1 + M2 = $34,698.00	Cost-reduction alternatives: If insurance reimbursement is available, reduce cost accordingly. If using age-related exam, schedule as recommended in NFPA 1582. Adjust annual cost to reflect the schedule. See Example A, below.
Level 3 Exam, All Members. Basic exam, resting EKG. Basic exam plus graded exercise test. Total number of department members.	A = $253.00 B = $380.00 N = 260	A × M = $65,780.00	If using a once-every-three-year schedule for all members, regardless of age, the actual cost will be 33% of the estimates determined on this work sheet.
Level 3A Exam + GXT for All Members. Total number of department members.	N = 260	Level 3A cost = $98,800. B × M = $98,800.	

Key

M	=	Total number of members.
M1	=	Ten percent of members.
M2	=	Number of members ≥ 40 male and ≥ 50 female.
A	=	Basic exam cost.
B	=	Basic exam cost including GXT (stress test).

Note: As shown, this department has a variety of cost options, ranging from $0 to $98,800, depending on LOP choice and whatever cost-reducing options are available.

Example A: Three-Year-Schedule Cost Estimate at Level 3A

Age Group	Year 1	Year 2	Year 3	Year 4
20–29	$28,880	$ 0	$ 0	$28,880
30–39	$41,800	$ 0	$41,800	$ 0
40+	$28,120	$28,120	$28,120	$28,120
Total	$98,880	$28,120	$69,620	$57,000

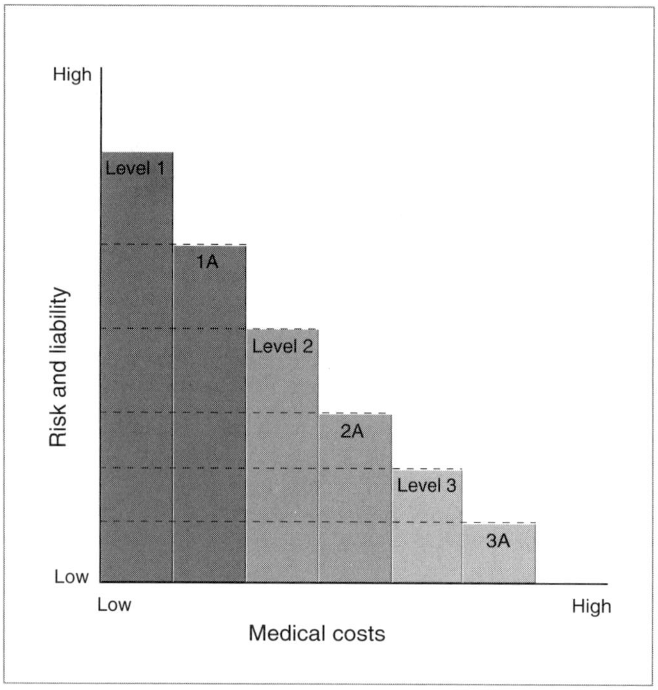

Fig. 7–2 Cost/Risk Comparison: LOP Medical Screen Levels

Once you have a cost estimate of the options at all three levels, you can compare these figures with your staff options. If the total cost of your medical exams exceeds the cost of employing a full-time physician, then you should consider employing the physician for pure economic reasons. If the cost is comparable or less, don't discount the option, since there are added incentives and benefits associated with having a full-time physician that may outweigh the additional cost. Infectious-disease control, wellness, and injury rehabilitation would all benefit from the services of a full-time, on-duty physician. The physician agreement (whether by referral, on a part-time basis, or by contract) that you made as a staffing option should prove influential when identifying medical alternatives and their cost.

Once you have completed the medical worksheet, there is one final cost adjustment to be made, based on your schedule and three important questions.

(1) Should I provide exams every year, two years, or three years?

(2) Should the exams be based on an age schedule, as identified in the NFPA standards?

(3) If the program includes a phase-in period, do I need to examine everyone the first year?

If you decide on using a voluntary-to-mandatory phase-in program, your costs may be spread out over a two-year period. My recommendation is that you use a phase-in period with exams scheduled every three years, based on need and regardless of age.

A basic exam should include the following.

1. A complete physical exam.
2. A complete medical history.
3. Lab hemoculture/V.A./biochemical analysis.
4. EKG (resting).
5. Health risk appraisal.
6. Pulmonary function.
7. Lumbar/sacral evaluation.
8. Graded exercise test (stress test) as an option.

EQUIPMENT AND FACILITIES

Estimating the cost of equipment and facilities is again a matter of collecting information and developing three cost options. The task should be assigned to a fitness coordinator, preferably someone knowledgeable about exercise equipment. He should use the equipment list provided in the LOP matrix.

The cost of equipment can range from a few dollars to an astronomical sum. Following the option criteria presented in Chapter 6, attach a dollar amount on an itemized basis. This can be done by way of direct contact with vendors or through a bid process, using the information given in the matrix. An example of equipment costs appears in Table 7–3. The figures are based on actual bids submitted by five equipment manufacturers in 1999.

TABLE 7–3

SAMPLE EQUIPMENT COSTS

(BASED ON 1999 MANUFACTURER BID PRICES)

Free-Weight Equipment	
Essential	
(a) Olympic weight bench	$ 95.00
(b) Olympic weight set (300 lbs.)	$199.00
(c) Weight tree	$ 35.00
(d) Roman chair	$145.00
(e) Power rack	$250.00
Totol	$724.00
Optional	
(a) Set of dumbbells 10-50lbs.	$ 216.00
(b) Dumbbell rack	$ 75.00
(c) Inclined bench	$ 145.00
(d) Curl bar	$ 30.00
(e) Lat-pulldown (plate load)	$ 285.00
(f) Leg-press (hip sled)	$ 795.00
(g) Two-100 lb. plates	$ 80.00
Total	$1626.00

Aerobic Exercise Equipment	
Essential	
None required. Can be done without equipment (e.g., walk/jog/run).	
Optional	
(a) Exercise bike	$ 600.00
(b) Stair machine	$1500.00 and up
(c) Rowing machine	$ 500.00 and up
(d) Ski machine	$ 450.00 and up

The same principle of phasing in the medical program can be applied to equipment and facilities. It isn't necessary to acquire everything that you potentially need or want in a startup program. To ask for the world is to ask for a delay. It's an excuse, not a reason. Start with the basics, and acquire additional equipment over time. I have found equipment purchasing to be an ongoing project, especially since it also entails maintenance and replacements.

When investigating resources, don't overlook the possibility of using secondhand equipment, or even of fabricating equipment in-house. Health and fitness clubs periodically update their inventory and often offer discounts on their old equipment. I know of one department that outfitted all of its firehouses to a Level 3 at bargain basement prices using secondhand machines from a local gym.

When determining what equipment you wish to purchase for the program, don't forget about test equipment. Dynamometers, calipers, stopwatches, and other such equipment may not be used for exercising per se, but they are essential to the fitness program. Consult the next chapter on test selection before listing your equipment options. You may want to select an ergometer or a variable resistance machine for testing purposes only; still, it must be included in your budget estimate. The equipment will dictate the test, not vice versa. You might even adopt a calisthenic program and purchase equipment solely for testing. Although the mode of exercise should determine the test, alternative modes of testing will at least be able to document a change in one's performance. Gather your figures, then plug in the numbers on the work sheet.

At Level 3, a fitness program would be based around a variety of exercise machines. When it comes to pricing, the sky's the limit, so take a reasonable approach and determine what you actually need. A seven-station unit consisting of a bench press, lat-pulldown, military press, leg press, leg curl/extension, and incline sit-up features is acceptable. As stated previously, if you're outfitting one central facility, you should buy top-quality devices.

RESOURCE REQUIREMENTS

Once the resource cost requirements have been determined for each category, the Level 1, 2, and 3 options for each area should be transferred to the budget estimate work sheet, shown in Table 7–4. You should provide an estimate for each category at each level. In this manner, you may be able to juggle your options and arrive at a cost that is within your budget.

TABLE 7–4

BUDGET ESTIMATE WORKSHEET

Category	*Level 1*	*Level 2*	*Level 3*
Direct Operating Costs			
Staff Physician Program director Program manager Fitness coordinator(s)			
Exams First level Second level (a)			
Training Fitness coordinator			
Materials Texts and publications—resources			
Direct Capital Costs			
Facilities (new or remodeled) Exercise equipment Test equipment			
Indirect Costs			
Administration Overhead			
Total Program Cost			
For each level For a combination of levels			

Adapted and reprinted from the National Fire Academy student manual *Training Program Management,* 1997, Emmittsburg, MD.

BUDGET APPROVAL

The last step is to submit the budget estimate for approval. I use the term *approval* optimistically, on the belief that a department that is committed to pursuing a fitness program will commit *some* funds toward accomplishing that goal.

Since a variety of cost combinations can be assembled (*e.g.*, Level 1 for equipment, Level 2 for medical, and Level 3 for staff), you can create a budget range from high to low, and thus gain some bargaining room based on your priorities. You also have some hard data to justify the budget request. The proposal should be submitted to the health and safety committee to prioritize the options. A notation of positive and negative factors should accompany the alternatives, expressed in both economic and social terms. Evaluating the positive and negative aspects of each option will ensure that the most effective alternative is chosen. If you are awarded a set dollar amount, you may choose to lower the expectations in one area and upgrade another. If one of the less desirable options is chosen, you may be able to qualify it as conditional, subject to change. The members may be more receptive to an unpopular choice if there are plans to upgrade that particular area as soon as funds become available.

The final proposal is then submitted to the chief for approval. Although it shouldn't be necessary to sell the program at this point (since the decision to proceed originated at the top level anyway), every attempt should be made to convince management to increase the level of services and to pursue upgrades, pending the results of the program.

RECOMMENDED READING

NFA course *Training Program Management, 1997*. Emmitsburg, MD: National Fire Academy.

CHAPTER 8

TEST SELECTION AND DEVELOPMENT

Once the fitness coordinators have been trained and certified, and the type of equipment has been determined, the next step is to develop or select tests. Since the choice depends on the equipment that you select, you should consider your testing purchases early, particularly if you intend to use equipment dedicated solely to testing.

To assess an individual's level of fitness, a specific test instrument designed to evaluate each fitness component must be selected and applied. A variety of tests are available for assessing all fitness dimensions. In this chapter, I will provide a menu of tests from which to choose, offering several options for each component. The menu does not include all of the tests that are available but, rather, a select few that I believe represent the most commonly used instruments. Deciding which test to use depends on a number of factors, including validity, reliability, objectivity, economy, and application. A test is deemed valid if it measures what it purports to measure. It is reliable if the test results are consistent with repeated measures, and it is objective when the results aren't affected by the administrator. Since all of the tests that I have listed are valid and reliable instruments, choosing between them will simply be a matter of preference.

Each fitness component has a separate menu, with the test listed by type or name, a coefficient of correlation when available, and a resource for additional information on each procedure. Just in case you decide not to investigate your options, a sample test battery with protocols is supplied in Appendix I for your convenience.

Since there are almost unlimited potential combinations of test batteries, I have intentionally excluded some test protocols in favor of references, hopefully to encourage a responsible test administrator to investigate the options. It is only through reading the literature that one can determine the most appropriate selection for his intended test group. One of the most important requirements of an effective program manager is to stay informed, since he may have to explain or even defend his choice.

The test lists aren't exhaustive, and if you presently use one that isn't listed, don't worry. If you have a valid test that works well within your current program, then continue using it, particularly if you have collected a significant quantity of raw test scores. The data will be useful in developing your test standards. Remember that the only requirement is that you evaluate each fitness component using a valid test instrument.

VALIDITY

There are three types of validity described in the *Uniform Guidelines on Employee Selection Procedures* (1978) that are accepted by the American Psychological Association (APA); namely, criterion, construct, and content. A test that appears to measure what it is supposed to in a somewhat obvious manner may be said to have content validity. Often face validity, a term not commonly appreciated, is used in the same context. A push-up, for example, can be said to possess face validity, for it measures the ability to perform push-ups. A hose drag included in a test battery for firefighters would be considered content valid in the sense that firefighters drag hose and the event looks like a firefighter task.

The best example of a construct is physical fitness as defined earlier, which included the four components of muscular strength and endurance; aerobic capacity; flexibility; and body composition. It can be said that these four underlying traits form the construct of physical fitness. A fitness test that includes assessment of all of these components can be said to be construct valid.

An example of criterion validity would be in the use of a 1.5-mile run as a test of aerobic capacity. The criterion ($\dot{V}O_2max$) is significantly correlated to the predictor (the 1.5-mile run).

RELIABILITY

Reliability refers to repeatability. For example, if you measured your waist circumference with a cloth tape, then repeated the measure again five minutes later with different results, the measure would not be considered reliable. Reliability and validity are related in that a measure that is unreliable cannot be deemed valid. However, reliability does not necessarily ensure validity. A test may reliably produce an invalid measure. Errors in the administration of a test often affect reliability.

OBJECTIVITY

A test is objective when the results are independent of the administrator. The instructions or directions given to the candidates should be clear and provided in the same manner to all of the candidates. A test is objective when there are no outside influences that can positively or negatively affect the results.

ECONOMY AND APPLICATION

Economy refers to the expense or cost associated with a test and the ease of use. The 1.5-mile run is a highly economical test, since it only requires a stopwatch and a measured track. An example from the opposite extreme would be the maximal graded exercise test, which requires expensive equipment and highly trained professionals to administer it.

Application refers to the type of test or population being tested. It would be inappropriate to assess firefighters using a test that was designed for children or one that was validated using elderly cardiac patients. Similarly, selecting an anaerobic test to assess aerobic capacity would be an example of improper application.

ENERGY FITNESS TESTS

In Chapter 4, we defined fitness and determined that energy fitness has two pathways, one being aerobic and the other anaerobic. The selection of only one aerobic test is necessary. Although others can be offered as options, this is generally impractical from an administrative perspective. Several anaerobic tests are provided in a separate menu for those who like using this type of test, although using them isn't considered necessary in a standard fitness test battery. There is no gold standard for anaerobic tests, and since there are no general anaerobic tests and anaerobic capacity is highly correlated with strength and evaluated with some endurance-type testing, I advise selecting one aerobic measure from this category.

Aerobic capacity refers to the body's ability to take in, transport, and use oxygen. The maximal value is expressed as $\dot{V}O_2max$ and reported in liters per minute ($L \bullet min^{-1}$ = aerobic capacity) or relative to body weight in milliliters per kilogram of body weight per minute ($mL \cdot kg^{-1} \cdot min^{-1}$ = aerobic power). The maximal graded exercise test ($\dot{V}O_2max$) is considered the gold standard for aerobic measures. The direct measure of $\dot{V}O_2max$ involves the use of a treadmill or cycle ergometer and sophisticated equipment to measure respiratory gases. The process is time consuming, expensive, and impractical for routine fitness programs, particularly when a large number of people have to be tested. Therefore, alternatives have been developed that can estimate aerobic capacity from submaximal field-test protocols. The test menu consists of field tests that have been statistically correlated to the $\dot{V}O_2max$ for validation. A coefficient of correlation is included, with reference to at least one resource or research study. The coefficient is a numerical value that expresses the relationship between two variables. The relationship may vary from $^-1$ to $^+1$. The closer to plus or minus 1, the higher the correlation. Don't get hung up on the numbers, however, since any of the tests listed will get the job done.

The aerobic tests identified in the menu are valid and reliable, yet they vary with respect to predictive ability, ease of use, economy, and application. The results will vary between tests to the extent that using more than one type to provide an individual option is not advised. To illustrate the point, I administered five different tests to the same firefighter over a two-week period. The result, as shown in Table 8–1, was five different scores.

TABLE 8–1

VARIANCE OF RESULTS OF TESTS ADMINISTERED TO A SINGLE SUBJECT

Treadmill	Stair Machine	1.0-Mile Walk	Ergometer	Step Test
45	32	41	28	36

Values per $\dot{V}O_2$, $mL \cdot kg^{-1} \cdot min^{-1}$

The test scores range from a low of 28 to a high of 45. Assigning a fitness classification to this firefighter depends on which test is selected. If the cycle ergometer is used, then the firefighter would be classified as having poor aerobic capacity. On the other hand, if the treadmill or the walk were selected, he would be classified as average to slightly above average. So which should we choose?

Ideally, the test mode should match the training mode. Stated simply, the best test for a runner is a run test, and the best test for a cyclist is the cycle ergometer. Since exercisers use a wide variety of training modes, providing all possible alternatives is impractical. It is therefore necessary to select the single best mode that would provide a fair estimate for the majority of participants. That mode should also be compatible with program procedures—group or individual testing, whichever is desired.

The treadmill, cycle ergometry, stair machine, and other similar tests that restrict testing to one individual at a time are less desirable and practical than those that can be administered to multiple subjects simultaneously. If adequate time and staffing are available to test one at a time, then selecting one of these formats is recommended. However, if limited resources require that you test multiple subjects, then your options are reduced to either a walk, a run, or a step test.

The walk, run, and step tests are inexpensive and easily adapted to groups. The ability to predict $\dot{V}O_2max$ accurately varies considerably with these tests. Specificity again plays a role. Step tests involve stepping up and down on a box or step that can be anywhere from twelve inches to eighteen inches in height, depending on protocol. Since even the lowest height of twelve inches is taller than a standard stair or walkway step, the test may result in early termination due to local muscle fatigue. Unless the subjects specifically train using a high-step routine, the aerobic prediction could be lower than actual capacity. The Modified Canadian step test (mCAFT), if administered properly, can provide good results. Step tests are inexpensive, easy to administer, and a good method for testing groups.

A running test, such as the 1.5-mile run, will provide a good estimation for runners, but it tends to underestimate larger individuals who are unaccustomed to running. It also requires a maximal effort, which may not be desirable for older members. Also, pacing has a significant effect on the result, a problem that isn't found in walk tests. The run test is the least expensive and easiest to administer, and it can be offered in a range of distances. It's also easy to score and can accommodate large groups.

Everyone walks, and walk tests compare favorably, biomechanically speaking, with the $\dot{V}O_2max$ test, treadmill version. If the walk test is performed on a track, an optional run test can be offered as a substitute and administered simultaneously. This is the only combination of alternative tests that I recommend for field testing a group (Table 8–2).

Choose *one* aerobic test method for the fitness test battery. Alternative modes may be selected for entry-level tests or for use in the annual medical exam. Choose an anaerobic test *if desired.* Anaerobic tests are appropriate for public safety officers. Although they typically aren't found in fitness test batteries, they are performance related and have been shown to correlate well with some job tasks.

As long as a test is a valid measure of aerobic capacity, it will serve the purposes of the program. A number of factors need to be considered in selecting a test, and preferences will vary. My own preference is the Rockport Walking Test.

ROCKPORT WALKING TEST

(The following is adapted from LeCuyer, J.E. (1998), Assessing firefighter aerobic capacity: the Rockport field test. *Fire Engineering,* February, pp. 49–54.)

The Rockport one-mile walking test was developed at the University of Massachusetts Medical School and funded by the Rockport Walking Institute in Marlboro, Massachusetts. The institute commissioned the medical school to develop an alternative field test to estimate maximal oxygen consumption ($\dot{V}O_2max$) using a one-mile walk. Over a period of eighteen months, 343 healthy adults (165 males, 178 females) with an age range of thirty to sixty-nine years were evaluated. The investigation generated six equations for estimating the $\dot{V}O_2max$ using a treadmill protocol, and the one-mile walk test was r = 0.93 (aerobic capacity expressed as $L \bullet min^{-1}$), indicating an accurate test for an individual's aerobic capacity (Kline *et al.,* 1987).

To evaluate the effectiveness of using a one-mile walk to assess the aerobic capacity of firefighters, I conducted a study with two Colorado career fire departments. This study deviates from the original conducted in 1987 by Kline *et al.* in several ways. First, the firefighters in Group A were instructed to walk as fast as possible per Kline *et al.,* and those in the Group B were instructed to walk briskly but not necessarily as fast as they could. Emphasis was placed on performing four even laps. The purpose of this was to test a theory deduced from the case study, that variable paces can estimate the $\dot{V}O_2max$ accurately.

The second deviation from the original study was that the best equation (aerobic capacity) reported by Kline *et al.* and presented in the article abstract wasn't used in this study. An optional equation (aerobic power) provided by Dr. Ann Ward, one of the original investigators, with a lower correlation (r = 0.88) than the best equation (r = 0.93) was selected. The purpose of this was to evaluate one alternative formula within a career firefighter population.

The third deviation is that the firefighters performed only one trial versus the two or more performed in the original study. The purpose of this was to limit the exercise to a single trial, as in an actual field test.

The fourth deviation is that a case study was added, using one volunteer performing eight walk tests at varying paces over a period of one week. The purpose of this was to assess the accuracy of the formula based on the theory that varying walk paces would produce heart rate adjustments that would preserve the accuracy of the equation.

TABLE 8–2

ENERGY FITNESS MENU

Test: Aerobic	Reference	Textual Mention	Fitness/Work-Task Relationship
Direct measure, maximal (clinic)			
1. Treadmill	Pollack *et al.* (1976); Dickinson (1997)	Chapter 8	Abeles *et al.* (1973)
2. Cycle ergometer	Bruce *et al.* (1973); ACSM (1993)		Barnard & Duncan (1975)
			Barnard (1974)
Indirect measure,			Burgess (1997)
maximal (clinic)			Cady et al. (1985)
1. Treadmill	Heyward (1998)		Collingwood
			Davis & Dotson (1987)
			Davis & Santa Maria (1975)
Indirect measure,			Davis & Santa Maria (1982)
submaximal (field test)			Davis *et al.* (1982)
1. Cycle ergometer			Davis *et al.* (1979)
a. Astrand-			Gardner et al. (1974)
Rhyming (m)	Legge & Bannister (1986), r = 0.98		Gilman (1993)
b. YMCA	Golding *et al.* (1989)		Gledhill & Jamnik (1992a)
c. Siconolfi	Siconolfi *et al.* (1982), r = 0.86 male,		Hanson (1985)
	r = 0.97 female		Hurley (1987)
			Jacobs (1976)
2. Step tests			Johnson (1976)
a. Siconolfi	Siconolfi *et al.* (1985), r = 0.92		Kilbom (1980)
b. mCAFT	CSEP (1997), r = 0.88		Lemon & Hermiston (1997a)
c. YMCA	Golding *et al.* (1989)		Louhevaara *et al.* (1979)
d. USFS	IAFF/IAFC (1997)		Lusa *et al.* (1994)
			Misner *et al.* (1987)
3. Run/Walk Tests			O'Connell *et al.* (1986)
a. 1.5-mile run	Cooper, K. (1982)	Appendix I	Puterbaugh & Lawyer (1983)
b. 12-minute	Cooper, K. (1977)		Raven (1977)
walk/run	Kline, G.M., *et al.* (1987), r = 0.88,		Sharkey (1997)
c. 1-mile walk	r = 0.90		Sothman *et al.* (1992)
(Rockport)	LeCuyer, J.E. (1998), r = 0.88	Appendix I	Sparks (1987)
4. Stair machine			
a. StairMaster	Vanderburgh & Considine (1995)		
5. Nonexercise test			
a. Paper/pencil	Jackson *et al.* (1990)	Appendix I	

Test: Anaerobic			
1. Margaria-Kalamen	Margaria *et al.* (1966)		Gledhill & Jamnik (1992)
2. Wingate anaerobic			Jacobs (1979)
test (WAT)	Inbar *et al.* (1996)		Lemon & Hermiston (1977b)
3. Vertical jump	CSEP (1997)		Misner (1987)
4. Arm-crank dynamometer	CDF (1987)		O'Connell (1986)
5. 200-meter run	Fox & Matthews (1974)		
6. Paper/pencil test	Jackson *et al.* (1990)		

Eighteen career firefighters volunteered to participate in the study. Each firefighter's age, weight, and vitals were recorded, in addition to a health risk appraisal and activity index prior to participating.

Each firefighter performed a treadmill test at the University of Colorado Human Performance Laboratory to determine $\dot{V}O_2$max. A warm-up period (HR: 98–120 bpm) was conducted for a minimum of four minutes at a 0-percent grade, with treadmill speed judged by activity history and heart-rate response to walking speed. If the firefighter appeared average or above in work capacity, based on warm-up heart rate and what a firefighter ought to be able to perform for a minimum of six minutes and a maximum of twelve to fourteen minutes, the grade was started at 5 percent. If the firefighter appeared average in work capacity, the test was started at 0-percent grade. The treadmill grade was raised 2.5 percent each two minutes until volitional maximum was reached by the firefighter. Each firefighter was encouraged to go all out and challenge the test. If a firefighter reached a 15 percent grade and completed the stage, the treadmill grade was lowered to 10 percent, and the firefighter was allowed to run at a comfortable speed of six to eight miles an hour. Metabolic measurements were taken each twenty seconds, heart rate each sixty seconds, and blood pressure every two minutes.

The walk test was performed on two separate days (one for each department) on a standard quarter-mile track. Only one test trial was performed by each firefighter. The original study required each subject to perform a minimum of two trials to account for possible learning effects. The results reveled that one trial was sufficient to produce an accurate estimate of $\dot{V}O_2$max, and since repeated trials would reduce the utility of the test, a single trial was performed in this study. The test protocol for Group A was to walk as fast as possible. For Group B, it was to walk briskly, but not necessarily as fast as possible. The emphasis was to maintain even laps. Each firefighter was equipped with a heart rate monitor, and the heart rate was recorded immediately on completion of the fourth lap. The total walk time was also recorded. The firefighter's age, body weight, final heart rate, and walk time were logged and entered into the formula to calculate $\dot{V}O_2$max. The formula selected was as follows.

$$(0.0769 \times \text{body weight}) + (0.3877 \times \text{age}) + (3.2649 \times \text{time}) + (0.1565 \times \text{heart rate}) - (132.853 + 6.315 \text{ male only}) = \dot{V}O_2\text{max mL/kg}^{-1}.\text{min}^{-1}.$$

The correlation between actual $\dot{V}O_2$ (treadmill) and estimated $\dot{V}O_2$ (walk) for Group A was $r = 0.88$, which is identical to the correlation found using this formula in the original study. The correlation for Group B was $r = 0.62$, considerably lower than Group A. The poorer correlation for this group can be attributed to a smaller group ($n = 6$) and slower test times, which for two members represented only 20 percent and 30 percent of their $\dot{V}O_2$max, an intensity well below the minimum threshold for an accurate estimation.

A case study was conducted to evaluate the potential for identifying a submaximal pace that could determine aerobic capacity with accuracy. One fifty-year-old career firefighter of average aerobic capacity volunteered to perform a series of one-mile walk tests over a period of seven days. During the test trials, both heart rate and time were recorded for each lap, thus providing four pairs for each one-mile walk test. Of the eight test trials, two were thrown out as a result of duplicate or near-duplicate times with other trials. This case study poses two questions. First, can varying paces in lieu of a maximal effort produce an accurate score? If yes, is there an accurate submaximal effort that falls within the American College of Sports Medicine guidelines of 40 to 60 percent of $\dot{V}O_2$max, thereby providing the option of testing higher-risk individuals without physician clearance?

In addressing the first question, remember that the walk test protocol instructs the subject to walk as fast as possible. The firefighter in this case study was instructed to walk slowly on the first walk test, then to increase his pace for each subsequent test until attaining his maximal speed. The reason for varying the pace was to identify the range of paces and intensity of effort that are possible without affecting the accuracy of the equation. As is typical of most firefighter groups, there are always a few who won't put forth their best effort, and if lower intensity levels can produce an accurate estimation, retesting the less motivated may not be necessary.

In addressing the second question, if lower intensities can produce an accurate estimation, then higher-risk individuals who would normally require a medical exam prior to testing could be safely evaluated without a pretest exam. The ACSM guidelines (*Guidelines for Exercise Testing and Prescription, Fourth Edition,* Philadelphia: Lea and Feibiger, 1991) state that "It is unnecessary for asymptomatic, apparently healthy men and women, irrespective of their age or CAD risk factor status, to have a medical evaluation by a physician before embarking on a program of moderate (*i.e.,* exercise intensity 40 to 60 percent of $\dot{V}O_2$max) exercise training."

In this case study, the fifty-year-old firefighter's corresponding heart rates for a 40 to 60 percent $\dot{V}O_2$max were as follows.

For this subject, a submaximal test that produces a heart rate of 134 or less is safe to perform based on ACSM guidelines. For members who are considered high risk, a heart rate monitor could be worn by the firefighter to ensure that the heart rate limit isn't exceeded.

The results presented in Table 8–4 support the conclusion that varying paces can be performed with good results. The first trial at 30 percent $\dot{V}O_2$max is too slow to be effective. The second trial, although an improvement, is still 3 ml. below the most accurate score. The third trial at 40 percent appears to represent the minimum level for a good estimation. Therefore, efforts ranging from 40 to 80 percent for this individual all represent acceptable intensity levels that will produce good results. Trial 5 produced the same score as the maximal effort in Trial 6, at a lower intensity, without undue stress or muscular discomfort for the subject, supporting the theory that a maximal effort isn't essential. The last trial was the subject's maximal effort, representing 80 percent $\dot{V}O_2$max and approximately 90 percent maximum predicted heart rate, the upper limit recommended by ACSM. Based on the ACSM guidelines, this subject could be safely evaluated without prior medical clearance and with reasonable accuracy (to within 1 mL of his max effort) in the 40- to 60-percent range.

TABLE 8–3

HEART RATES PER $\dot{V}O_2$MAX IN A FIFTY-YEAR-OLD SUBJECT

% $\dot{V}O_2$max	Heart rate	% max heart rate
40	116	68%
45	120	71%
50	125	74%
55	130	76%
60	134	79%

TABLE 8–4

TRIAL RESULTS AT VARYING INTENSITIES

Test trial	Ending heart rate	Test time completion	$\dot{V}O_2$max estimate	% $\dot{V}O_2$max
1	107	21:22	19	30%
2	114	15:43	37	38%
3	117	14:30	40	40%
4	125	14:06	40	50%
5	136	13:15	41	62%
6	151	12:43	41	80%

Note: To calculate the corresponding heart rate with percent $\dot{V}O_2$, use the Karvonen Formula provided in Chapter 12.

Having used this test for thirteen years, I've had the opportunity to assess both the advantages and the limitations of this method. The conclusion that I've reached is that the advantages far outweigh the disadvantages, and that there are few field tests that can compete with the Rockport Walk Test. The test is easily administered, and it was validated over a broad age range of men and woman who were heterogeneous in $\dot{V}O_2max$. This enables generalization to a broad population, including firefighters. Also, the correlation of the walk test to the $\dot{V}O_2max$ test is quite high (r = 0.88), with a standard error of estimate (SEE) of -4.98. Cross-validation reported r = -0.92. Most tests aren't cross-validated nor provided a measure of the SEE, which describes the accuracy of prediction. Also, the test doesn't require maximal effort, as demonstrated in the case study. Unlike running tests, where pacing significantly affects the outcome, walking is fairly easy to pace. The even pace was reflected in the lap-time splits for heart rate, which ranged from 2 to 8 bpm. Also unlike running tests, the time factor alone doesn't determine the $\dot{V}O_2$ score. This eliminates a competitive environment.

Walking is virtually injury free. The landing force of each foot plant is about one to one and a half times one's body weight, as compared with running forces of three to four times body weight. This is particularly important to firefighters who weigh 180 pounds or more.

Since the test can be performed with good results at varying paces and intensities, it is a good alternative for overweight or high-risk individuals. The results can then be used to prescribe a walking program, which is an excellent mode of training for such individuals. It's easy for such a person to test himself periodically to assess his own aerobic condition. The concept of specificity applies to the test protocols for a walk test. Since not all modes of training can be accommodated within a specific test, the walk test is the best alternative, since everyone does it.

There are several limitations that I have found. First, the faster paces require the development of a technique. Also, the anterior tibialis (shin) muscle becomes irritated during fast walking. A third problem is that the participants won't always follow directions. Since an accurate assessment can be achieved at a comfortable pace, the first two problems are easily rectified. The anterior tibialis is typically a weak muscle, and fast-walk training will help to strengthen it. Thus, the problem will eventually disappear. The third problem is present with all types of tests. To minimize adverse responses, the administrator should emphasize the goal.

MUSCULAR FITNESS TESTS

The assessment of muscular fitness includes muscular strength, muscular endurance, and flexibility. Both strength and endurance measures can be assessed using either static or dynamic techniques or equipment. The static (isometric) mode implies that no observable movement takes place, and the dynamic mode refers to movement in which the muscle or body part does exhibit motion.

STATIC TESTS

Static tests employ equipment such as cable tensiometers or dynamometers. Cable-tension devices are typically used to assess leg and back strength, although virtually all muscle groups can be assessed with the equipment. The subject is required to apply maximal force, usually for a period of about three to five seconds. The force is transferred by the compression of a spring device that records the force in pounds or kilograms.

The most familiar dynamometer, and the easiest one to use, is the hand-grip type. Results vary among investigators, with strong relationships (r = 0.90) established between grip strength and other strength measures, such as bench and leg presses. There is a less desirable correlation with firefighter work simulations (r = 0.56). The exception in the relationship between grip and work simulation was with female firefighters, who exhibited a strong relationship (r = 0.85) (LeCuyer, 1998).

A study using an arm dynamometer to assess the upper-body static strength of steelworkers reported an average correlation of r = ⁻0.84 with twelve simulated work tasks (Arnold et al, 1982), while a similar assessment of one hundred firefighters reported only r = ⁻0.42 (Wollack and Assoc., 1986). Varying test protocols and work tasks may account for the difference in results. In the same two studies, a half-squat dynamometer test using the same protocol but different work tasks produced similar results (firefighters at r = ⁻0.52; steelworkers at r = ⁻0.61).

Since the physical demands of steelworkers, based on the DOT coding, are comparable to the strength requirements of firefighters (V, very heavy), it may be assumed that such measures would be acceptable in a fitness test battery. The tasks used in both studies were generically similar, such as lifting, carrying, and so on, with comparable weight loads expected of both occupations.

The Arnold study concluded that the single measure of static arm strength was sufficient (r = ⁻0.84) to predict total body performance, and it therefore appears to be the best choice of the static dynamometer test protocols. The study implied that various types of strength are sufficiently related to produce a general strength assessment using one test.

Isometric measures of muscular strength and endurance don't require equipment, and they're scored on the basis of time as opposed to force applied. The wall sit is one example, in which the subject sits unsupported with his legs in a right-angle position, his back against the wall for a specified period, usually one to three minutes. The flexed arm hang is another isometric endurance test that requires the subject to hang from a chin bar, his arms flexed in the up position. A sit-up crunch may also be used, in which the individual is timed while holding a sit-up position at an angle of about 45 degrees. Some of the static test options are illustrated in Appendix I.

The advantages of using tensiometer and dynamometer equipment for field tests are their ease of use, standardization, and the accuracy of recording forces. The isometric tests are also easy to use and accurate if performed correctly.

The disadvantages are that the test protocols require maximal effort, forcing the subjects to strain themselves in attempting to move an immovable object. Typically an individual will cease straining within one and a half to two seconds when attempting to lift an object that won't move. This action, when sustained for several seconds, will produce elevated blood pressures, which can be dangerous to high-risk individuals. Most experienced weight lifters know within a second or two whether a load is too heavy, and will consequently stop applying force to avoid injury. In the static test, one is required to continue applying force, sometimes for as long as five seconds. In the assessment of leg and back strength, the lower back is the weak link. Subjecting it to an excessive load can easily result in injury. Since the lower back is the limiting factor in determining strength level, leg strength may not be properly evaluated. Strength is also highly specific to the angle of the joint; therefore, repeated accuracy in determining the correct angle is essential for proper evaluation. The final disadvantage is related to specificity, a term mentioned above with respect to training. Most work and activity involves motion. Work is inherently dynamic, and tests designed to measure strength should reflect this observation. The use of static tests is effective and necessary in the context of injury rehabilitation, but it's of questionable use in a fitness test battery and can be dangerous, particularly when administered by inexperienced testers.

DYNAMIC TESTS

Since the real world is a place of movement, dynamic tests are the preferred choice. Most dynamic tests involve a variety of equipment, usually free weights and variable-resistance machines, although some tests, such as calisthenic types, require no equipment at all.

Earlier in the text, I expressed a preference for using free weights in an exercise program. Understanding the concept of specificity, which states that the test mode should match the training mode, free weights would appear to be the logical choice for strength-test equipment; however, with regard to testing, I make an exception. There are several reasons not to use free weights as a test instrument. The most important is safety. When attempting to assess maximal strength (I recommend sub-max efforts later), there is an element of risk in losing control with free weights, particularly for the novice. The problem can be resolved by using spotters or a Smith machine, in which a free-weight bar is attached to a power rack-type cage. This controls the movement while providing a safety lockout that can be engaged anytime during the lift. The safety problem is basically nonexistent with variable-resistance machines, making them the preferred choice on this issue alone. If you choose to use more than one mode-free weight and variable-resistance machine, you'll have to adjust the scores to ensure that no one device provides an advantage. For example, a plate-loaded hip sled would require an adjustment factor of 0.71 to correlate with a free-weight lift.

A second problem is related to the variations in training modes (*e.g.*, free weights, calisthenics, variable-resistance machines, isometrics) as mentioned earlier with respect to aerobic training. It is impractical to offer all variations of test alternatives. Therefore, we again must decide on the best test, or one mode of preference that will provide a fair assessment for the majority of participants. The mode that I recommend is a variable-resistance machine (VRM). Either the plate-load or select-stack type will work. The choice is a matter of cost. Anyone, regardless of training preference, can be accurately evaluated without prior experience. If your training program is based on calisthenics, then calisthenics are the obvious choice for a test; however, you can effectively employ a VRM for test purposes in a calisthenic-based program. If an individual's strength has improved as a result of a strength-training program that doesn't employ a VRM, his strength gains will still be detectable using a VRM test. Although a test that has a constant resistance (*e.g.*, free weights) is different from one that has a variable resistance, the last few inches of bar travel to lockout is the same, and that's what's scored. The purpose of testing is to document change, and although an accurate assessment is desired, precision is unnecessary and best left to the research environment. The VRM gets the job done, and it serves our purposes well.

UPPER BODY

Any strength assessment should incorporate one test for each of the three main body segments; namely, the upper body, the lower body, and the midsection.

The bench press has historically been regarded as the gold standard for strength testing. It has been highly correlated with job performance, particularly in occupations that require heavy lifting. The bench press is therefore recommended for assessing upper-body strength. A VRM is the preferred mode to assess this dimension. One alternative is the push-up. Although push-ups are viewed as a test of muscular endurance, their relationship to absolute strength is significant enough to warrant using them. The relationship between absolute strength and absolute muscular endurance is positive (Biddle, 1986). If you're using a program of calisthenics, then push-ups are your best option for measuring upper-body strength. Although the relationship between strength and relative endurance is considered negative, I have found a significant relationship between job performance tasks and a sixty-second leg-press test of maximum repetitions, 100 pounds above the firefighter's body weight (LeCuyer, 1988). Other upper-body measures include seated military presses, lat pull-downs, and arm curls. The last two correlate well ($r = 0.72$, $r = 0.77$) with job-task simulations. Additional calisthenic options include dips, chin-ups, and a flexed-arm hang, to name a few (Johnson and Nelson, 1979).

Lower Body

The single best measure of lower-body strength is the leg press. The plate-loaded leg-and-hip sled is preferred over the standard VRM leg press for reasons related to the start position. The legs are flexed on most VRMs in the start position, thus requiring substantial hamstring involvement to begin each repetition. The plate-load version starts with the legs extended, a much more desirable position, and the exercise can be performed in a horizontal (leg press) or vertical (squat) position. The same problem exists with the VRM bench press, but it can easily be corrected by providing assistance to raise the bar, thus helping the subject to start in the extended position (Maud and Foster, 1995). A free-weight dead lift is another alternative that can be performed in a power rack to assess leg and back strength. Full-squat dead lifts should be discouraged in favor of partial lifts starting above a leg angle of 90 degrees.

The calisthenic or body-weight options are the one- or two-legged squat, squat thrusts, and the isometric wall sit. The one-legged squat requires an assistant or rail to help balance the subject, and the measures are repeated for each leg. This is a high-intensity test. If it is too intense, then a two-legged version for repetitions can be substituted. Again, you should begin with a leg angle of greater than 90 degrees to avoid overstressing the knee.

Although squat thrusts are considered to measure a combination of agility and overall body endurance, they have also been identified as an excellent predictor of general strength among steelworkers, and they do sufficiently tax the leg muscles (Arnold *et al.*, 1982). Used by the U.S. Navy Seals, the squat thrust is an excellent exercise that can assess the anaerobic energy system quite well if performed in two-minute timed reps.

Midsection

The obvious test for the torso is the standard sit-up. The best method is a one- or two-minute maximal repetition. I recommend the crunch-style version, in which the lower back remains in contact with the floor as the subject curls up enough to raise the scapula off the floor. This is the safest version, since it isolates the abdominal muscles while reducing assistance from the hip flexors. Thus, it provides a quality assessment of abdominal strength. It can be performed dynamically or statically using the isometric sit-and-hold variation. A further comparison of these exercises appears in Table 8–5.

Flexibility Tests

There is no single test of flexibility, since range of motion (ROM) is specific to each body segment or part. Direct measures of flexibility are done using a variety of equipment, such as dual-bubble inclinometry, the Leighton flexometer, and the goniometer. The direct assessment of flexibility using these devices is best left to the professional physical or occupational therapist. Direct assessments should be included in the annual physical exam so that a baseline status can be recorded. This information will be useful in the clinical setting, providing baseline data for comparison in the event of a postinjury evaluation. For field tests, an indirect method, such as the standard Wells-Dillon sit-and-reach test, is easy to administer and highly recommended. Increased flexibility is often associated with reduced risk of injury, although no conclusive evidence is available to support this theory. It is generally believed to be true, since tight muscles would logically be prone to injury if overextended. The principal reason to evaluate ROM is more related to establishing a baseline for later comparison, as mentioned above. The lower-back region of the body is the most frequently injured; therefore, assessment of the lower back and hamstring group is advisable. The sit-and-reach test is

TABLE 8–5

MUSCULAR FITNESS MENU—STRENGTH AND ENDURANCE

Test: Upper Body	Reference	Textual Mention	Fitness/Work-Task Relationship
Calisthenic (body weight)			
Push-ups	Johnson & Nelson (1979)		Arnold *et al.* (1987)
Dips	Cicciantelli (1991)		Ben-Ezra & Verstraete (1988)
Chin-ups	Johnson & Nelson (1979)		Blankenship (1987)
Pull-ups (modified)	Johnson & Nelson (1979)	Appendix I	Cady *et al.* (1979)
Flexed arm hang	Terra (1987)	Appendix I	CDF (1987)
			Clark (1966)
Free Weights			Clark (1986)
(Constant Resistance)			Davis (1982)
Bench press	Jackson *et al.* (1980)		Dawson & Hurley (1989)
Upright row	NSCA Journal		Dickinson (1997)
Bar curl	Johnson & Nelson (1979)		Doolittle (1979)
			Gledhill & Jamnik (1992)
Machine (VRM, resistance,			Hoffman (1979)
static, other)			Humphrey (1988)
Bench press	CSEP (1997)		Kroemer (1985)
Lat pulldown	CSEP (1997)		LeCuyer (1989)
Seated military press	CSEP (1997)		Lemon & Mermiston (1977a)
Arm curl	CSEP (1997)		Morrow (1981)
Hand-grip dynamometer	Heyward (1998)		Terra (1987)
Arm-curl dynamometer	Blankenship (1987)		
Test: Midsection			
Sit-ups (crunch style)	Juker *et al.* (1998)	Appendix I	
Sit-ups (isometric)	CDF (1987)	Appendix I	
Sit-ups (bent-leg variations)	CSEP (1997)		
	Golding *et al.* (1989)		
Test: Lower Body			
Calisthenic (Body Weight)			
Wall sit (isometric)	CDF (1987)	Appendix I	
One-leg squat	Phoenix Cardinals, NFL	Appendix I	
Burpee (squat thrust)	Johnson & Nelson (1979)	Appendix I	
Free Weight			
Squat	Johnson & Nelson (1979)		
Deadlift (partial)	NSCA Journal		
Machine (VRM, dynamometer)			
Leg press	Altug (1987); Jackson (1980)		
Hack squat	Sigman & Duncan (1990)		
Back-leg dynamometer	Heyward (1998)		
Leg extension/curl	Altug (1987); CSEP (1997)		

considered a reliable method of assessing hamstring flexibility, but somewhat less effective in assessing low-back flexion. Since both are interrelated, the sit-and-reach test is the best bet for assessment in the field.

Performance on the Wells-Dillon sit-and-reach test appears to be affected by body type, gender, and limb length. Therefore, a modified version was developed to correct for limb length. Women typically perform better on both versions due to anatomical features. Their pelvic structure and greater range of motion in the elbow allow them to extend further when tested (Gelabert, 1966). The modified sit-and-reach test was designed to adjust for the limb length of individuals who aren't proportional, thereby affecting their scores and the accuracy of assessment. Since the test protocol is adjusted on an individual basis, I refer to this modified version as a relative measure and the nonadjusted version as an absolute measure. I prefer the absolute measure, since it is more closely related to the real world. An officer seated in a vehicle who drops his radio and is forced to reach down and forward to retrieve it won't experience the cab floor rising to adjust for his limb length. He'll either reach it or he won't. I have documented several injuries associated with similar moves.

A third method of performing the sit-and-reach test involves isolation and measurement of the hamstring muscles, both alone and in conjunction with the lower back. This evaluation produces two scores, isolating each muscle group and therefore identifying the ROM culprit. This technique is often used by physical therapists. The final choice is a matter of preference, and any variation will serve our purposes (Appendix I). For entry-level candidates, the inclinometry method is preferred and recommended. Direct and indirect measures of flexibility are shown in Table 8–6.

BODY COMPOSITION

Although body composition isn't considered a performance measure, it is assessed in a fitness test battery because it has considerable influence on performance ability (Wilmore, 1976; Wilmore and Haskell, 1972; Davis, 1980). In addition to performance, the health risks are also of primary concern and reason enough to assess this important dimension.

The hydrostatic test is the gold standard for determining body composition. This test is expen-

TABLE 8–6

MUSCULAR FITNESS MENU—FLEXIBILITY

	Reference	Textual Mention	Fitness/Work-Task Relationship
Direct Measures (Clinical)			
Dual-bubble inclinometry	Alter (1996)		Cady *et al.* (1985)
Goniometer	Heyward (1997)		Hilyer *et al.* (1990)
Flexometer	CSEP (1997)		Khalil *et al.* (1980)
			Russell & Weld (1992)
Indirect Measures (Field Test)			Sharkey (1997)
Wells-Dillon sit and reach	Wells & Dillon (1952)	Appendix I	Williams & Sperryn (1976)
	Golding *et al.* (1989)		
Modified sit and reach	Hoeger (1987, 1990)	Appendix I	
	Alter (1996)		

sive and time-consuming, and it requires an elaborate setup that isn't practical for routine use. The alternatives for field testing are many, but the most frequently used are bioelectric impedance, skin-fold caliper, and anthropometric measures. Anthropometric techniques are easy to use and relatively accurate; however, the degree of accuracy varies greatly between techniques, and the resultant scores should be accepted as ballpark estimates only. Table 8–7 illustrates this point clearly. One firefighter was evaluated using five different techniques, with the results shown below. In this case, he would certainly express a preference for the hydrostatic test, if offered a choice.

TABLE 8–7

COMPARISON OF METHODS FOR DETERMINING BODY COMPOSITION

Bioelectric impedance	Skin-fold Caliper	Hydrostatic	Circumference	Anthropomorphic measures
12%	15%	7%	11%	18%

The routine use of body-fat measures should be confined to documenting before-and-after measures to determine whether any change has occurred. The hydrostatic test should be reserved for individuals with serious weight problems, since they could profit from a more accurate analysis. Since few individuals in any given organization would require such special attention, the added cost would be justified to help these individuals.

Choose one method for the test battery. Since I don't recommend scoring body composition, the accuracy of the method isn't critical. Ease of use and the ability to detect change are the main criteria. For individuals who have a weight problem and are enrolled in a specific weight program, the hydrostatic test may be the best choice. Measures of body composition are further addressed in Table 8–8.

TEST DIMENSIONS

EQUIPMENT

A menu of tests has been provided for each fitness component. The choice of test depends on four essential variables: validity, reliability, objectivity,

TABLE 8–8

BODY COMPOSITION MENU

	Reference	Textual Mention	Fitness/Work-Task Relationship
Clinical Test			
Hydrostatic weighing (underwater)	Heyward (1997)		Davis & Starck (1980)
Plethysmograph	McCrory *et al.* (1995), r = 0.96		Davis *et al.* (1982)
			Williams & Evenson (1988a)
Field Tests			Wilmore & Haskell (1972)
Skin-fold calipers	Jackson & Pollack (1985), r = 0.91 male, r = 0.85 female		Wilmore (1976)
Body circumference	Hodgdon (1992), r = 0.88	Appendix I	
	Heyward (1997)		
Bioelectric impedance	Lukasi *et al.* (1985), r = 0.86		

and ease of administration. One other important aspect of testing, independent of these variables, is the type of equipment that we select. Unless a department can afford equipment for testing purposes only, the equipment and type of program will dictate the choice of tests. Don't be overly concerned with having the exact same test equipment as exercise equipment. Although this is the ideal, it may be impractical. A variable-resistance machine is an excellent test instrument, but it would be very expensive to provide it for everyone to use on a daily basis. Individuals who have improved their strength through alternative modes of exercise can demonstrate their progress on a VRM even if they're unfamiliar with the equipment. The stronger of two individuals will score higher on any given test of strength. The ultimate goal is to document and track a member's physical status during the course of a training program and throughout his career.

TEST PROTOCOLS

Once the tests have been selected, you should review the literature to verify the application, as well as to obtain test procedures and protocols. Complete the test battery table by listing your choices for each dimension. Then, write a protocol for each.

TEST SCORES

Once the tests have been selected for each component, a method of scoring should be determined. All tests will produce raw data, some of which can be used as a final score. For instance, calisthenic tests provide a final score, the number of repetitions. The sit-and-reach test records inches, another final score. Aerobic and strength tests, however, may produce raw data that requires further manipulation to derive a final score. The walk test mandates that heart rate and time be recorded, then age and weight are factored in to produce a final $\dot{V}O_2$ score. Strength tests may require estimates of the one-repetition maximum by using a factor multiplied by weight lifted, then a division process to arrive at a relative score. The relative score would be a ratio of the body weight to the weight lifted. This would be considered the final score. If points are used in the scoring process, another step must be added for that calculation.

Energy Fitness Test

1. Aerobic ______________________________
   Alternate ______________________________
2. Anaerobic ______________________________
   Alternate ______________________________

Muscular Fitness Test

1. Strength ______________________________
   A. ______________ Alternate ______________
   B. ______________ Alternate ______________
   C. ______________ Alternate ______________
2. Endurance ______________________________
3. Flexibility ______________________________

Body Composition

1. Body Fat Measure ______________________________

Fig. 8–1 Fitness Test Battery

TEST

Title: ____________________

Version: ____________________

Purpose: To define the dimension that you're measuring.

Equipment: Identify all equipment necessary for the assessment. Don't forget items that aren't specific to the test but are necessary, such as a weight scale.

Procedure: Provide a step-by-step procedure as to how the test is to be administered and performed. A practice session is recommended to ensure that all steps are identified. Don't forget the alternatives if you choose to allow any.

Illustration: A drawing of the start and end position may be desirable for test administrators.

Scoring: Identify how the data is to be collected and used. If there is a formula involved, write it down. Include all information needed to compute the score, such as weight, age, gender, and the like. Note whether you're using relative or absolute scores.

Standard: Note the criterion or minimal standard.

Reference: Cite the test source and attach any literature that is appropriate for future reference, particularly studies relevant to the occupation.

Fig. 8–2 Test Protocol

The simplicity of scoring therefore depends on the type of test and the desired type of score. Some standards don't provide any numerical score, but rather, a categorical score, such as "good" or "excellent." Other tests require the use of formulas or reference tables. Therefore, the ease of scoring should be considered in the selection of tests, particularly if many subjects are to be evaluated. If a complex procedure is selected, the only efficient alternative is to develop a computerized method for calculating the formulas and scores. Computerization is the preferred method, since it can speed the process and ensure accurate computation.

TEST STANDARDS

To complete the scoring process, the raw scores must be converted into final scores, then applied to a standard. In the development of your program, it may not be possible to predetermine your standards. If your standards are based on the statistical computation of incumbent scores, then your standards cannot be developed until the pilot program tests are administered and completed. If you intend to use available norms or standards, then you should determine your scoring procedures at this time. Those who elect alternative methods need only determine, at this stage of the program, the type and method of collecting raw scores and converting them to final scores. Standards are best left for development following the pilot study.

RECOMMENDED READING

Alter, Michael (1996). *Science of Flexibility, Second Edition*. Champaign, IL: Human Kinetics.

Heyward, Vivian (1998). *Advanced Fitness Assessment and Exercise Prescription, Third Edition.* Champaign, IL: Human Kinetics.

Golding, L.A.; Myers, C.R.; and Sinning, W.E. (editors) (1989) *The Y's Way to Physical Fitness: The Complete Guide to Fitness Testing and Instruction.* Champaign, IL: Human Kinetics.

Maud, P.J.; and Foster, C. (1995) *Physiological Assessment of Human Fitness.* Champaign, IL: Human Kinetics.

Chapter 9

Developing the Pilot Study

The basic program plan has been decided, the staff has been selected, and the necessary equipment has been acquired. A certified fitness coordinator, assigned to the task of selecting and developing tests, has completed his research and assembled a pretrial test battery. Now a plan for implementing the program must be developed, beginning with a pilot presentation. There are two presentations that need to be written: a lesson plan to introduce the departmental program and test battery, and a second lesson plan for the individual exercise and training program. The two presentations would be delivered on separate days when implementing the program departmentwide; however, for the pilot program, it is recommended that both be presented on the same day, divided into a morning and afternoon session. In addition to the lesson plans, all necessary resource materials must be developed or acquired prior to presentation.

Developing the Lesson Plan

The first lesson plan is designed to introduce the program to the membership; describing its origin, policy, and purpose; outlining the steps for implementation; and generally selling the product. The second lesson plan addresses the actual exercise program that is to be performed by individual members while on duty. A good method for writing these is to follow the precepts of the National Fire Academy's lesson plan format, as promulgated in its *Instructional Methodology and Course Development* course. The five steps of the program are Motivation, Presentation, Application, Evaluation, and Conclusion.

Motivation Step

Lesson Plan No. 1. Two questions must be answered in the first step: "What is the purpose of this program?" and "Why do we need it?" The first three chapters of this book should serve as a guide toward answering these.

Lesson Plan No. 2. To accomplish the goals of the department regarding the fitness of its members, each member must participate in a structured program that adheres to the fundamentals of exercise training. Departmentwide improvement is only possible through application of these principles.

Presentation Step

Lesson Plan No. 1. The decisions as to how the program has been designed can be explained using the levels of participation matrix and by discussing the options and the reasons for selecting a particular format. Constructive feedback from the pilot study can also be used to support decisions once the pilot is complete and the members have been introduced to the program.

Lesson Plan No. 2. This should consist of an explanation of the training effect, as well as the four

FITT principles; namely, frequency, intensity, time, and type. A handout should be developed for instructional purposes.

APPLICATION STEP

Lesson Plan No. 1. The application step is the activity segment. It should include a health risk appraisal (HRA), the collection of basic personal information, and a description of the test protocols and procedures. The HRA is evaluated, and medical referral is given as needed. Each test is explained and demonstrated.

Lesson Plan No. 2. The different types of exercises for strength, flexibility, and aerobic conditioning are explained, discussed, and demonstrated. The basics of exercise programming are presented with examples of common workout routines. Individuals are then asked to compute their beginning exercise intensity levels based on information provided in Chapter 12.

EVALUATION STEP

Lesson Plan No. 1. The actual test is administered at this stage. You should also include a description of the scoring methods and standards. For more information on these, consult Chapter 11.

Lesson Plan No. 2. The evaluation portion of this lesson plan involves cross-checking individual calculations, such as heart rate estimates, and reviewing individual exercise routines to determine whether the basic training principles have been applied. After this paper-and-pencil activity has been completed, direct each member to perform his individual exercise routine, with the fitness coordinator observing to ensure proper form and technique.

CONCLUSION

Lesson Plan No. 1. Following the administration of the test, describe how the results will be tabulated and distributed. Schedule any retests or individual counseling at this time. You should also distribute a survey at this stage of the process.

Lesson Plan No. 2. Provide instruction with regard to exercise safety, and inform the members that their exercise routines will be periodically observed. Offer additional information and assistance as requested.

RESOURCE MATERIALS

You should provide an exercise log and diary to each member. Exercise logs are commercially available in a variety of formats, or they can be designed by a fitness coordinator and reproduced in-house. A log should note the frequency, intensity, and duration of each exercise. Entries in the diary are more subjective, expressing how one feels during and after a given workout. Both, if kept accurately, are important to the participant and the fitness coordinator. The information from the log and diary can be used to identify problems and make adjustments in the exercise prescription.

Each member should also maintain a health and fitness logbook. This is a confidential personal file best kept in a three-ring binder. A health and fitness file should include information on injury and accident reports; occupational exposures; fitness test results; performance test results; injury rehabilitation log; medical exam reports; exercise log summaries; and other health and fitness information.

You should also create a reference guide for the station, containing comprehensive information on the exercise program. The reference guide can be in the form of a three-ring binder. You can develop it from numerous sources and include information on all aspects of fitness and training. The guide should also include policies and procedures relevant to the fitness program; the organization and administration of the program; rules; tests; standards; and any other relevant information. Combining all of the information in one accessible source can eliminate much confusion and many redundant questions that will arise if the essentials aren't publicized.

Post exercise charts and illustrations in the workout area. Many types are available commercially from fitness organizations and equipment manufacturers. Some basic charts can be made in-house and laminated for durability.

RECOMMENDED READING

Canadian Society for Exercise Physiology (1997). *The Canadian Physical Activity, Fitness, and Lifestyle Appraisal: CSEP's Guide to Health and Active Living*. Ottawa, Ontario.

Gledhill, N., and Jamnik, V.K. (1992) Characterization of the physical demands of fire fighting. *Canadian Journal of Sports Medicine*. 17 (3): 207–213.

Golding, L.A.; Meyers, C.R.; and Sinning, W.E. (1989) *The Y's Way to Physical Fitness: The Complete Guide to Fitness Testing and Instruction, Third Edition*. Champaign, IL: Human Kinetics.

Heyward, Vivian (1998). *Advanced Fitness Assessment and Exercise Prescription, Third Edition*. Champaign, IL: Human Kinetics.

NFA student manual: *Instructional Methodology* (1987). Emmitsburg, MD: National Fire Academy.

PART III REFERENCES

Abeles, F.J.; DelVecchio, R.J.; and Himel, V.H. (1973) *A Fire Fighter's Integrated Life Protection System, Phase I: Design and Performance Requirements*. New York: Grumman Aerospace Corp.

American College of Sports Medicine (1993). *ACSM Resource Manual Guide for Exercise Testing and Prescription, Second Edition*. Philadelphia: Lea and Febiger.

Adams, T.D.; Yanowitz, F.G.; Chandler, S.; Specht, P.; Lockwood, R.; and Yeh, M.P. (1986) A study to evaluate and promote total fitness among fire fighters. *Journal of Sports Medicine*. 26: 337–345.

Alter, Michael (1996). *Science of Flexibility, Second Edition*. Champaign IL: Human Kinetics.

Altug, Z.; Hoffman, J.L.; Slave, S.M.; Farabaugh, E.J.; Truschel, D.W.; and Bemesderfer, L.L. (1990) Work circuit training. *Industrial Physical Therapy*. 10 (5): 41–48.

Arnold, J.D.; Rauschenberger, J.M.; Soubel, W.G.; and Guion, R.M. (1983) Validation and utility of a strength test for selecting steelworkers. *Journal of Applied Physiology*. 67(5): 588–603.

Bahrke, M.S. (1982) Voluntary and mandatory fitness programs for fire fighters. *Physician in Sports Medicine*. 10: 126–132.

Barnard, R.J., and Duncan, H.W. (1975) Heart rate and ECG responses of firefighters. *Journal of Occupational Medicine*. 17: 247–250.

Barnard, R.J. (1974) *Electrocardiographic and Heart Rate Responses to Sudden Exercise and to the Fire Alarm*. Proceedings of Second Symposium on the Occupational Health and Hazards of the Fire Service. New York: International Association of Fire Fighters.

Ben-Ezra, V., and Verstraete, R. (1988) Stair climbing: an alternative exercise modality for fire fighters. *Journal of Occupational Medicine*. 30: 103–105.

Biddle, S.J. (1986) Personal beliefs and mental preparation in strength and muscular endurance tasks: a review. *Physical Education Review,* 8 (2): 90–103.

Blankenship, K. (1987) *Preemployment Screening With Isometric Strength Testing*. Macon, GA: American Therapeutics Inc.

Bruce, R.A.; Kusumi, F.; and Hosmer, D.(1973) Maximal oxygen intake and nomographic assessment of functional aerobic impairment in coronary disease. *American Heart Journal*. 85: 545–562.

Cady, L.D.; Bischoff, D.P.; O'Connel, E.R.; Thomas, P.C.; and Allan, J.H. (1979) Strength and fitness and subsequent back injuries in fire fighters. *Journal of Occupational Medicine*. 21: 269–272.

Cady, L.D.; Thomas, P.C.; and Kawasky, M.S. (1985) Program for increasing health and fitness in fire fighters. *Journal of Occupational Medicine*. 27: 110–114.

CDF (1987). *Physical Fitness Guidelines, No. 1839*. Sacramento, CA: California Department of Forestry, pp. 16–17.

Canadian Society for Exercise Physiology (1997). *The Canadian Physical Activity, Fitness, and Lifestyle Appraisal: CSEP's Guide to Healthy Active Living*. Ottawa, Ontario.

Cicciantelli, P. (1991) The dip. *National Strength and Conditioning Association Journal*. 13 (6): 53.

Clarke, D.H. (1966) *Muscular Strength and Endurance in Man*. Englewood Cliffs: Prentice–Hall.

Clarke, D.H. (1986) Sex differences in strength and fatigability. *Research Quarterly Exercise and Sport*. 57: 144–149.

Collingwood, T. (1995) Physical fitness standards: measuring job relatedness. *The Police Chief*. February, pp. 31–34, 46–47.

Cooper, K.H. (1977) *The Aerobics Way*. New York: Bantam.

Cooper, K.H. (1982) *The Aerobics Program for Total Well-Being*. New York: Bantam.

Craig, F.N.; Blevins, W.V.; and Roehlich, H. (1971) *Training to Improve Endurance in Exhausting Work of Men Wearing Protective Masks: A Review and Some Preliminary Experiments*. AD 729787, Aberdeen Proving Ground, MD.

Davis, P.O.; Biersner, R.J.; Barnard, R.J.; and Schamadan, J. (1982) Medical evaluation of fire fighters: how fit are they for duty? *Postgraduate Medicine*. 72 (2): 241–248.

Davis, P.O., and Dotson, C.O. (1987) Physiological aspects of fire fighting. *Fire Technology*. 23: 280–291.

Davis, P.O.; Dotson, C.O.; and Sanata Maria, D.L. (1978) The physical requirements of fire fighters. *Fire Command*. January, pp. 36–38.

Davis, P.O.; Dotson, C.O.; and Santa Maria, D.L. (1982) Relationship between simulated fire fighting tasks and physical performance measures. *Medicine and Science in Sports and Exercise*. 14: 65–71.

Davis, P.O., and Starck, M.A. (1980) Excess body fat—not age—viewed as greater culprit in fitness decline. *Fire Engineering*. June, pp. 33–37.

Dawson, P., and Hurley, G.F. (1989) Strength training improves endurance performance and lactate threshold without altering $\dot{V}O_2max$. *Medicine and Science in Sports and Exercise.* Supplement 21 (2): 523.

deVries, Herbert (1974). *Physiology of Exercise, Second Edition.* Dubuque, IA: Wm. C. Brown, p. 436.

DOT (1993). *Selected Characteristics of Occupations Defined in the Revised Dictionary of Occupational Titles*. U.S. Dept. of Labor: Washington, DC: U.S. Govt. Printing Office.

Duncan, H.W.; Gardner, G.W.; and Barnard, R.J. (1979) Physiological responses of men working in fire fighting equipment in the heat. *Ergonomics*. 40: 482–489.

Dickinson, A. (1996) *Human Factors in Organizational Design and Management V: Combining Kinesiology, Aging Science, and Research Design to Produce Valid and Defensible Job Analyses*. Proceedings of the Fifth International Symposium on Human Factors in Organizational Design and Management. Held in Breckenridge, CO, July 31–August 3.

Doolittle, T.L. (1979) *Validation of Physical Requirements of Firefighters*. Seattle: Seattle Fire Department.

Dotson, C.O.; Santa Matia, D.L.; Schwartz, R.A.; and Davis, P.O. (1976) *Development of Job-Related Physical Performance Tasks for Firefighters*. Washington, DC: U.S. Fire Administration.

Ebbeling, C.; Ward, A.; Puleo, E.; Widrik, J.; and Rippe, J. (1991) Development of a single-stage submaximal treadmill walking test. *Medicine and Science in Sports and Exercise*. 23: 966–973.

Espich, J., and Williams, B. (1967) *Developing Programmed Instructional Materials*. Palo Alto, California: Fearon Publishers.

Foster, C.; Pollock, M.L.; Rod, J.L.; Dymond, D.S.; Wible, G.; and Schmidt, D.H. (1983) Evaluation of functional capacity during exercise radionucleide angiography. *Cardiology,* 70: 85–93.

Gagne, R.; and Briggs, Leslie J. (1974) *Principles of Instructional Design*. New York: Holt, Rinehart, and Winston Inc.

Gardner, G.W.; Barnard, R.J.; Disco, N.V.; and Kattus, A.A. (1974) *Near-Maximal ECG Stress Testing and CHD Risk Factor Analysis in Los Angeles City Firemen*. A paper presented at the 21st Annual Meeting of the American College of Sports Medicine. Knoxville, TN.

Gelabert, R. (1966) *Raul Gelabert's Anatomy for the Dancer.* New York: Danad.

Gerkin, R.; Kelley, P.; Perry, R. (1998) *Correlation of $\dot{V}O_2max$ During Maximal Treadmill Stress Testing With $\dot{V}O_2max$ at 85% PMHR: A Retrospective Review of the Phoenix Fire Department Treadmill Protocol.* Unpublished paper. Phoenix Fire Department.

Gledhill, N.; and Jamnik, V.K. (1992) Characterization of the physical demands of fire fighting. *Canadian Journal of Sports Medicine.* 17 (3): 207–213.

Golding, L.A.; Meyers, C.R.; and Sinning, W.E. (1989). *The Y's Way to Physical Fitness: The Complete Guide to Fitness Testing and Instruction, Third Edition.* Champaign, IL: Human Kinetics.

Hanson, R.I. (1985) *A Comparison of a Stair Treadmill Ergometer and Actual Stair Climbing Based on Oxygen Consumption.* University of Southern California Graduate School.

Heyward, Vivian, H. (1984) *Designs for Fitness: A Guide to Physical Appraisal and Exercise Prescription.* Macmillan Publishing Company.

Heyward, Vivian (1998). *Advanced Fitness Assessment and Exercise Prescription, Third Edition.* Champaign IL: Human Kinetics.

Hilyer, J.C.; Brown, K.C.; Sirles, A.T.; and Peoples, L. (1990) A flexibility intervention to reduce the incidence and severity of joint injuries among municipal firefighters. *Journal of Occupational Medicine.* 32 (7) 631–637.

Hodgdon, J. (1992) Body composition in the military services: standards and methods. *Body Composition and Physical Performance, Applications for the Military Services.* Marriott, B.; and Grumstrop-Scott, J. (editors), Washington, DC: National Academy Press, p. 65.

Hoeger, W.K. (1987) *The Complete Guide for the Development and Implementation of Health Promotion Programs.* Englewood, CO: Morton.

Hoeger, W.W.K.; Hopkins, D.R.; Button, S.; and Palmer, T.A. (1990) Comparing the sit-and-reach with the modified sit-and-reach in measuring flexibility in adolescents. *Pediatric Exercise Science,* 2(2)156–162.

Hoffman, T.; Stauffer, R.W.; and Jackson, A.S. (1979) Sex differences in strength. *American Journal of Sports Medicine.* 7: 265–267.

Humphrey, D. (1988) Abdominal muscle strength and endurance. *The Physician and Sports Medicine.* 16 (2): 201–202.

Hurley, B.; Hagberg, J.; and Goldberg, A. (1987) Resistance training can reduce coronary risk factors without altering $\dot{V}O_2$max or percent body fat. *Medicine and Science in Sports and Exercise,* 20: 150–154.

IAFF/IAFC (1997). *Fire Service Joint Labor Management Wellness-Fitness Initiative.* Washington, DC: International Association of Fire Fighters.

Inbar, O.; Bar-Or, O.; and Skinner, J.S. (1996). *The Wingate Anaerobic Test.* Champaign IL: Human Kinetics.

Jackson, A.S.; Blair, S.N.; Mahar, M.T.; Wier, L.T.; Ross, M.; and Stuteville, J.E. (1990) Prediction of functional aerobic capacity without exercise testing. *Medicine and Science in Sports and Exercise,* 22: 863–870.

Jackson, A.W.; and Langford, N.J. (1989) The criterion-related validity of the sit-and-reach test: replication and extension of previous findings. *Research Quarterly for Exercise and Sport,* 60(4): 384–385.

Jackson, A.S.; and Pollock, M.L. (1986) Practical assessment of body composition. *The Physician in Sports Medicine,* 13 (5): 76–90.

Jacobs, D.T. (1976) *Physical Fitness and the Fire Services.* Boston, MA: National Fire Protection Association.

Juker, D.; McGill, S.; Kropf, P.; and Steffen, T. (1998) Quantitative intramuscular myoelectric activity of lumbar portions of psoas and the abdominal wall during a wide variety of tasks. *Medicine and Science in Sports and Exercise,* 30 (2): 301–310.

Johnson, B.L.; and Nelson, J.K. (1979) *Practical Measurement for Evaluation in Physical Education, Third Edition.* Minneapolis: Burgess.

Khalil, T.M.; Asfour, S.S.; Martinez, L.M.; Waly, S.M.; Rosomoff, R.S.; and Rosomoff, H.L. (1992) Stretching in rehabilitation of low-back-pain patients. *Spine,* 17(3): 311–317.

Kilbom, A. (1980) Physical work capacity of firemen. *Scandinavian Journal of Work-Environment Health,* 6: 48–57.

Kline, G.M.; Pocari, J.P.; Hintermeister, R.; Freedson, P.S.; Ward, A.; McCarron, R.F.; Ross, J.; and Rippe, J.M. (1987) Estimation of $\dot{V}O_2$max from a one-mile track walk, gender, age, and body weight. *Medicine and Science in Sports and Exercise,* 19: 253–259.

Kroemer, K.H.E. (1985) Testing individual capability to lift material: repeatability of a dynamic test compared with static testing. *Journal of Safety Research,* 16: 1–7.

LeCuyer, J.E. (1988) *Recruit Training Program.* Unpublished data. Denver, CO: Rocky Mountain Fire Academy.

LeCuyer, J.E. (1989) *The Relationship Between Fitness Profiles and Job Performance.* Unpublished report. Denver, CO: Rocky Mountain Fire Academy.

LeCuyer, J.E. (1998) Assessing aerobic capacity: the Rockport field test. *Fire Engineering*, February, pp. 49–52, 54, 56.

Legge, B.J.; and Banister, E.W. (1986) The Astrand-Rhyming nomogram revisited. *Journal of Applied Physiology*, 61: 513–515.

Lemon, P.W.R.; and Hermiston, R.T. (1977a) The human energy cost of fire fighting. *Journal of Occupational Medicine*, 19: 558–562.

Lemon, P.W.R.; and Hermiston, R.T. (1977b) Physiological profile of professional firefighters. *Journal of Occupational Medicine*, 19: 337–340.

Louhevaara, V.; Smolander, T.; and Tuomi, T. (1985) Effects of an SCBA on breathing pattern, gas exchange, and heart rate during exercise. *Journal of Occupational Medicine*, 27 (3): 213–216.

Lukasi, H.C.; Johnson, P.E.; Bolonchuk, W.W.; and Lykken, G.I. (1985) Assessment of fat-free mass using bioelectrical impedance measurements of the human body. *American Journal of Clinical Nutrition*, 41: 810–817.

Lukaski, H.C.; Bolonchuk, W.W.; Hall, C.B.; and Siders, W.A. (1986) Validation of tetrapolar bioelectrical impedance method to assess human body composition. *Journal of Applied Physiology*, 60: 1327–1331.

Lusa, S.; Louhevaara, V.; and Kinnunen, K. (1994) Are job demands on physical work capacity equal for young and aging firefighters? *Journal of Occupational Medicine*, 36 (1): 70–74.

Margaria, R.; Aghemo, P.; and Rovelli, E. (1966) Measurement of muscular power (anaerobic) in man. *Journal of Applied Physiology*, 21: 1662–1664.

Maud, P.J.; and Shutz, B.B. (1989) Norms for wingate anaerobic test with comparison to another similar test. *Research Quarterly for Exercise and Sport*, 60 (2): 144–151.

Maud, Peter; and Foster, Carl (1995) *Physiological Assessment of Human Fitness*. Champaign IL: Human Kinetics, p. 121.

Matthews, D.K. (1973) *Measurement in Physical Education, Fourth Edition*. Philadelphia: W.B. Saunders Co.

McCrory, A.; Gomez, T.D.; Bernauer, E.M.; and Mole, P.A. (1995) Evaluation of a new air-displacement plethysmograph for measuring human body composition. *Medicine and Science in Sports and Exercise*, 27 (12): 1686–1691.

Mealy, M. (1979) New fitness for police and fire fighters. *Physician in Sports Medicine*, 7: 96–100.

Misner, J.E.; Plowman, S.A.; and Boileau, R. (1987) Performance differences between males and females on simulated firefighting tasks. *Journal of Occupational Medicine*. 29: 801–805.

Morrow, J.R.; and Hosler, W.W. (1981) Strength comparisons in untrained men and trained women athletes. *Medicine and Science in Sports and Exercise,* 29: 801–805.

NSCA (1988). *Strength Training and Conditioning for Speed Development: A Coaches Guide.* Colorado Springs, CO: National Strength and Conditioning Association, pp. 42, 44, 54–55.

NFA student manual *Course Development* (1988). Emittsburg, MD: National Fire Academy.

NFA student manual *Instructional Methodology* (1987). Emittsburg, MD: National Fire Academy.

NFA student manual *Training Program Management* (1997). Emittsburg, MD: National Fire Academy.

O'Connell, E.R.; Thomas, P.C.; Cady, L.D.; and Karwasky, R.J. (1986) Energy costs of simulated stair climbing as a job-related task in fire fighting. *Journal of Occupational Medicine,* 28: 282–284.

Pauletto, B. (1988) Basic program design in *Strength Training and Conditioning for Speed Development: A Coaches Guide.* National Strength and Conditioning Association, pp. 61–77.

Pollack, M.L.; Bohannon, R.L.; Cooper, K.H.; Ayers, J.J.; Ward, A.; White, S.R.; and Linnerud, A.C. (1976) A comparative analysis of four protocols for maximal treadmill stress testing. *American Heart Journal,* 92: 39–46.

Puterbaugh, J.S.; and Lawyer, C.K. (1983) Cardiovascular effects of an exercise program: a controlled study among firemen. *Journal of Occupational Medicine,* 25 (8): 581–586.

Raven, P.B.; Davis, P.O.; Shafer, C.L.; and Linnebur, A.C. (1977) Maximal stress test performance while wearing a self-contained breathing apparatus. *Journal of Occupational Medicine,* 19: 802–806.

Romet, T.T.; and Frim, J. (1987) Physiological responses to fire fighting activities. *European Journal of Applied Physiology,* 56: 633–638.

Russel, P.; Weld, A.; Peary, M.J.; Hogg, R.; and Unsworth, A. (1992) Variations in lumbar spine mobility measured over a 24-hour period. *British Journal of Rheumatology,* 31 (5): 329–332.

Sharkey, B.J. (1997) *Fitness and Work Capacity, Second Edition.* USDA Forest Service, Technology and Development Program. Missoula, MT. 4E42P30.

Siconolfi, S.F.; Garber, C.E.; Laster, T.M.; and Carleton, R.A. (1985) A simple, valid step test for estimating maximal oxygen uptake in epidemiologic studies. *American Journal of Epidemiology,* 121: 382–390.

Siconolfi, S.F.; Cullinane, E.M.; Carleton, R.A.; and Thompson, P.D. (1982) Assessing $\dot{V}O_2$max in epidemiologic studies: modification of the Astrand-Rhyming test. *Medicine and Science in Sports and Exercise,* 14: 335–338.

Sigmon, C.; and Duncan, D. (1990) The hack squat. *National Strength and Conditioning Association Journal,* August/September, 12 (4): 28–31.

Sothman, M.S.; Saupe, K.; Jasenof, D.J.; and Blaney, J. (1992) Heart rate response of fire fighters to actual emergencies: implications for cardiorespiratory fitness. *Journal of Occupational Medicine,* 34 (8): 797–800.

Sparks, P.J. (1987) *Minimum Medical Standards for Fire Fighters.* Olympia, WA: Association of Washington Chiefs.

Terra, D. (1987) *Validation of Physical Ability Tests for Selection of Firefighter II and Fire Apparatus Engineer.* Criterion Validation Unit, California State Personnel Board, Sacramento, CA.

Vanderburgh, P.M.; and Considine, W.J. (editors) (1995) *The Fitness Handbook, Second Edition,* pp.131–156. St Louis, MO: Wellness Bookshelf.

Wells, K.F.; and Dillon, E.K. (1952) Sit-and-reach, a test of back and leg flexibility. *Research Quarterly,* 23: 115–118.

Williams, T.; and Evenson, S. (1988) Physically fit for duty? by whose standards? part I: mandatory physical fitness programs. *Fire Chief,* pp. 43–46.

Williams, J.C.P.; and Sperryn, G. (1976) Stiffness is a disease of the unfit. *Sportsmedicine, Second Edition*. Baltimore, MD: Williams and Wilkins.

Wilmore, J.H. (1976) *Athletic Training and Physical Fitness: Physiological Principles and Practices of the Conditioning Process.* Boston: Allyn and Bacon Inc.

Wilmore, J.H.; and Costill, D.L. (1994) *Physiology of Sport and Exercise.* Champaign, IL: Human Kinetics.

Wilmore, J.H.; and Haskell, W.L. (1972) Body composition and endurance capacity of professional football players. *Journal of Applied Physiology,* 33: 564–567.

PART IV
IMPLEMENTATION

CHAPTER 10

CONDUCTING THE PILOT STUDY

The preliminary program has been developed and we are now prepared for implementation, which will begin with a pilot study. Depending on how well prepared we are, one or more pilot studies may be required, and they may consume the better part of a year. Don't become overanxious about getting the program off the ground. The extra time spent up front will be well worth the cost of a few months' delay.

The purpose of the pilot is to test the program, to work out the bugs, and to serve as a dress rehearsal prior to full implementation (DuBois, 1984). The pilot group should consist of the same number of members that you expect to schedule with incumbents in one test session.

Selecting subjects for the pilot program isn't the same as sampling for a research study or validation study. Although a fair cross section of the department is desirable, you probably won't be able to get one on a voluntary basis, nor is it necessary. A group of volunteers is preferable for several reasons. The volunteers that you get will typically be from among the higher-fit members, and they are most likely to be avid supporters of the program. They tend to be receptive to experimentation and welcome the opportunity to participate. They're often knowledgeable and can provide valuable suggestions. Since they're fit, any health concerns regarding their participation and the cost of screening them will be minimized. Not the least important benefit is that such people can help sell the program to the membership.

While conducting the pilot, make note of the time required to test an individual. You should also get a feel for the number of members that can be handled smoothly at one time. If possible, opt for smaller groups. Approximately ten to twenty people can be handled by a single administrator. If there are more test administrators or equipment test stations available, then the number of participants can be increased. Still, I recommend rotating small groups, with a short break in between. Splitting a group and conducting more test sessions provides greater control over the process.

In experimenting with schedules, test arrangements, and procedures, the goal is to find maximal efficiency with minimal downtime and disruption. If members are on duty but out of service, you'll want to return them to active status as soon as possible to avoid unnecessary delays. Don't expect to refine the process overnight. You'll continue to make adjustments for several years.

ADMINISTERING THE TESTS

Pay attention to the details. Score sheets and paperwork sound simple, but they can create a problem. The score sheets should be easy to use and allow you to track several subjects on a single page. Flipping through pages for a name can be irritating. The score sheet should also be arranged in a logical order, following the sequence of tests. Make sure that you have enough copies, and don't forget basic information sheets, health risk appraisals, and an attendance roster.

The tests should be arranged in some logical order so that unnecessary activity will neither lengthen the process nor inhibit performance. Don't conduct a leg-press exercise then ask the members to run immediately afterward. Alternate your focus on muscle groups, and provide adequate rest by rotating the subjects through an event. For example, if four members are performing a bench press, the rest period between individual attempts will provide adequate time to rest, and no downtime will be incurred. If only two members are using the bench-press station, however, the rest period will be too short for adequate recovery, and extra rest will be required between attempts, thereby increasing the total test time. In other words, having four members share the equipment is more efficient than letting two trade off. If free weights are used on a plate-loaded machine, gradually increase the weight and have everyone rotate through the exercise until they reach their individual limit. You can continue adding weight until each participant has reached his limit. Make sure that you start with a weight that is appropriate for the weakest subject. This way, you won't have to continually add or subtract weight based on an individual's ability. Load it once and unload it once.

Insist on proper warm-up trials. Repeated trials with lifting exercises produce better scores. Such warm-up trials won't wear people out, provided that the subjects are given sufficient time to rest in between.

The instructions should be standardized, and all test administrators should be familiar with them. Be prepared for the creative alternative. There will be someone who wants to do something differently, and you'll need an answer. Sometimes the suggestions you receive will be good, so feedback during the pilot is encouraged. You should also use a reliability check. Reliability is assessed by way of a test-retest procedure, by which each administrator tests the same individuals. The scores are then statistically correlated to derive a coefficient of reliability. The relationship should be at least 90 percent, meaning that the testers are getting the same results. If the correlation is low, then the reliability of the test is poor. If you don't use a test-retest procedure, the program manager should observe each test administrator separately to ensure that the instruction and administration of each test is consistent with each individual. If one test administrator is always producing high scores as compared to the other administrators, check that individual's procedures and correct them. The test-retest shouldn't produce objections from participants during the pilot; however, retesting should not be mandatory. The retest can be scheduled on a different day, if necessary.

Gather all of the equipment that you'll need ahead of time, and have a contingency plan ready in case the equipment breaks. Check each item prior to conducting the test. Don't wait until you say "go" on a run test to discover that your stopwatch is nonoperational. It makes people unhappy to have to repeat a test due to administrator error.

Schedule the tests prior to lunch or dinner. Get your subjects when they're hungry, not full after a meal. A day of rest before test day is advisable. Inform people in advance so they can abstain from working out. Weatherwise, schedule the tests for mild weather, fall or spring, if they are to be performed outdoors. Use a standard heat-humidity and cold chart to decide whether it'll be necessary to reschedule a given event. Precipitation means an automatic cancellation.

Make safety a priority, and always explain the scoring procedure and standards. Include the standards in the program policy so that all members will know the requirements. I am in the minority of opinion in that I don't believe in telling someone how much weight he needs to lift or how fast he has to run to achieve a specific score. I advise test subjects to do their best. A standard is irrelevant, since one cannot exceed his own physical capabilities. Also, most tests do require some manipulation of the raw data. The walk test, for example, requires a formula, so it's almost impossible to compute on test day for each individual, since there are four variables and an unlimited number of possible combinations (LeCuyer, 1998). Instruct the subjects to do their best, and you'll receive your best score. Providing the protocols and standards prior to testing may help the members calculate their own scores in advance, if they choose to test themselves.

TEST PROCEDURE

The fitness test procedure is depicted in the flow chart in Figure 10–1. The first step is to collect all necessary demographics: name, gender, height, weight, and ID or assignment number, if necessary. Create a basic form to record this information and check it for completeness.

The basic assessment of blood pressure, resting pulse, and HRA values should be recorded during the pretest period. Simple tests, such as the handgrip dynamometer or body composition analysis, should also be conducted at this time (Golding *et al.*, 1989; Heyward, 1998). One word of caution: Determining body composition is a highly personal process. If possible, this assessment should be performed with all due respect for privacy.

Once the preliminary data has been collected, verify that each subject has been cleared to take the fitness test. Direct anyone who requires physician clearance to the appropriate medical authority. Preparing a simple step-by-step procedure in graphic form will serve in the decision process and help you avoid confusion at test time. An example of a health-risk screening process for level of participation 1A appears in Figure 10–2.

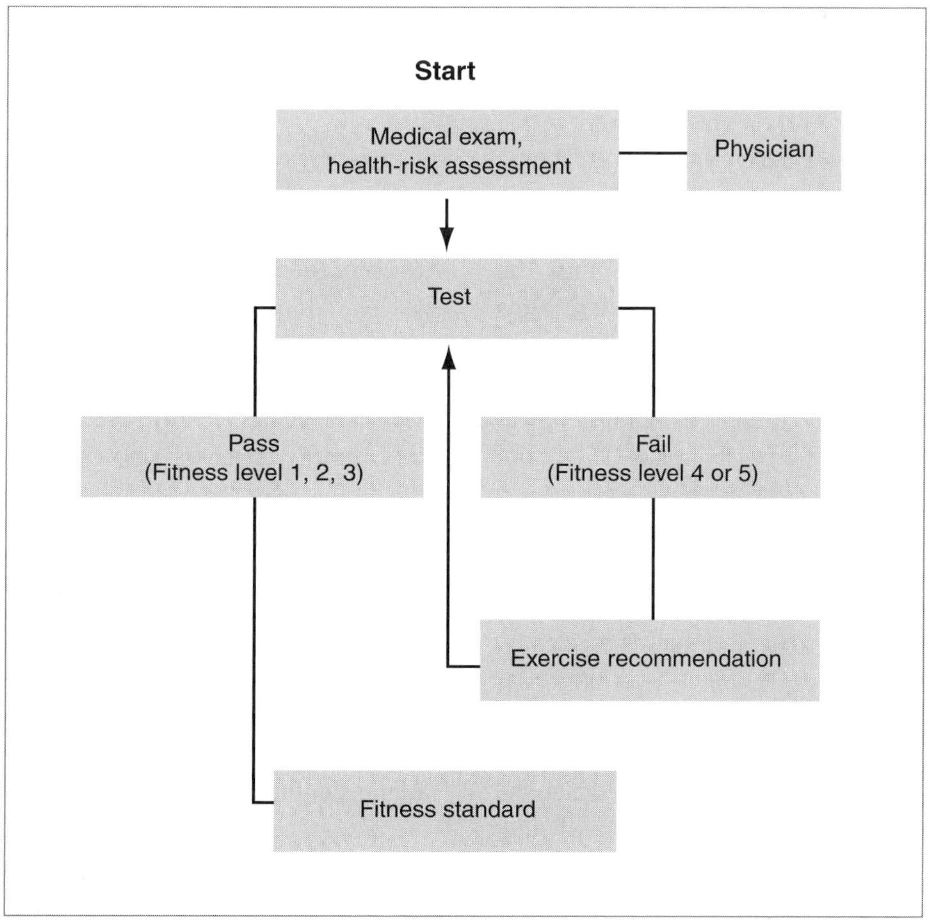

Fig. 10–1 Physical Fitness Test Flow Chart

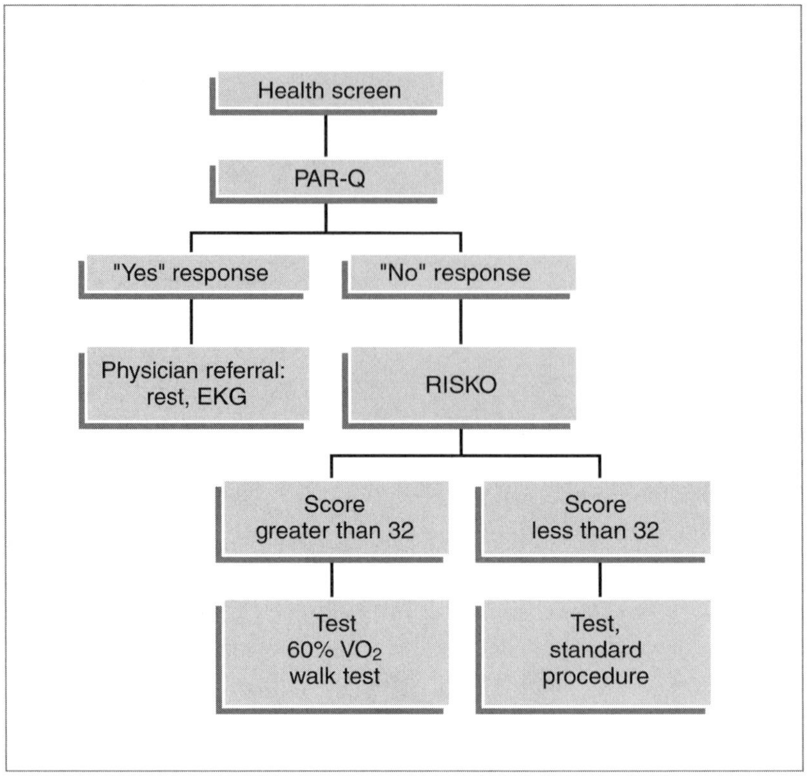

Fig. 10–2 Medical Exam Flow Chart for LOP 1A

Members who have been cleared to test should be escorted to the test area and given instruction as to the proper technique for each exercise. Allow them to practice the technique during warm-up. The test administrator should provide advice during this phase. Once everyone is comfortable with the technique, the test may begin. It is best to describe and demonstrate each test just prior to administering it. If you explain five tests at once, it may be an hour before the last test is administered and everyone will have forgotten the instructions by then.

Emphasize proper form and reinforce the pursuit of accuracy. Emphasize to everyone also that the test is for their benefit. The goal is to identify their physical status as it relates to a particular component, provide an exercise recommendation, then reevaluate them to check their progress. If they cheat or perform incorrectly, the results will have little meaning, and they won't know whether their physical status has improved or declined. There is a tendency for people to work with too much weight, which results in poor exercise form. It also leads to disappointment when, during a test, they are required to display correct form and so must reduce the weight to do so. Providing clear test procedures to the members prior to test day will provide advance notice as to how they'll be expected to perform.

Aerobic tests that require determination of the heart rate are particularly susceptible to miscalculation, though members who provide EMS can be expected to take pulses accurately. The best method is to use a heart-rate monitor, but these are expensive. A free-standing thumb-pulse monitor may be an alternative. Emphasize that there is nothing to gain by trying to beat the system by cheating, for they would only be deceiving themselves.

The most important duty of the fitness coordinator during the implementation phase is simply to administer the test. A valid and reliable test instrument is only as good as its administrator. If the test is administered improperly, the procedure will be invalid and the resulting data will be skewed (Johnson and Nelson, 1979). If you're using a dynamometer to assess isometric leg strength, for example, you must identify the proper angle for the test. This can be a difficulty with such devices, but you can construct a wooden template that duplicates the angle so as to check for proper positioning. Retest scores using different angles can vary considerably, and this will affect the overall reliability of the test. Similarly, skin calipers must be used in a precise manner. To become proficient with this device requires considerable practice.

If you didn't practice the test procedures prior to test day, you may discover your selection to be less than desirable (Maud and Foster, 1995). Aerobic tests, for example, require more time to conduct than other types of tests. When a group is scheduled to meet at a central test location and individuals have to wait an hour or more to be tested, idle hands and minds can become problematic. Also, out-of-service time is unproductive. When using a test mode such as a stair machine, only one person can be tested at a time. In such a case, it's best if the administrator conducts the regimen at the duty station so that the members not being tested at any given moment can remain occupied with their regular duties. Allowing alternative test modes can create havoc, particularly if they require additional equipment or extra time. Substituting an exercise such as pull-ups for an endurance test is flawed, since this is more of a strength test than a measure of endurance. Some quality tests might be desirable but prove impractical because of administrative difficulties. Work it out beforehand.

Don't allow variations in the performance of a test. A pull-up can be performed with either an underhand or an overhand grip. The underhand grip is easier, and substitution will yield different results. Stick to your protocol and choose one or the other. A bent-leg sit-up versus a crunch-style sit-up will also give different results, as will male versus female test procedures for push-ups. A push-up can be performed with a wide or narrow hand stance, and this will affect the range of motion and the extent to which the triceps are taxed. Specify a position—hands at shoulder width, for example—and pay attention to form. A seated military press compared to the standing version is more difficult but less likely to cause injury. Beware the fluctuating results that are inherent to aerobic tests. Restrict a run test to active runners only, and always test them using the same protocol.

Make sure that the equipment is adjusted properly. Seat heights and bar heights should suit the participant. Most equipment is designed to fit an average person, so you must accommodate the very short and the very tall, even if it means fashioning an auxiliary seat or making some other modification. Provide a range of sizes for weight-lifting belts. This is a must if the test includes free-weight lifts such as dead lifts or squats. Don't forget whom you're testing, and make sure that you have enough weight. Some VRMs don't have adequate weight stacks for larger, stronger people. As always, use spotters for safety.

Treat everyone alike. Don't cheer on one individual then stand by in silence for the next. Give encouragement and pay attention to the exertion level of the test subject. Don't allow someone to strain for ten seconds attempting to lock out a lift. If he gets stuck and can't move the weight for more than two seconds, stop the lift. The only exception to this is in the case of dynamometers that require about five seconds of strain per protocol. There is an inherent safety disadvantage to such static tests. Recognizing overexertion is important and only comes with experience. It's also important to be able to recognize a lack of effort.

Beware the competitive challenge, and don't allow someone to attempt a lift that he can't handle safely. You can avoid this by conducting multiple trials with a gradual weight increase. To deemphasize competition, avoid making comparisons. Each individual should compare his performance with his previous performance or expected ability, not with someone else. Explain why you're using submaximal tests and how they're designed to estimate a maximal effort without the subject having to perform at a high level. A person can be evaluated effectively without overexerting himself, though any given test should be challenging and not so easy that it becomes meaningless.

Underachievement can be a problem. Some individuals set personal goals to perform only to the minimum. Although this standard of performance may be acceptable, setting out to meet only the minimum while being capable of more is nonproductive, and it defeats the purpose of testing. If a member habitually performs to the minimum, he'll never know his capability or his true physical status. This is important because performance declines with age, and an individual who always performs to the minimum may be unaware of his rate of decline. One day he may be unpleasantly surprised to find that he is no longer able to do what he once took for granted. The goal is to document where you are today, physically, as compared to where you were during the previous battery of tests. Sometimes underachievers are highly fit and avoid demonstrating their true ability for a variety of reasons. Encourage them to do their best, or even offer a private assessment if it will render what you need—an honest effort.

Typically, preexercise screenings don't include questions about the use of over-the-counter treatments or prescription drugs. You should ask, however, since many of these do affect performance. Refer to the *ACSM Resource Manual* (ACSM, 1993) for specific information. If a participant states that he cannot perform because he has a cold or some other minor complaint, you should tell them that his options are either to take sick time or to take the test. If he is too ill to perform a submaximal test, then he is too ill to perform at maximum level if the alarm sounds. Confronting him with these options may sound harsh, but it is necessary, since makeup tests can be a nightmare to reschedule. In some unique situations, I may allow a retest, but in general, I try to avoid it.

Test Schedule

NFPA 1500 recommends that the fitness test be conducted on an annual basis (NFPA 1500, 1997). The performance test and medical exam, however, are also recommended on an annual schedule. Therefore, you need to consider how many tests and of what type you intend to provide. I believe that you should provide all three, though not necessarily on an annual basis.

One recommendation is to provide a complete medical evaluation and exam annually, but most physicians would agree that such a schedule is far too frequent. Since cost is an important factor, a schedule based on age group, as recommended by the NFPA, is one viable option. Another option is to evaluate everyone once every three years, regardless of age. On this sort of schedule, you can evaluate one-third of the department each year. A third alternative is to schedule the over-forty group every three years and the under-forty group every five years. A fourth option is to schedule personnel as needed based on a health risk appraisal or review of an individual's health history. For all practical purposes, an annual exam is desirable but fundamentally unnecessary.

The performance test can be done annually, or it can trade off with the fitness test on alternate years. This involves a policy decision. Since training time is always limited, substituting company standards training for the assessment may be the preferred option.

The fitness test should be conducted annually, as recommended by NFPA 1500. An annual fitness test includes the basic health screen, as well as the fundamental assessments of blood pressure, height, and weight as identified in NFPA 1582, *Medical Requirements for Fire Fighters*. The fitness test can serve as a substitute for the annual medical evaluation. Even though it doesn't qualify as a medical test, there is sufficient overlap to justify substitution. Although a review of one's medical and occupational history is considered part of the annual medical evaluation, any significant changes, such as exposures or job-related injuries, will be handled at the time that they occur, making additional review during the fitness test unnecessary. The annual fitness test can also provide certification for the use of respiratory protection as required by OSHA (1998 final rule) and NFPA 1404, *Fire Department Self-Contained Breathing Apparatus Program, 1996 Edition*. The aerobic test instrument, preferably the one-mile walk test, can adequately serve as a qualifier in place of the recommended spirometry test (LeCuyer, 1998). Therefore, an annual test schedule is advised, not only for reasons of standard compliance, but also because it can effectively satisfy three requirements in one test session: respiratory protection certification, the annual fitness test, and an annual medical evaluation.

It is commonly recommended that fitness tests be held twice a year. This frequency is ideal; however, it can present a serious logistical problem for many departments, especially larger ones. If adequate time and staff are available, test twice a year.

The most common frequency for the fitness test is once a year. Fitness levels can drop considerably within such a time frame, particularly the aerobic component, and some people have a tendency to avoid exercise until a month before they're scheduled to take the test. Testing more frequently is one solution, but mandatory on-duty workouts are the best way to keep personnel fit.

Another alternative is to test above-average personnel on an annual basis and below-average personnel twice a year. This provides incentive for the poorer performers to improve and helps reduce the number of marginal scores. You can also test parts of the test battery. For instance, you may conduct the entire test annually, then concentrate on a given member's weakest area a second time that year. This will serve to encourage each member to focus on improving his area of weakness, knowing that he'll be reevaluated on that particular component. Another option is to perform the complete test battery annually, then retest one dimension of it again that year. The aerobic test is a good choice for this semiannual test, since aerobic capability declines rapidly with a layoff. Still another possibility is to test annually, then retest all low scores on a three- or six-month schedule. This can cause scheduling problems, but it can also help you focus on weak areas, provide incentive, and monitor the compliance of those who repeatedly fail.

If you test everyone annually, then subsequently test the top performers every other year, you'll reduce the total number of people to test, but only by a small margin. Those whom you will exclude, the most fit, are the ones who commonly enjoy testing, so there may be little incentive to give this group a year off, considering their small number.

The best choice of scheduling combinations will vary between departments. Remember that the goal is to close the gap between the level of fitness and the demands of the job. This requires a focus on the marginal group. The fitness coordinator will discover that ninety percent of his time will be spent working with ten percent of the department—the target group. Since the members who fail the fitness test are prescribed an exercise routine and reevaluated periodically, there is probably little need to administer the fitness test more than once a year. Still, once a year is a minimum frequency. The exception to this is the introductory year. When you first start the testing program, you should arrange two follow-up test dates at three-month intervals. A sound exercise program should produce a measurable gain within twelve weeks. Thus, a retest at twelve-week intervals is appropriate. This will help establish good baseline data and familiarize the members with the process.

EVALUATING THE TEST

Once testing is complete, you should have a raw score for each individual. In the pilot program, you may or may not have established a scoring system. If you elected to use available normative data, you can convert the raw scores to a final score and inform the members of their standing on test day. If you have elected to develop standards based on your population scores, you'll only be able to state generally whether a given member performed above or below the criterion for any particular dimension. This is acceptable for the pilot if you make the members aware up front that you'll be developing standards and offering feedback at a later date.

If you've already defined the standards and a scoring system, you can provide each member's results shortly after his test. Those who don't meet the standard should be counseled and provided an exercise recommendation following the test battery. Reschedule them in approximately eight to twelve weeks, depending on the score. If a member's score was borderline, I often allow a retest on the same day or shortly thereafter. Everyone is subject to daily fluctuations, and someone caught on a bad day should be allowed to repeat the test battery in part or in whole, if requested.

Once the pilot testing is finished, give the participants a survey to complete. Encourage them to comment on the process. Review the program and consider all suggestions. Don't expect to have to restructure the entire program. If you did your planning well, the program shouldn't require significant modifications. If you do anticipate significant changes, however, you should conduct a second pilot.

RECOMMENDED READING

ACSM (1993), *Resource Manual for Guidelines for Exercise Testing and Prescription, Second Edition*. Philadelphia: Lea & Febiger.

Part IV References

Dubois, Paul (1984). Promoting physical fitness in the fire service: a statewide effort. *Fire Journal,* May, pp. 51-55.

Golding, L.A.; Myers, C.R.; and Sinning, W.E. (1989) *The Y's Way to Physical Fitness, Third Edition.* Champaign, IL: Human Kinetics.

Heyward, Vivian (1998). *Advanced Fitness Assessment and Exercise Prescription.* Champaign, IL: Human Kinetics.

Johnson, B.L.; and Nelson, J.K. (1979) *Practical Measurement for Evaluation in Physical Education, Third Edition.* Minneapolis, MN: Burgess.

LeCuyer, John (1998). Assessing firefighter aerobic capacity: the Rockport field test. *Fire Engineering,* February, pp. 49-56.

Maud, P.J.; and Foster, C. (1995) *Physiological Assessment of Human Fitness.* Champaign, IL: Human Kinetics.

NFPA 1404 (1996) *Fire Department Self-Contained Breathing Apparatus Program, 1996 Edition.* Quincy, MA: National Fire Protection Association.

NFPA 1500 (1997) *Fire Department Occupational Safety and Health Program, 1997 Edition.* Quincy, MA: National Fire Protection Association.

NFPA 1582 (1997) *Medical Requirements for Fire Fighters, 1997 Edition.* Quincy. MA: National Fire Protection Association.

PART V EVALUATION

Chapter 11

Standards and Statistics

The organization theorist Mason Haire once remarked, "What gets measured gets done." Although the message is clear, the terminology is somewhat inaccurate. A better phraseology might be, "What gets evaluated gets done." Measurement entails using tools and various techniques to collect data. We use a sphygmomanometer to measure a person's blood pressure, a weight scale to measure body weight, and so on. The term *measurement* in relation to our fitness program simply connotes testing and recording a score.

Administering a fitness test for the purpose of documenting a score is of no value to anyone, however, including the participant, unless the scores are subsequently interpreted. The term *evaluation* implies measurement, as well as some method of judging or interpreting the result. In order to interpret a result, we must have some standard to use as a comparison.

Suppose that you're the fitness coordinator and that you are administering a $1\frac{1}{2}$-mile run test to several firefighters. At the end of the six laps, one of the test subjects asks, "How did I do?" You respond, "You ran well. Your time was 11 minutes 58 seconds." (Measurement.) The participant asks, "What does that mean?" (Evaluation.) You then pull a chart from your clipboard, look up the aerobic capacity score nearest the 11:58 mark, and state, "According to the norm chart, your $\dot{V}O_2$ is 42, and that's good." (Standard.) Unfortunately for you, that answer isn't good enough for him, and he inquires further. "What's a norm, who made that chart, what does 42 have to do with my job, and who says it's good?" he demands. The proper action for you to take at this point is to look at your watch and say, "Sorry, but I'm in a hurry. I have a meeting in ten minutes. We'll talk some other time. See ya."

Of course, you don't attend any meeting because you don't have one. Instead, you return to your office and pull out your old textbooks and search for some answers. What is a norm? What is a standard? How do we develop them? The answers to these questions aren't difficult, but they do require a basic understanding of statistics.

Statistics

The very word *statistics* is somewhat intimidating for many people, and if we look at some of the more complex computations, that fear may be justified. The purpose of this chapter is to present some of the fundamentals needed to evaluate test performances and develop standards. A discussion of the more advanced concepts is unnecessary for our purposes and therefore excluded. The basic calculations that you'll need to perform can be done using a handheld calculator. If you have a sizable number of raw scores, you should consider using an appropriate computer program.

Normal Curve

The *normal curve* is a term that many people have either heard or used without any prior knowledge of statistics. How many times have you heard a teacher state that he or she was grading on a curve? Perhaps you didn't know exactly what that meant, but you believed that it implied less stringent grading. In point of fact, the teacher was referring to the normal curve, which is a pattern formed by a distribution of scores and based on the assumption that they are distributed in a normal manner (Figure 11–1). This is also referred to as a *probability curve,* because it is based on chance. If we were to measure the height of a large group of public safety officers, then plot the results on a graph, chances are that the numerical values would form a normal curve pattern. There would be people represented at the extremes, both short and tall, with the majority clustered around the middle, close to the average height.

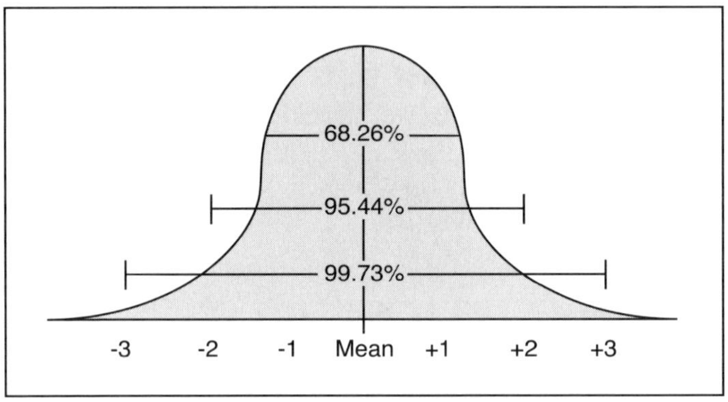

Fig. 11–1 Normal Curve Showing Percentage of Scores Expected Within Standard Deviation

MEASURES OF CENTRAL TENDENCY

There are three common average values, and each has a different meaning. Collectively referred to as measures of central tendency, they are the *mean,* the *mode,* and the *median.*

To illustrate the difference between them, let's use the following height measures, expressed in inches: 64, 67, 68, 69, 69, 70, 72. The mean is the arithmetic average of them, or 68.4. The mode is the value that appears most frequently, or 69 in this set. The median is the value in the middle of the set, also 69. Since this example has an odd number of values (seven in all), the median score is the fourth number. If there were an even number of values, you'd have to average the two closest to the middle; e.g., the fourth and fifth values in a set of eight. The scores must be numerically ranked, as in the example, before you can locate the middle score.

I have intentionally used a small number of values in this example. Realistically, this would be too small a sample for practical use, but it will serve for illustrative purposes. The distribution of these seven values would resemble a normal curve. The mean, mode, and median would line up close to the center of the curve, indicating a normal distribution.

We compute the three measures of central tendency so as to select the one that best represents the entire group. In this case, any of the three would suffice.

Now let's use a different set of scores for our example. Let 90, 39, 35, 35, and 27 be the number of push-ups that five test subjects can perform in one minute. The mean of this set is 45.2, the mode is 35, and the median is also 35.

In this example, the mode and median scores are of the same value, and the mean is over ten points higher. The reason the mean is higher is because of the one extreme score of 90. The mean is easily affected by extreme scores. Typically, the median would better represent a given group, since the mode can often be a weird value, especially when the set is small.

One way to correct this sample would be to drop the extreme score from the calculation. With the skew removed, the new set consists of 39, 35, 35, and 27. Accordingly, the mean is now 34, the mode is still 35, and the median also remains 35.

Thus, with the extreme score removed, the mean, mode, and median again form a fairly normal curve. In this case, dropping the high score proved useful. You can support doing so on the rationale that a person cannot perform 90 push-ups in one minute correctly. Anyone who scored this high would probably be using improper form. Still, this is a very small set of numbers, and the results might not have been so much to the extreme if the set had been larger. The practice of dropping extreme scores should not be habitual. In some cases, the extremes are good data. The best approach is to perform adequate sampling.

The extent to which the plotted values form a normal curve depends on the number of people represented. To measure all of the members of your department is to measure a population. To measure only a small group of firefighters is to measure a sample. The larger the sample, the more likely it is that the results will display a normal curve. The mean for a population is known as a parameter. The mean from a sample is called a statistic. The parameter is the real mean of the population. The sample statistic is an estimation. The closer the sample size approaches the population, the more accurate the data, assuming that the sampling procedure is unbiased. Normally we don't test an entire population because of logistical reasons. If the sample is a good cross-section of the population, and if our measurement is objective and accurate, we can infer that what we observed in the sample is true of the population and that the data is valid and reliable. When good sampling techniques are used, the preferred measure of central tendency is the mean.

STANDARD DEVIATION

The averaged distance of variance from the mean is known as the standard deviation. Using push-up scores of 32, 24, 39, 36, 28, 33, 35, 37, 36, and 40, the computed mean is 34. Computing the standard deviation for a small set of numbers is accomplished as follows (Table 11-1).

In this sample, 68.26 percent of the subjects scored within plus or minus one standard deviation of the mean (assuming that this is a normal distribution). The standard deviation and the mean are then used to develop standard scores. For all practical purposes, these two statistical calculations are all that you'll need to perform.

STANDARD SCORES

You have just completed administering a fitness test battery to a group. You now have a collection of scores for each fitness component for each participant. The scores are recorded as the number of repetitions, the number of pounds lifted, or expressed in terms of inches or time. What does thirty-five sit-ups mean, and how does that relate to a 200-pound bench press or a ten-minute $1\frac{1}{2}$-mile run? Are the scores high, low, average, good, or bad? To interpret

TABLE 11–1

COMPUTING STANDARD DEVIATION

Score	d	d^2
32	-2	4
24	-10	100
39	5	25
36	2	4
28	-6	36
33	-1	1
35	1	1
37	3	9
36	2	4
40	6	36
340		220

d = the difference between the score and the mean.
d^2 = the difference squared.
Σd^2 = the sum of d^2.
N = the number of scores.
s = the standard deviation.

$$s = \sqrt{\frac{\Sigma d^2}{N-1}}$$

340 ÷ 10 scores = 34 (mean)

$$s = \sqrt{\frac{220}{9}}$$

s = 4.94

$s = 4.94$

Note: A different formula is used for large groups, and the result is best found with a calculator or computer.

the scores, they must first be assembled or combined to create one standard score. There are two popular methods of standardizing scores; namely, percentage tables and scales. The four types of scales that we will consider are the T scale, the Hull scale, the Sigma scale, and Z scores.

To construct a scale, you must first compute the mean and the standard deviation from a given group of scores. Once you have determined these values, the procedure is similar for three of the scales except for the number of standard deviations that are used. The T scale uses 5 standard deviations, the Hull scale uses 3.5, and the Sigma scale uses 3. All of these scales are related to the Z score, which is basically a score expressed in standard deviations from the mean.

Table 11–2 shows the relationship between the scales using the following data as a sample. You administered a push-up test and collected a group of raw scores. The mean score is 25, and the standard deviation is 9.

First, develop a score column of numbers from 1 to 100, and place the mean score of the push-ups (25) next to the number 50. For the sake of clarity, only a portion of the chart is presented.

Multiply the standard deviation by 5 for the T scale, by 3.5 for Hull, and by 3 for Sigma, then divide by 50.

Then, add the resulting number to the mean (25) to get the corresponding values for 51 through 100. Subtract this number from the mean for the corresponding values of 49 through 1.

Round off the numbers as shown for ease of interpretation. You may also prefer to use larger point increments (*e.g.*, 50, 55, 60, etc.) when using the scales. As shown in Table 11–2, the T scale includes the greatest range of scores, and the Sigma scale includes the fewest scores. The T scale range

TABLE 11–2

COMPARISON OF SCALES USING SAMPLE DATA

Score	T scale	Hull scale	Sigma scale	Z score	
100	70.0	56.5 (56)	52.0		
60	34.0	31.3 (31)	30.4 (30)	+1.0	
59	33.1 (33)	30.67 (31)	29.86 (30)	+0.9	
58	32.2 (32)	30.04 (30)	29.32 (29)	+0.8	
57	31.3 (31)	29.41 (29)	28.78 (29)	+0.7	
56	30.4 (30)	28.78 (29)	28.24 (28)	+0.6	
55	29.5 (30)	28.15 (28)	27.70 (28)	+0.5	
54	28.8 (29)	27.52 (28)	27.16 (27)	+0.4	
53	27.7 (28)	26.89 (27)	26.62 (27)	+0.3	
52	26.8 (27)	26.26 (26)	26.08 (26)	+0.2	
51	25.9 (26)	25.63 (26)	25.54 (26)	+0.1	
50	**25**	**25**	**25**	**0.0**	**(Mean score)**
49	24.1 (24)	24.37 (24)	24.82 (25)	-0.1	
48	23.2 (23)	23.74 (24)	23.92 (24)	-0.2	
47	22.3 (22)	23.11 (23)	23.38 (23)	-0.3	
1	0	0	0		

of 5 standard deviations is considered an advantage, since virtually all possible scores can be included. On the other hand, it may be viewed as disadvantageous, depending on the data, because it will include very high scores at the top that would be difficult to achieve.

The Sigma and Hull scales could present the opposite problem and not include all of the raw scores due to the small increments. The highest scores in this example for the Sigma and Hull are 52 and 56 respectively. It is possible to perform more repetitions in the allotted time than the tables can score. The positive side is that a small improvement can result in a higher score. An increase of two push-ups on the Sigma scale can result in a four-point score increase, whereas the same two push-ups would only gain two extra points on the T scale.

At the low end of the scales, where values are subtracted, negative scores may appear. Since it's impossible to perform a negative push-up, however, record a zero instead.

The fourth method of scoring, shown in the far right column in Table 11–2, is the Z score. In fact, the other scales are all directly related to and based on the Z score. The Z scores are expressed in terms of standard deviation from the mean, and the mean is assigned a value of zero. The Z score is computed by subtracting the raw score from the mean and dividing by the standard deviation. In our push-up example, the mean was 25 and the standard deviation was 9. If an individual performed 22 push-ups, his score would be calculated as follows.

$$\text{Z score} = \frac{\text{score (22)} - \text{mean (25)}}{\text{Standard deviation (9)}}$$

$$\text{Z score} = \frac{^{-}3}{9}$$

$$\text{Z score} = {^{-}0.33}$$

Since Z scores are small and expressed in both positive and negative values, they are difficult to use and interpret. The solution is to convert them to T scores [T = 50 ± 10 (z)]. Of the scales presented, the T scale is the one that is most commonly used today, along with the Z scores. The relationship between T scores, Z scores, and the normal curve is presented in Figure 11–2.

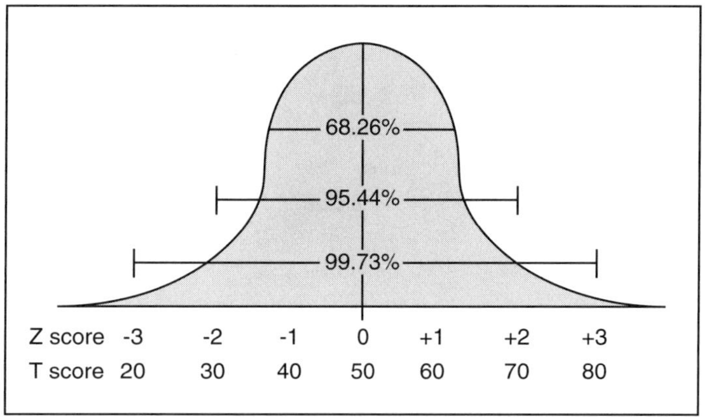

Fig. 11–2 Comparison of T Score, Z Score, and Normal Curve

PERCENTAGE TABLES

The last method of standardizing scores that needs mention is the percentage table. The use of percentiles is quite popular, particularly with physical educators. The American Alliance for Health, Physical Education, Recreation, and Dance (AAHPERD) makes use of percentiles, allowing teachers to compare the performance of their students with other students nationwide. The popularity of percentage tables can be attributed to the ease of interpretation by test participants. This is one objection to the T scale, since these scores are often misinterpreted as percentages. A score of 80 on the T scale is a high score, often erroneously mistaken for 80 percent, when in fact it equates to the 99th percentile.

Percentiles are popular since everyone knows the difference between a score of 90 percent versus one of 50 percent. A score that places someone in the 95th percentile clearly indicates that that person performed better than 95 percent of the rest.

If I were to perform 25 push-ups in a one-minute time span, and if that score correlates to the

TABLE 11–3

PREDICTED MAXIMAL OXYGEN CONSUMPTION FOR MALE AND FEMALE COLLEGE STUDENTS

Percentile Ranking	Predicted Maximal $\dot{V}O_2$ ($mL \cdot kg^{-1} \cdot min^{-1}$) Female	Predicted Maximal $\dot{V}O_2$ ($mL \cdot kg^{-1} \cdot min^{-1}$) Male
100	42.2	60.9
95	40.0	59.3
90	38.5	57.6
85	37.7	54.2
80	37.0	52.5
75	36.6	50.9
70	36.3	49.2
65	35.9	48.8
60	35.7	47.5
55	35.5	46.7
50	35.1	45.8
45	34.8	44.1
40	34.4	43.3
35	34.2	42.5
30	34.0	41.6
25	33.3	40.8
20	32.6	39.1
15	32.2	37.4
10	31.8	36.6
5	29.6	34.1

Note: When a minimum requirement for $\dot{V}O_2max$ is set at a specific value (*e.g.*, 42 $mL \cdot kg^{-1} \cdot min^{-1}$), a female would have to rank in the 100th percentile and a male in about the 35th percentile to qualify. However, if a percentile is selected, 50 percent of females would have to achieve a $\dot{V}O_2$ of 35.1 versus the higher value of 45.8 for males. Since $\dot{V}O_2$ varies with percentiles, the best choice is to select a specific value as the criterion standard.

80th percentile for a firefighter, should I be pleased? I can assume that 25 push-ups is a good score. The percentile rating certainly indicates that it's a good score, but is it really? (May and Nicewater, 1994.)

I can accept the results and be content, or I can question the results in the manner described earlier, when the firefighter confronted the test administrator following his run test. If this were an actual scenario and I had actually performed 25 push-ups, placing me in the 80th percentile, I would intuitively be disappointed. Although it reflects my position within the group, 25 push-ups is fair at best, which means that I didn't perform so well. I was simply being compared to a group of low performers! In fact, the score of 25 push-ups, based on the criterion standards that I will present later, is a minimum score for fire fighters. The 80th percentile actually means that I met the minimum standard, and 80 percent of the sample members are below standard.

This is an inherent problem with percentiles. They can be deceptive, and for this reason, I am no proponent of their use. Percentile tables don't take into consideration the mean or the standard deviation. They simply report how many scores there are above and below yours, and the intervals between the percentiles imply that they're equal, but they aren't.

Consider Table 11–3, which employs percentile ratings for $\dot{V}O_2$ scores. A female whose $\dot{V}O_2$ score is reported to be 33.3 receives a ranking in the 25th percentile. If she participates in an aerobic program and improves her score by 2 $mL \cdot kg^{-1} \cdot min^{-1}$, her ranking will jump from the 25th percentile to the 55th percentile, or 30 points. On the other hand, if a female with a $\dot{V}O_2$ of 40 and ranked in the 95th percentile were to make a 2 $mL \cdot kg^{-1} \cdot min^{-1}$ improvement, she would move to the 100th percentile, a jump of only 5 percentile points for the same degree of improvement (which, I might add, is more difficult to achieve at the high end of the spectrum).

A second problem evident in this table is the percentile ranking between males and females. A female is ranked in the 100 percentile category if she has a $\dot{V}O_2$ of 42.2 mL • kg^{-1} • min^{-1}. A male with the same score is ranked in about the 35th percentile. The table poses the question, "Is the 35th percentile good?" In terms of percentages, the answer is no; in terms of a $\dot{V}O_2$ score, the answer is yes. Before we can answer the question properly, we need to turn our attention to the subject of standards.

STANDARDS

Earlier in the chapter, we defined evaluation as an interpretation of measurement. That interpretation required comparison to a standard. Some of the problems associated with the use of scales and tables are in part related to their method of construction. As stated earlier, the T scale, Hull scale, Sigma scale, and Z scores make use of a probability curve, the mean and standard deviation of the sample. A percentage table doesn't use the mean or standard deviation, but instead is calculated from the frequency distribution that shapes the curve. From this normal curve, norms are derived and standards developed. There are two types of standards: norm-referenced and criterion-referenced.

The norm-referenced standard is developed from the statistical procedures previously described. The scales and percentage tables are norm-referenced standards. These are used to compare an individual's performance to the group (individual to group).

A criterion-referenced standard is based on some predetermined criterion, such as "A firefighter must be able to perform forty correct push-ups in one minute." Each individual's performance is then compared to that standard (individual to task). If the goal is to close the gap between job requirements and one's level of fitness, which it should be, then the use of the criterion approach is mandatory, and the criterion must reflect the job requirements to be of any value.

A general discussion of criterion standards is necessary before we can begin to construct a standard table. There is no consensus as to what should constitute a criterion standard for all fitness dimensions relevant to public safety. Most researchers generally agree on what should be accepted as the criterion for aerobic capacity for firefighters and police, but other measures, such as strength, have yet to be determined. There is also debate as to whether to use relative or absolute values and sliding scales when scoring. Therefore, the following discussion is an attempt to arrive at some logical conclusion as to what criterion should be established for each of the fitness components, as well as what measures or scales are the most desirable toward achieving the goals of the program. The values recommended in this section are intended to apply to all public safety officers; specifically, to firefighters and police.

The component of fitness most frequently studied, relative to public safety occupations, is aerobic capacity. The measurement of this dimension is popular because it fits the science process well and serves as the foundation for exercise physiology. The occupation of firefighting predominates in the research literature, perhaps due to the unique nature of the work. The earliest research conducted by Dr. Brian Sharkey for the U.S. Forest Service in the 1970s determined an aerobic capacity of 45 mL • kg^{-1} • min^{-1} to be the recommended minimum (Sharkey, 1978). Twenty years later, the same value was recommended by researchers at York University, Ontario, Canada (Gledhill and Jamnik, 1992). Numerous studies conducted within that twenty-year time frame have estimated values as low as 33.5 mL • kg^{-1} • min^{-1} to a high of 51 mL • kg^{-1} • min^{-1}, with the 42 and 45 mL • kg^{-1} • min^{-1} figures being the most frequently cited (Southman *et al.*, 1992; Davis *et al.*, 1982; Jacobs, 1976; Davis and Dotson, 1986; Lemon and Hermiston, 1977; Schoenfeld *et al.*, 1990; Gledhill and Jamnik, 1992; and Sparks, 1987). There is sufficient data available to establish a minimum of 42 mL • kg^{-1} • min^{-1}, with a recommended level of 45 mL • kg^{-1} • min^{-1}. These are the most frequently accepted values reported by research (Davis and Dotson, 1987; Gledhill and Jamnik, 1992; Doolittle, 1979; and Sharkey, 1997).

As mentioned previously in the text, aerobic measures are expressed in both relative terms (aerobic power equals $mL \cdot kg^{-1} \cdot min^{-1}$) and absolute terms (aerobic capacity equals L/min^{-1}). The relative measure allows for a direct comparison between individuals without consideration of body weight. The absolute measure provides information as to one's total capacity (Sharkey, 1990). The relative value is the most common expression; however, the absolute measure will show a stronger correlation to public safety work and is therefore recommended. The choice is a matter of preference, and departments may elect to report both values when scoring members, or perhaps to provide an option in terms of standards. Having an option will satisfy everyone, since smaller members score better with relative scores, whereas larger members do better with absolute expression. From my experience, individuals between the extremes—weighing more than 150 pounds but less than 185 pounds—tend to score well using either method of reporting.

Relative value is calculated from a formula or derived from a table, based on test protocol. The standard is 42 $mL \cdot kg^{-1} \cdot min^{-1}$ minimum and 45 $mL \cdot kg^{-1} \cdot min^{-1}$ recommended.

Absolute value is also calculated from a formula or derived from a table, based on test protocol. The standard is 3.0 $L \bullet min^{-1}$ minimum, 3.2 $L \bullet min^{-1}$ recommended.

To convert $mL \cdot kg^{-1} \cdot min^{-1}$ to $L \bullet min^{-1}$, use the following formula. Consider a person with a body weight of 176 pounds to have a relative value of $\dot{V}O_2 = 45\ mL \cdot kg^{-1} \cdot min^{-1}$.

(1) First, convert pounds to kilograms by dividing 176 by 2.2.

$$\frac{176}{2.2} = 80 \text{ kg.}$$

(2) Then, divide the result by 1,000.

$$\frac{80}{1000} \div 1{,}000 = 0.08.$$

(3) Multiply that result by 45 ($mL \cdot kg^{-1} \cdot min^{-1}$).

$$0.08 \times 45 = 3.6\ L \bullet min^{-1}.$$

$$\text{Absolute} = 3.6\ L \bullet min^{-1}.$$

A conversion table is provided in Appendix J.

AEROBIC COMPONENT

Although the research suggests that a high aerobic level is essential in determining performance capability (Davis *et al.*, 1982), other physical factors, such as strength, age, body composition, and physical stature, also play a significant role. It has been reported that a firefighter with an aerobic capacity of less than 33.5 $mL \cdot kg^{-1} \cdot min^{-1}$ is not likely to perform (Southman *et al.*, 1992). However, one cannot draw a line and state that 34 is acceptable and 33.5 is not, particularly if the $L \bullet min^{-1}$ value is greater than 3.0. The predictive ability of $\dot{V}O_2$ measures to job task performances will be marginal, in part due to the high anaerobic component of firefighting (Lemon and Hermiston, 1977; and Gledhill and Jamnik, 1992). Additionally, the potential error of prediction can also be compounded depending on the test mode.

The correlation between aerobic capacity and a physical performance test using the relative measure was ($r = 0.32$), and with the absolute measure it was ($r = 0.56$). The L•min is the better predictor, but not significantly so (LeCuyer, 1994). A more recent investigation using <3.0 L•min as a criterion to identify low-scoring performers and potential test failures on a job performance test was 94 percent effective in identifying these two groups. The relative measure identified approximately 69 percent (LeCuyer, 1998).

One thing that can be concluded is that the chance of failure to perform increases as the aerobic capacity of the individual declines. Since the risk of failure for public safety officers can have serious consequences, a value of 42 $mL \cdot kg^{-1} \cdot min^{-1}$ (or 3.0 $L \cdot min^{-1}$) is a reasonable minimum, and a value of 45 (3.2 $L \cdot min^{-1}$) is more desirable to provide a margin of safety (Astrand and Rodahl, 1970). The average of one department was 3.7, based on the testing of 225 firefighters. Researchers have determined that the most demanding firefighters tasks (10 percent of all operations) require a mean $\dot{V}O_2$ of 41.5 $mL \cdot kg^{-1} \cdot min^{-1}$ as a recommended minimum, and the remaining tasks (90 percent of operations) require a mean $\dot{V}O_2$ of 23.4 $mL \cdot kg^{-1} \cdot min^{-1}$ (Gledhill and Jamnik, 1992). Therefore, establishing a minimum criterion of 42 would provide a reasonable objective, with a level of 45 as the goal. This approach is consistent with the fitness program purpose statement.

Fitness standards should be based on the individual's job description and reflect the physical requirements of their assigned duties. One example is the U.S. Forest Service, which has established different strength and aerobic requirements for wildland firefighters ($\dot{V}O_2$ = 45 $mL \cdot kg^{-1} \cdot min^{-1}$) and smoke jumpers ($\dot{V}O_2$ = 48 $mL \cdot kg^{-1} \cdot min^{-1}$) based on job demands. This approach is appropriate, since it reflects the different requirements of each assignment, as shown in Table 11–4 (Sharkey, 1997). A public safety officer assigned to desk duties may have different requirements that would change the physical standard. The norms for the general population might be acceptable, since desk officers wouldn't be performing life safety tasks. The exception is the police officer or firefighter who, based on job description, is expected to be on call and to perform critical tasks both on and off duty, regardless of assignment.

TABLE 11–4

U.S. FOREST SERVICE STANDARDS: REQUIRED AND RECOMMENDED FITNESS LEVELS

	Pack Test (in minutes, 45-lb. pack)		Aerobic Fitness ($mL \cdot kg^{-1} \cdot min^{-1}$)		Muscular Fitness (Recommendations)*				
Job	Req'd	Recmmnd	Req'd	Recmmnd	Leg Press lbs. x Body Weight	Bench Press lbs. x Body Weight	Pull-Ups	Push-Ups	Sit-Ups
Field worker	—	45	—	45	1.5x	0.7x	3	15	25
Wildland firefighter	45	—	45	—	2.0x	0.8x	5	20	30
Hot-shot	45	42.5	45	50	2.5x	1.0x	6	25	40
Smoke jumper**	—	42.5	48	50	2.5x	1.0x	7	25	45

* Recommended fitness levels indicate a capacity to do the job safely and well, plus a reserve to meet unforeseen emergencies.
** Smoke jumper requirements: 1.5-mile run in 11 minutes, 7 pull-ups, 25 push-ups, 45 sit-ups, pack-out of three miles with 110 pounds in less than 90 minutes.

Reference: Sharkey, B.J., *Fitness and Work Capacity, Second Edition.* Missoula, MT: USDA Forest Service, April 1997, p. 51.

ANAEROBIC COMPONENT

Structure firefighting involves a high anaerobic component, as reported by Gledhill and Jamnik. Although important, there is no single test available that demonstrates a significant relationship to all essential job performance measures, due to the highly specific nature of such tests. The Margaria-Kalamen test was used in a study of incumbent firefighters. This revealed a fair correlation to a hose-drag task ($r = {}^{-}0.56$) and a good correlation to a one-minute leg press ($r = 0.78$). The one-minute leg press also exhibited a good correlation with the hose drag ($r = 0.72$), and would therefore be preferable to the Margaria test (LeCuyer and Kakos, 1992). Anaerobic capacity is directly related to muscle mass and strength measures, and since strength is measured in our test battery, the need to assess the anaerobic parameter independently is unnecessary. Additionally, anaerobic tests require high-intensity efforts, and since there is no submaximal test version available, it is suggested that they not be routinely included in a fitness test battery for safety reasons. Some tests, such as timed push-ups, burpees, and sit-ups, do tax the anaerobic system well, and can therefore be regarded as substitutes for testing this dimension.

If an anaerobic test is required, however, the preferred test mode is the arm-crank ergometer. In an extensive study conducted by the California Department of Forestry, a significant relationship was established between the arm-crank ergometer and structure firefighting (Terra, 1987). The absolute values survived statistical analysis and correlated well with all firefighter tasks, while the relative values were rejected as insignificant. The test was developed for the purpose of selecting firefighter candidates.

MUSCULAR FITNESS

There is relatively little research available that can accurately estimate the desired strength level for public safety officers. This is surprising, since I have found strength, in several independent studies, to be the single best predictor of a firefighter's job performance. The assessment of upper-body strength using the bench press was the best predictor of five different fitness tests in a study conducted using recruit firefighters, reporting a correlation coefficient of $r = {}^{-}0.86$ to the performance test. The criterion standard recommended for strength measures will vary with the type of test selected in contrast to the single aerobic criterion ($\dot{V}O_2$) produced by all aerobic tests.

The recommended minimum criterion for the bench press is 100 pounds (absolute) free weight, or a ratio of 1:1 weight lifted to body weight (relative). The selection of 100 pounds is based on the DOT classification for very heavy work, which requires the ability to lift at least 100 pounds or more. The relative measure would ensure that the minimum absolute of 100 pounds is achieved, since it is unlikely that a public safety officer would weigh less than 100 pounds. It is reasonable to expect a firefighter or police officer to be capable of handling his own weight. A sample table with tests and weight recommendations for free weights and variable resistance machines is provided in Appendix C.

The most common assessment of leg strength is the leg press or squat. An expectation in relative terms would be 2.0 times one's body weight, using a variable resistance machine to remain consistent with the bench press recommendation (Heyward, 1984). An absolute score for a leg press should be 200 pounds if using free-weight lifts. When using a machine for testing, a correction factor should be applied to achieve the equivalent free-weight lift. For instance, one hip sled correction factor is about 0.71, or 71 percent of a free-weight lift. Therefore, a higher weight requirement would be necessary for a comparable lift.

Recommendations for the arm curl are one-half one's body weight (0.50), or a free-weight lift of 75 pounds. For the lat pulldown, it's one's body weight (1.0 or 100 pounds). For the chin-up, it's six repetitions. This data is based on the testing of incumbent firefighters, and the criteria determined from work and training observations over a period of ten years.

Static strength standards are based on two studies: one with steelworkers, and the other using firefighters (Wollack and Assoc, 1986; and Arnold *et al.*, 1982). Until sufficient data is available, it is recommended that individual departments choosing to use leg or back dynamometers employ the criteria provided as a starting point only. Each department should develop its own standards following the evaluation of its incumbents.

The exception to this is the hand-grip dynamometer, which has been highly correlated (r = 0.95) to other strength measures, such as the bench press and leg press, in a study of 203 career firefighters (LeCuyer and Kakos, 1992). The grip-strength scores ranged from 35 kg to 85 kg. The minimum value recommended in the California Department of Forestry (CDF) study was 39 kg (86 pounds), which is consistent with the minimum estimate that I derived from performance test observations, in which firefighters scoring <35 kg did not meet the minimum performance standard.

The most frequent muscular endurance measures are sit-ups and push-ups. A high muscular endurance requirement for firefighters has been recommended by some researchers. Sit-ups were shown to correlate well with some firefighter tasks (Davis *et al.*, 1982; and LeCuyer, 1994) and poorly with some police tasks (Spitler *et al.*, 1987). Push-ups correlated significantly (r = 0.90) with bench-press measures in twenty career firefighters, providing support for the option of substituting push-ups for bench presses if so desired (LeCuyer, 1989). A frequency distribution showing the relationship between bench presses and push-ups using 240 career firefighters appears in Figure 11–3.

The suggested criterion for a minute's worth of push-ups is twenty-five. For a minute's worth of sit-ups, it's thirty. Another test that may be classified as an endurance measure—although it is sometimes described as a strength and agility test—is the squat thrust, or burpee. This test has been highly correlated with other static-strength tests (Arnold *et al.*, 1982). The recommended criterion is twenty-five repetitions in one minute, or forty-five in two minutes. The test is sometimes performed either for several minutes or for maximum repetitions to exhaustion. Isometric tests, such as the 30-degree sit-up and hold, as well as the wall sit, may be viewed as

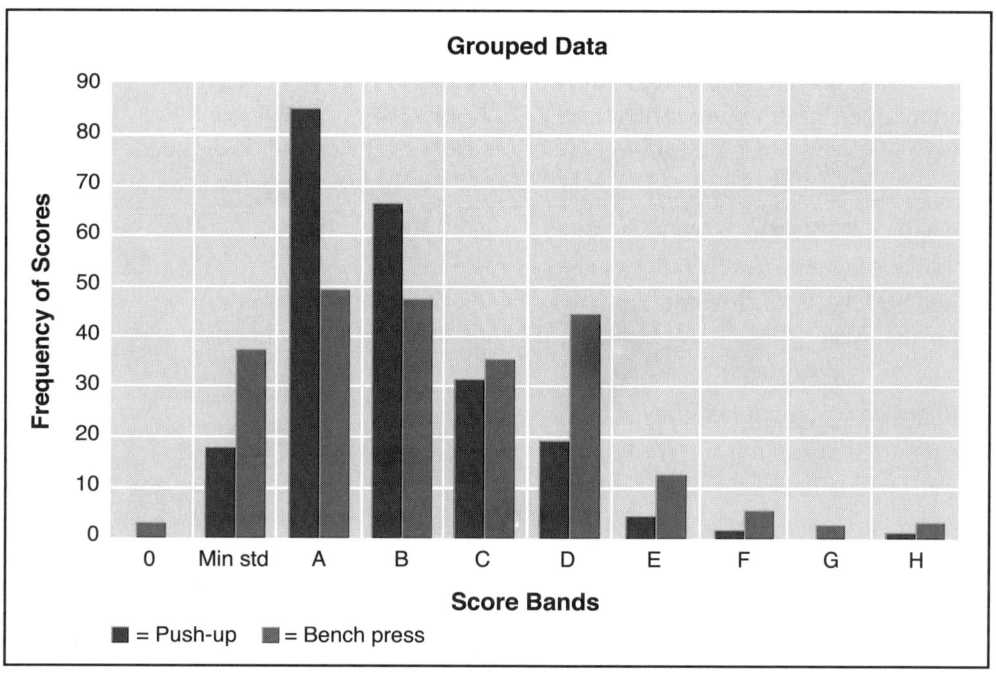

Fig. 11–3 Frequency Distribution of Bench Press and Push-Up Scores

either strength or endurance tests. They are good isometric options when no equipment is available.

The most commonly applied field-test measure of flexibility is the sit-and-reach test. This is a measure of low-back and hamstring flexibility. Increased flexibility is believed to reduce the risk of low-back pain (Caillet, R, 1988; and Khalil, T.M. *et al.*, 1992). The ballpark standard is the ability to touch one's toes. Using the standard sit-and-reach test, the criterion, based on one protocol, is fifteen inches—touching your toes. The modified sit-and-reach would be viewed as a relative measure, since the start point of zero is adjusted to the length of the individual's limbs.

BODY COMPOSITION

The measure of body composition results in a single value similar to the $\dot{V}O_2$ measure, regardless of the choice of tests. The assessment of body composition is the only measure in which different standards based on gender are acceptable. Since body composition isn't a performance measure, dual standards are recommended, particularly when the use of a single male standard would create a health risk for women.

Although not considered a performance measure, body composition does have a significant effect on job performance, and it should therefore be evaluated. A correlation of $r = 0.81$ was reported between body-fat measures and performance on a stair-climb firefighter task (LeCuyer and Kakos, 1992). Similar results were reported by other investigators (Davis and Starck, 1980). The criterion recommendation for males is 15 to 18 percent. For females, it's 21 to 25 percent. Body composition measures should not be scored; rather, they should be recorded and reported to the individual when he is being counseled on exercise recommendations. Since improvements in other fitness dimensions will positively affect body composition, progress in total fitness may be viewed as a positive sign of changes in body composition.

RELATIVE VERSUS ABSOLUTE SCORES

In terms of strength measures, absolute scores are preferred, since they relate directly to the real world. Relative scores are acceptable if one wishes to compare individuals to one another (norm-referenced), but they're of little value when the comparison is intended to match the job demands (criteria-referenced). When the objective of the program is to close the gap between fitness and performance, the comparison should only be made with the job. We have all seen people whom we consider strong for their size. If a small person can lift a heavy weight, the next question becomes whether he can drag a hose or subdue a suspect. In terms of relative measures, smaller people tend to be stronger than larger people. This is clearly demonstrated in the use of weight divisions for lifting competitions. However, if you want to pick up a heavy object, get the strongest person, because simply being strong for your size won't help if you can't perform the task. Public safety work is largely a matter of size and strength in the absolute, regardless of relative capability (Doolittle, 1979).

The choice is personal and will depend on the questions that you want answered regarding your program: How does a firefighter or police officer compare with other incumbents or the general public, and how does a firefighter or police officer compare with the demands of the job?

Endurance measures such as sit-ups, push-ups, and so on are reported as absolute scores, since there is only one score, and each such measure is fundamentally relative, since the resistance is the individual's own body weight (Table 11–5).

CONSTRUCTING A STANDARD SCALE OR TABLE

Now that we have some idea as to what scales and standards are all about, and armed with a list of cri-

TABLE 11–5

RECOMMENDED MINIMUM STANDARDS FOR PUBLIC SAFETY OFFICERS

Dimension	Relative score	Absolute score
1. Bench press	1.00 (0.70 FW)*	100 lbs. (145 lbs. VR machine)
2. Push-up	25 reps (1 min.)	25 reps (1 min.)
3. Leg press	2.00 (1.40 FW)*	200 lbs. (290 lbs. VR machine)
4. Arm curl	50% body wt.	75 lbs.
5. Lat pull	100% body wt.	100 lbs.
6. Dynamometers:		
Hand grip		39 kg. (86 lbs.)
Arm		39 kg.
Leg		160 kg.
Back		125 kg.
7. Sit-up	30 reps (1 min.)	30 reps (1 min.)
8. Sit and reach	6 in.	15 in.
9. Squat thrust	25 reps (1 min.)	25 reps (1 min.) or 45 (2 min.)
10. Chin-up	6	6
11. Aerobic—$\dot{V}O_2$	42 $mL \cdot kg^{-1} \cdot min^{-1}$	3.0 $L \bullet min^{-1}$
12. Wall sit (90°)	60 seconds	60 seconds
13. Body composition	Male: 15–18% Female: 21–25%	

* Free-weight equivalents to a variable-resistance machine. Equivalents may vary with equipment.

terion standards, we can return to our discussion of scales and begin to plug in the numbers to create a working table.

POMPIER SCALE

Although the T scale and percentage tables are most frequently used, I propose using yet another scale. Like the others, it too is based on the Z score. Being a full-blooded Frenchman, I shall use the term *pompier* to describe this scale. Translated into English, pompier means firefighter. Most firefighters are familiar with the term pompier ladder, which has a long history within our industry, though it has been relegated to antique status by most departments in the U.S.

A pompier ladder is a single-beam scaling ladder that was used to scale the side of a building, floor to floor, by means of a hook inserted into a window well. The terms *scale* and *ladder,* in conjunction with *pompier,* seem more than appropriate for a firefighter fitness scale.

The Pompier scale uses the standard deviation, as does the Z score, but it replaces the zero value with a criterion value. The scales constructed earlier placed the mean score at the 50-point value, or 0 for the Z score. The Pompier scale places the selected criterion score at the 50, then adds and subtracts the standard deviation from that point in the same manner as a T scale or Z score (Table 11-6).

To construct the Pompier scale, begin by computing the mean and standard deviation of the test that you've selected. The mean is used to determine the standard deviation only and is thereafter disregarded.

To describe the process of constructing the scale, I will use the aerobic criterion of $\dot{V}O_2 = 42\ mL \cdot kg^{-1} \cdot min^{-1}$ and a standard deviation of $3\ mL \cdot kg^{-1} \cdot min^{-1}$.

TABLE 11–6

BASIC POMPIER SCALE

Fitness Level	*Aerobic Capacity*	*Points*	*Standard Deviation*
1	48	10	plus 2
2	45	8	plus 1
3	**42**	**6**	**criterion**
4	39	4	minus 1
5	36	2	minus 2

Basic Table (1 std. dev.)

Fitness Level	*Aerobic Capacity*	*Points*
1	48	10
2	45	8
3	**42**	**6**
4	39	4
5	36	2

The basic table can be used in the simple format shown or modified by subdividing each category using a fraction of the standard deviation as shown in Options 1 and 1A.

	Option 1 (½ std. dev.)		**Option 1A (⅓ std. dev.)**		**Option 2**	**Categories**
Fitness Level	*Score*	*Points*	*Score*	*Points*	*Score*	*Category*
1	48	10	49	10.8	48	Excellent
			48	10		
2	46.5	9	47	9		
	45	8	46	8.5	45	Good
			45	8		
3	43.5	7	44	7		
	42	6	43	6.5	42	Average
			42	6		
4	40.5	5	41	5		
	39	4	40	4.5	39	Fair
			39	4		
5	37.5	3	38	3		
	36	2	37	2.5	36	Poor

(1) First, place the criterion value on the baseline to represent the minimum standard. This will also establish the minimum fitness level category.

(2) Next, add the standard deviation (3) to the criterion (42) for each box above the criterion. Subtract the standard deviation from the criterion for each box below the criterion.

(3) Assign points or some value to correspond with each of the raw scores. Any system is acceptable. You may choose a numerical value of 1 to 100, with a score of 50 as the criterion baseline as in the T scales presented earlier. You may assign categories (poor, fair, below average, average, above average, excellent), or you may perhaps keep it simple and use a scale of 1 to 10, with an average score of 5 at the criterion baseline (Table 11–6).

You now have a standard for each of the basic categories and a score assigned to each box. Since there are five boxes and ten values, I elected to assign points in increments of two. There are no rules for assigning points.

Now that the basic table has been constructed, you may choose to subdivide the box into smaller increments using half a standard deviation, or perhaps one-third a standard deviation, or some other fraction thereof. Remember, the corresponding points also have to be divided. With small increments, small improvements will receive extra points, as they do on the Sigma scale. Please note that although the boxes are subdivided, the basic category doesn't change. The $\dot{V}O_2$ score of 42 is still in the same fitness level category. So whether you opt to subdivide or not, it won't alter the classification or the minimum standard. In this example, it is 6.0. Once you have plugged in the values, transfer the data to your standard table.

Repeat the same procedure for all of your fitness test measures. Table 11–7 shows the aerobic score that we just completed, as well as a strength

TABLE 11–7

SAMPLE POMPIER TABLE WITH A STRENGTH SCORE ADDED

Fitness Level	*Test 1* Score (Aerobic)	*Test 2* Score (Strength)	*Test 3* Score (Strength)	*Test 4* Score (Strength)	*Test 5* Score (Endurance)	*Test 6* Score (Flexion)	*Test 7* Score (Body comp)	Points	Total
1	48	**1.5**	A bench press score is added to the standard table using the criterion of 1.00 and the standard deviation of 0.25. The score intervals above and below are determined by adding or subtracting the standard deviation in the same manner as the aerobic score.					10	
2	45	**1.25**						8	
3	42	1						6	
4	39	**0.75**						4	
5	36	**0.5**						2	
						Total points			
						Average			
						Fitness Score			

score example where the criterion was 1.00 and the standard deviation was 0.25. Fitness classification box three will contain all scores, regardless of test, that fall within a +1 standard deviation. All scores that are two standard deviations above fall into fitness class two, and three standard deviations above in fitness class one. The same is true in the negative direction. In Figure 11–4, sample fitness levels are shown on the curve. Fitness level ratings are sometimes divided in increments of 0.5 of the standard deviation (ACSM, 1993), while others recommend a full standard deviation between categories (Gledhill and Jamnik, 1992). The curve in Figure 11–4 uses 1.0 standard deviation between ratings to ensure higher standards. In the same figure, there are six categories: three positive and three negative. I don't show the negative three category or use it in the Pompier table, since it represents three standard deviations below the criterion—a score so low that no police officer or firefighter could possibly record it and remain employed.

Each score will have a corresponding point value, and the final score can be totaled, then averaged for a composite score.

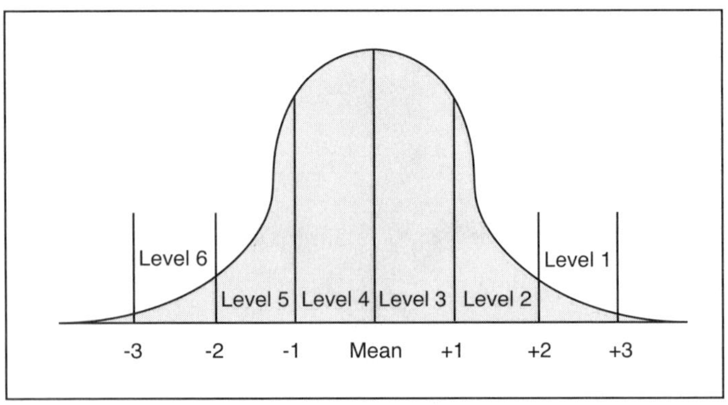

Fig. 11–4 Levels as Defined in the Text Relative to the Standard Deviation

If the distribution of scores doesn't fall along a normal curve, you should use a normalizing method. To normalize the data, first convert the raw scores to percentile ranks, then convert the result into a standard scale, usually a T scale. Consult a statistician for more details on this option.

COMPOSITE SCORES

There are a variety of methods for constructing a scoring system, not to mention any that you may dream up yourself. Almost any method will serve the purpose of attaching some value to the test scores. Whether you elect to use a numerical system or a category system is a matter of preference, as long as the scoring system and the corresponding values are understood by the membership.

The raw score should be provided with its corresponding value. Each raw score will have a value assigned that may be presented individually or added to arrive at a cumulative score. One method is to total all the raw scores, such as the total weight lifted in the bench press, plus that of the leg press, plus that of the curl, and so on. Total strength correlated well in two studies that I performed with two different departments (r = −0.95, r. = −0.84). Another method is to total all score points and provide an average or composite score. Variations of the composite method appear in the literature (Gledhill and Jamnik, 1992; Davis et al, 1982).

I have a preference for using composite scores to identify fitness levels. Since fitness is defined as having five components, all of which are interrelated and combine to form a numerical description of "fitness," it appears logical that the resultant scores of each dimension should be joined to form a composite score. Since we cannot draw a line on any specific dimension, the averaging method eliminates test failure due to the presence of a single marginal score, for which other strengths might compensate. By averaging the scores, weak areas may be offset by stronger areas and thereby produce a more accurate total picture.

In Table 11–8, a composite score method is shown using the Pompier scale. The raw scores from an individual's test are identified on the table, and the point value for each is totaled in the far right column. The column points are then added and divided by the total number of tests. The result is the composite fitness score. The fitness level or classification is then determined by identifying:

TABLE 11–8

SAMPLE COMPOSITE POMPIER SCORE

Fitness Level	Test 1 Score (Aerobic)	Test 2 Score (Strength)	Test 3 Score (Strength)	Test 4 Score (Strength)	Test 5 Score (Endurance)	Test 6 Score (Flexion)	Test 7 Score (Body comp)	Points	Total
1	48	2.25	3.7		**50**	19		10	10
2	**45**	1.75	**2.9**		40	17		8	16
3	42	**1.25**	2.1		30	15		6	6
4	39	0.75	1.3		20	**13**		4	4
5	36	0	0.5		10	11		2	0

Total points	36
Average	7.2
Fitness Score	3

where the composite point score is located within the fitness levels. In the example, the average score is 7.2, which is less than the 8.0 points required for Fitness Level 2, and more than the 5.0 points in Level 4. Therefore, this individual is categorized as a Fitness Level 3.

SLIDING SCALE

A sliding scale is a standard adjusted for age or gender using normative data. The premise behind using a sliding scale is related to two widely held beliefs: that women aren't physically comparable to men, and that fitness declines with age. The differences between men and women in regard to fitness measures are considered by some to be related more to physical size than gender (Wilmore and Costill, 1994). However, other studies report morphological differences other than size alone that account for differences (Shepard, 2000). There is an age-associated decline in fitness that can affect performance, but the rate of decline is highly individualistic and directly related to activity level, (Kilborn, A., 1980). Also, job demands are independent of who performs them and cannot be adjusted; therefore, the individual must possess the requisite fitness to match those demands regardless of age or gender (Lusa *et al.,* 1994). The literature supports the opinion that there is little justification for sliding scales (Williams, T. and Evenson, S., 1988).

Still, I contend that there *is* justification for a sliding standard, based on three arguments. First, since the fitness program test and standards are neither designed nor intended to make job decisions, different standards don't negatively affect the members as long as they're used properly. Second, if the goal is to determine whether the members are physically capable of performing, and if the standards match the job, then gender is irrelevant. Third, knowing that fitness declines with age, the logical solution is to adjust for the decline by increasing the entry-level standard. I refer to this as a positive slide rather than a negative slide. The typical negative-slide approach is set to the minimum requirements for the youngest age group. Subtractions are made from it to account for the decline of the older groups. By a positive slide, the minimum standard is assigned to the oldest age group. Adding the associated decline to this minimum raises the standards for the youngest group.

The requirement of a higher fitness standard for the youngest age group will offset any declines, ensuring a fitness level that will match the job demands throughout one's career. This may be interpreted as a dual standard, but in fact, it's a way of integrating an acceptable level of performance and a desired level into the fitness standards (Williams and Evenson, 1988a). The younger members should easily be capable of meeting the higher recommended level, and the older members, through a properly designed exercise program, should be capable of arresting physical decline and achieving the minimum standard commensurate with their assignment.

If you decide to use a sliding scale, per Figure 11–5, simply slide the scores upward by age group; however, don't move the point values! You may divide the age groups by any interval you choose. The ten-year time frame is the most common. In this example, a twenty-five-year-old with a raw score of 42 will receive fewer points than a forty-year-old with the same score.

STANDARD CHOICE

Using the concept of levels of participation (LOP), as presented in Parts II and III, you have three LOP choices for developing standards. This is a matter of preference, and the decision about it should be included in the program policy. Any of the options will work; however, the criterion method will be the most effective. No matter what method you use, the criterion will ensure the minimum standard.

LEVEL 1

Choose available general population normative data. The Institute of Aerobic Research has the largest general-population database, and it can

STANDARD AEROBIC CAPACITY BY AGE

	40+	30–39	20–29
Fitness Level 1	48		
Fitness Level 2	45	48	
Fitness Level 3	**42**	45	48
Fitness Level 4	39	**42**	45
Fitness Level 5	36	39	**42**

Slip the scale one standard deviation (three points in this example) to correspond with the age-associated decline in aerobic capacity. The minimum (42) is set for the oldest age group, and the scores are adjusted upward for the younger groups. The same procedure is used to develop the sliding scale for strength and flexibility scores.

Age-Corrected Score for the Minimum Standard of a $\dot{V}O_2$ of 42 mL·kg^{-1}·min^{-1}

Age	20	25	30	35	40	45	50
Mets	13.7	13.4	13.1	12.8	12.6	12.3	12.0
$\dot{V}O_2$*	48	47	46	45	44	43	42

* $\dot{V}O_2$ rounded off to whole number.

Fig. 11–5 Constructing a Slip Scale

serve the purpose for six basic tests: bench press, aerobic test, leg press, sit-ups, push-ups, and body composition/flexibility. For additional test options, normative data is available from the Canadian Society for Exercise Physiology and in referenced texts by Heyward (1997) and Johnson and Nelson (1986). Additional normative data on a variety of other tests is included in Appendix C.

The advantage of using available data is that your standards have already been developed for you, although you will have to devise a scoring system. There are also some disadvantages. For one, the standards may not be applicable to your population, and none may be available for the tests you choose. More importantly, they may not reflect the physical demands or requirements of your occupation, depending on the source.

LEVEL 2

The second-level option is to develop your own normative data using your population. This is done by testing your department members and computing the mean and standard deviation for each fitness test. Then, select one of the scales or develop a percentage table to construct your own standards. The advantage is that the resultant standard reflects your specific group. The disadvantage is related to the physical status of your members. If they are in excellent or poor condition, the standards may be too high or too low. Again, they may not reflect the job requirements. A second option may be to use a hybrid of Level One and Level Two by selecting the Pompier scale provided in Appendix C.

LEVEL 3

Use the criterion standards provided in Table 11–5, and construct a Pompier scale as described above. This is the preferred method, since the standards are based on job requirements and your specific department members. Sample tables are constructed from public safety officer test data and provided in Appendix C.

Although there are a variety of methods available for developing standards, the Pompier scale offers several advantages. First, it provides a minimum standard that can be used by all agencies. The Pompier table requires the establishment of the criterion at Fitness Level Three. The standard for all departments is the same, which allows for cross-comparison and analysis.

Another advantage is that it reflects an individual department's profile. The minimum standard remains constant, while allowing for variations between departments in determining categories using the standard deviation. The standard deviation computed from each department's test score is used to identify the interval above and below the minimum standard to establish the remaining fitness categories for internal use only. In Table 11–9, four departments are compared using actual test data plugged into the Pompier table. Note that the minimum requirement is the same ($\dot{V}O_2$ = 42, Fitness Level 3), but the other categories vary, based on the results of each department's test scores. A $\dot{V}O_2$ of 48 is a Fitness Level 2 in Department D, whereas the same score is a Fitness Level 1 in Department A. This variance means that the results won't necessarily translate between departments, but what's important is the ability to identify members as meeting, exceeding, or falling short of the standard.

The Pompier scale establishes a minimum standard. The standard is job-related and not determined by the personal physical condition of department members. Also, the method of testing can vary between departments, but it is the end product of the test mode that provides one standardized value for scoring and comparison.

If all departments used the same table; *i.e.*, the same minimum standard, comparisons could be made with specific injury data. For example, NFPA 901, *Standard Classification for Incident Reporting and Fire Protection Data, 1995 Edition,* mandates a description of the physical condition of a member both at the time and prior to the time of injury. Per an NFPA proposal, and following the precepts of 12–10.3, a person's aerobic fitness prior to the time of injury could be used to classify data; e.g.:

TABLE 11–9

POMPIER SCALE COMPARISON

Aerobic Scores in $mL \cdot kg^{-1} \cdot min^{-1}$

Department	Mean Score	Standard Deviation
A	45	3.2
B	39	4
C	41	2
D	40	6

Fitness Classification	Dept. A	Dept. B	Dept. C	Dept. D
1	48.4	50	48	54
2	45.2	46	44	48
3	**42**	**42**	**42**	**42**
4	38.8	38	40	36
5	35.6	34	38	30

The minimum standard is set at 42 $mL \cdot kg^{-1} \cdot min^{-1}$. The standard deviation for each department is used to determine the intervals above and below the standard for the five fitness classifications.

1. Above standard.

2. Meets the minimum of the standard.

3. Below standard.

9. Status of the injured person not measured.

0. Status of the injured person undetermined or not reported.

Using these codes in conjunction with the Pompier scale would identify the physical status of the injured member relative to the minimum standard. The injury-fitness relationship between specific fitness dimensions and specific injuries could easily be analyzed. An example of this would be the percentage of those reporting cardiac symptoms (Code 11) and those who are below standard in aerobic capacity. The system also allows for departments that don't use the Pompier to exclude analysis if they don't use the table (Code 0) or don't have a program (Code 9).

The recommended approach is to start with the aerobic value of 42, which has been well defined and has been relatively stable in research reports for the past twenty years. If the data proves meaningful, other fitness dimensions, such as strength, could be added and correlated to other symptoms, such as the sprain/strain code.

RECOMMENDED READING

Johnson, B.L.; and Nelson, J.K. (1986) *Practical Measurements for Evaluation in Physical Education, Fourth Edition*. Minneapolis, MN: Burgess Publishing Co.

Heyward, V. (1998) *Advanced Fitness Assessment and Exercise Prescription, Third Edition.* Champaign, IL: Human Kinetics.

Chapter 12

Exercise Recommendations and Training

The exercise prescription is the application of a treatment (exercise plan) based on the diagnosis (fitness profile). The term *prescription* suggests a medical procedure or activity reserved for the physician. In fact, the guidelines of the American Heart Association place exercise prescription exclusively in the physician's domain. This precaution is appropriate for high-risk or diseased (CHD) patients, but it's unnecessary for the majority of participants in a departmental program. The American College of Sports Medicine is less restrictive and assigns exercise prescription to a certified exercise specialist or program director. The ACSM's approach is more realistic, particularly if the fitness coordinator is properly certified by one of the organizations listed in Appendix G.

The practice of making exercise recommendations (EXRX) can range from being a highly complex and scientific procedure to one that is simple and basic. The degree of complexity depends on the needs of the participant and the expertise of the exercise programmer. An exercise recommendation cannot be completed without a current physical profile of the individual. Therefore, the testing process must first be completed and scored, and the data must be recorded to produce a fitness profile. The exercise recommendation is predicated on analysis of the profile.

The effectiveness of a fitness coordinator in providing sound advice depends on his or her qualifications, which at minimum must include certification by an appropriate organization. The certifying agencies vary in curriculum, but they should all provide the essential information necessary to writing a basic exercise recommendation, or EXRX. For those who aren't comfortable writing recommendations, I offer alternatives, using the level of participation concept presented in Parts II and III.

If you constructed a fitness standard table as discussed in the previous chapter, you will have the basic format on which to base your EXRX. The majority of your department members will require no more than a simple, basic approach. The remaining few I place in four categories: (1) athletic, (2) high risk, (3) special need, and (4) focus group.

Your top performers are in the athletic group. The more members that you have in this group, the less work you'll have. Typically, the top performers represent about ten percent of a department. They are in the top group because whatever they're doing in terms of exercise works! For all practical purposes, you won't be required to provide exercise recommendations to this group. Depending on your personal background and level of expertise, any recommendations that you do give to this group will most likely have to do with training. The athletic group is interested in fitness, exercise, and training regimes that extend beyond the requirements of the job and into the competitive realm of sports. As a fitness specialist or coordinator, your concerns are occupational. If you have the expertise, then provide it. If not, then direct them to an organization, club, or journal that can provide the information that they desire.

The National Strength and Conditioning Association (NSCA) is a good resource for strength training specific to a variety of sports. The high-end performers may require a more complex training regime if they are to improve. Typically they have a solid understanding of their physical capabilities and will often profit from advanced techniques and routines offered by associations such as the NSCA.

The use of interval training for all types of activities is becoming increasingly popular, and it is effective for all levels of participants. The fundamental difference again is in complexity. An interval program designed for the high-end performer is very precise, being highly regimented and with performance expressed in terms of seconds. For the novice, it's more general, with inexact directives, such as "speed up" and "slow down." Some advanced techniques, such as interval training, are provided for both strength and running programs in a text by Fox and Mathews (1974).

The high-risk group is comprised of those members who require a medical referral and a physician's clearance before being allowed to participate in an exercise program. For all practical purposes, you wouldn't be required to prescribe exercise for such a person; rather, you should refer him for expert advice. A member referred to a physician for a specific medical problem would be directed to the proper professional for rehabilitation and conditioning recommendations prior to returning to duty. The physician/therapist would prescribe the activity, and the fitness coordinator would interact with the provider and perhaps monitor the member's activity. Once the case has been resolved, the member, if cleared for duty, would be reintroduced to the department fitness program to resume activities with the other members.

In the special-need group are those individuals who have some limitation as a result of injury or illness that doesn't affect their ability to perform as a firefighter, but that does require alternative exercise modes to participate. For example, a firefighter who has suffered a rotator-cuff injury wouldn't want to perform dips as part of his standard exercise program, nor should he be tested in this mode. Although the exercise develops good strength in the anterior deltoids, performing it does mimic a partial dislocation of the shoulder joint. The solution for such an individual would be to have a physical therapist provide alternative exercises that will strengthen the muscles or muscle group without increasing the chance of reinjury. Individuals who have suffered specific injuries that limit their ability to perform certain types of exercise are aware of the problem and have usually received some professional advice as to what activities they should avoid. They should also have been informed as to viable alternatives. If they haven't, it's best for the fitness coordinator and the member to consult with a physical or occupational therapist who can provide some direction. An alternative test mode, such as push-ups instead of a bench press, may be necessary for such an individual. However, beware the member who is capable yet seeks an alternative to avoid testing. If a member cannot participate in an exercise program, he certainly cannot be cleared for public safety work.

The focus group will demand a significant amount of attention and will occupy eighty percent of a fitness coordinator's time, even though the people in it generally represent less than ten percent of the membership. The current philosophy of many program directors in the private sector is to write off this group and focus on maintaining the fitness levels of the majority. Although this may be acceptable in the private sector in terms of time invested, it is unacceptable in the realm of public safety, where everyone must perform. A trait common to this group is their consistent performance on fitness tests, in that they repeatedly perform poorly.

The low performers fall into three basic categories. Those with medical contraindications cannot perform for legitimate reasons. The solution to this problem, if not correctable through medical treatment, is either retirement or reassignment to a sedentary job.

The second category consists of those with behavioral problems. Within this category are those who have problems related to stress, substance abuse, or psychological issues that require professional intervention. These members need referral to an appropriate authority by way of the employee assistance program, and the fitness coordinator would only become involved following treatment and a clearance to return to work. Within this category are also those who simply refuse to participate. If the program is mandatory, it becomes an insubordination issue if someone refuses to perform the assigned duty. In this case, disciplinary action is warranted (Williams and Evenson, 1988).

The third category has to do with exercise programming. When an individual fails a fitness test and is then prescribed an exercise regimen, then fails the retest, it may or may not be the individual's fault. The first question then becomes whether or not the participant did the work. Whether the participant followed the exercise plan or not can be ascertained simply by asking for and reviewing his fitness log. If the log is incomplete, it may become a disciplinary issue. If it is complete and reflects the exercise recommendation, one need only compare the information in it with the test results. It's easy to detect a false report by comparing the log to the test results. If it doesn't match, then you may have another disciplinary issue. If the log is accurate and the individual performed as assigned, then the fitness coordinator should rewrite the exercise prescription and monitor the individual's activities to ensure compliance with the proper procedures.

Any individual who is capable of participating, is provided an appropriate exercise recommendation, and who performs the exercise routine can pass a valid fitness test if he was capable of performing the job at the time of employment. Remember that all fitness test failures, barring a medical issue, involve fundamental, individual decisions that can be traced to a given member's attitude toward compliance.

The fitness specialist's role in exercise prescriptions isn't difficult if one adheres to the basic principles and employs good judgment. The four categories mentioned above are easily addressed by consulting with medical providers, fitness experts, and department administrators. Those outside of the four categories, or about eighty percent of the department, can be provided exercise recommendations without difficulty using one of the three alternatives presented.

LEVEL 1

At the most basic level, the exercise recommendation may be made by way of a handout that reflects the precepts of ACSM's "Position Statement on the Quality and Quantity of Training for Developing and Maintaining Fitness in a Healthy Adult." This ACSM statement provides only a general guide that incorporates all levels of fitness. Therefore, the fitness coordinator must provide some interpretation based on individual fitness levels when using this handout. For example, the aerobic intensity for a low-fit individual would be 60 percent of HR (heart rate) or less, versus a recommendation of closer to 90 percent maximum HR intensity for a high-fit individual. The fitness coordinator must be able to distinguish between the two when writing a recommendation, since the guide simply provides a range, such as 60 to 90 percent of HR. The condensed version of the exercise recommendation below, or the more informative position statement itself, may be distributed as a handout. Either way, the fitness coordinator will have to provide some sort of explanation or interpretation, such as how to calculate a training heart rate. The expertise required for such interpretation is acquired through a certified fitness coordinator program, the minimum prerequisite for program staff members.

The following recommendations are adapted from the American College of Sports Medicine, *The Recommended Quantity and Quality of Exercise for Developing and Maintaining Cardiorespiratory and Muscular Fitness and Flexibility in Healthy Adults* (1998).

Aerobic Training

Frequency: Three to five days a week.

Intensity: 55/65-90 percent of maximum heart rate (HRmax), or 40/50-85 percent of maximum oxygen uptake ($\dot{V}O_2max$) or HR reserve (HRR). The lower intensity values (i.e., 40-49 percent of $\dot{V}O_2$ or HRR and 55-64 percent of HRmax are most applicable to individuals who are unfit).

Duration: Twenty to sixty minutes of continuous aerobic activity. The duration depends on the level of intensity. Lower-intensity activity should be performed longer (thirty minutes or more), whereas high-intensity activity should be conducted for shorter periods (about twenty minutes).

Mode: Any activity that uses large-muscle groups, that can be maintained continuously, and that is rhythmical and aerobic in nature, such as walking, hiking, jogging, running, cycling, cross-country skiing, dancing, swimming, skating, or various games that require endurance. Cross-training may be employed to achieve a more well-rounded effect.

Resistance Training

Frequency: Two days a week at minimum, three days recommended.

Intensity: A minimum of one set of eight to twelve repetitions for each muscle group, performed to near fatigue for eight to ten exercises that condition the major muscle groups. Individuals past the age of fifty may select a higher rep routine, such as ten to fifteen repetitions.

Duration: The time it takes to complete a comprehensive, well-rounded program, with the exercises performed at moderate to slow speed, moving through a full range of motion and without impeding normal forced breathing. Since programs of more than sixty minutes per session are associated with higher dropout rates, sessions of less than an hour would appear preferable.

Mode: Muscular strength and endurance can be developed by means of static (isometric) or dynamic (isotonic or isokinetic) exercises. Although resistance training equipment may provide a better graduated and quantitative stimulus for overload than traditional calisthenics, calisthenics and other types of exercise can still be effective in improving and maintaining strength.

Flexibility

Frequency: Depending on the individual's degree of flexibility, a daily routine is best, particularly before and after an exercise session, and periodically throughout the day while at work.

Intensity: Stretch to a comfort level at or just below your limitations, but not to the point of pain.

Duration: Ten to forty-five seconds, determined by feel. About eight to ten exercises lasting fifteen to thirty minutes.

Mode: Preferably slow, static movements, although dynamic techniques are acceptable. Appropriate warm-up and cool-down periods are recommended.

Body Composition

If the primary purpose of the training program is weight loss, then regimes of greater frequency and duration of training, at low to moderate intensity, are recommended. Training should be combined with a dietary regime for effective weight loss and maintenance (ACSM, 1983).

If the total body mass and fat reduction aren't considerations, then shorter-duration, higher-intensity programs may be recommended for healthy individuals at low risk for cardiovascular disease and orthopedic injury.

LEVEL 2

The second-best choice would include the recommendations from Level 1, plus a more detailed guide for providing an exercise recommendation. There are two ways to provide an increased level of service without having to upgrade the fitness coordinator qualifications beyond the basic prerequisite. The first is to employ the services of an exercise specialist from outside the department. A local community college or university is the least expensive source. If an internship is available from a university exercise physiology program, your problem is solved. One assignment for the intern would be the development of a series of prepared exercise recommendation cards that could be handed out to members, based on their fitness profile and personal goals. The cards would encompass the broad spectrum of fitness levels: one card for each fitness level and each specific fitness component. The fitness coordinator would then simply match the member's fitness level with the appropriate card.

The second alternative is to use the following card that I have prepared. It is based on the fitness categories developed with regard to statistics and standards, as described in the preceding chapter. The chart boxes correspond to the standards box. For example, a member scoring a Fitness Level 2 on an aerobic test would be prescribed the routine found under the aerobic column, Fitness Level 2 on the corresponding EXRX card (Figure 12-1).

The concept behind using a predeveloped card is based on the same table format, presented with standards. The process becomes one of selecting the exercise recommendation card that corresponds with the same standard score box. Notice that the EXRX card chart is designed to match the Pompier table. An aerobic score that is classified as a Fitness Level 2 would receive the fitness Level 2 exercise recommendation, which would then be transferred to the program recommendation. Blank cards are provided in Appendix D. The exercise recommendations correspond to the fitness classification boxes on the standard table. The method may not be precise, but it will get an individual to the starting line. An exercise recommendation is only a start guide. It should always be adjusted by the individual, based on his personal response to the workload.

LEVEL 3

The preferred approach to exercise recommendation is to provide individual coun-

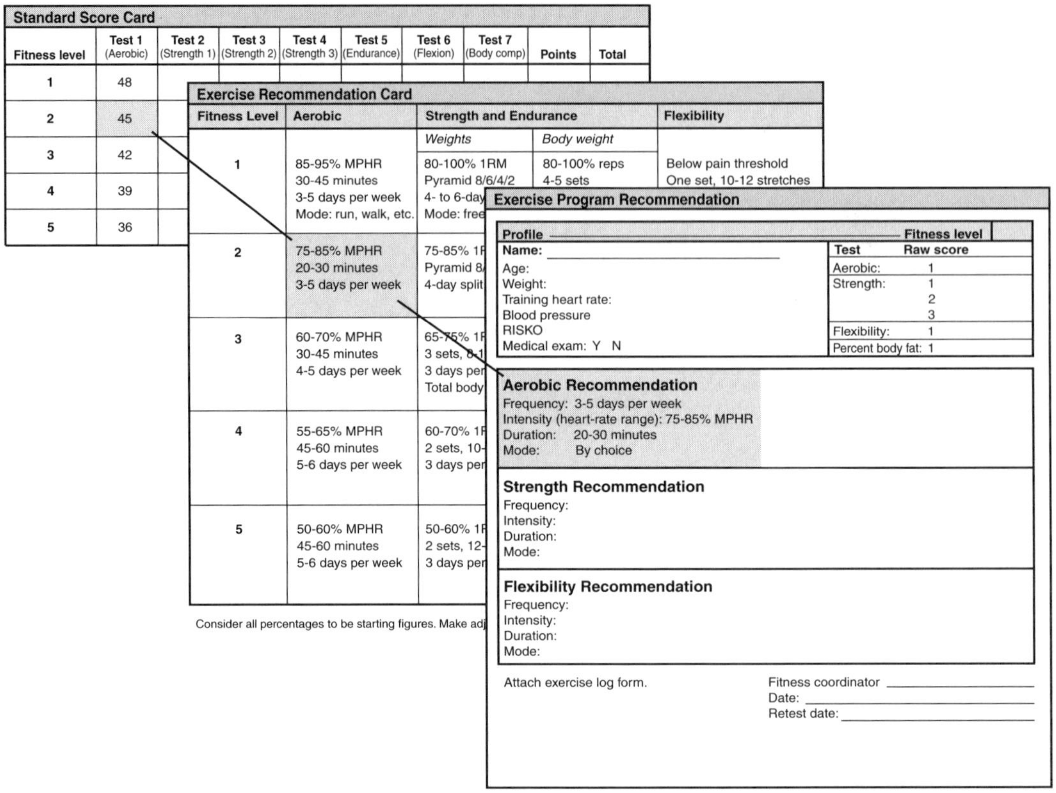

Fig. 12–1 Exercise Recommendation Cards.

Writing an EXRX card is a simple three-step process. First, identify the score on the standard score card. In this example, it's an aerobic score of 45, which correlates with Fitness Level 2. Then, find the corresponding box at Fitness Level 2 on the exercise recommendation card. Finally, transfer the information to the exercise program card.

Follow the same procedure for each component. Strength will have more than one recommendation, and the EXRX may differ for each.

seling and prescription based on the specific needs of the member. The ability to provide this level of service depends on the qualifications of the fitness specialist. The Level 3 classification specifies graduate-level education in exercise physiology, kinesiology, or a related field. The fitness coordinator who meets this level of qualification will have the necessary expertise to provide the best service. Decisions about staff and qualification level should be made carefully, since decisions in one area can have repercussions throughout the program.

TRAINING PRINCIPLES

There are several fundamental principles that you should follow when prescribing and participating in any exercise routine. These principles apply regardless of the type of program. Adherence to the general guidelines presented here is necessary to ensure success with any program or routine. The principles covered in this section can be reproduced and distributed in the form of a handout, along with the exercise recommendation.

THE TRAINING EFFECT

The training effect refers to the physiological changes that occur as a result of applying the training principles in an exercise program. A few of the changes that you can expect are as follows.

- Increase in maximal O2 uptake.
- Lower blood lactate levels at a given workload.
- Greater utilization of anaerobic energy reserves.
- Stroke volume increased.
- Increase in total work output at exhaustion.
- A more rapid heart-rate recovery and return of blood pressure to normal.
- A lower resting heart rate.
- Muscle hypertrophy.
- Muscle glycogen stores increased.

The four factors involved in achieving the desired effect are frequency, intensity, time, and type (mode), as referenced by the acronym FITT.

Frequency: This refers to how often the session is performed. Typically, three to five days per week is recommended for aerobic work, and two days minimum (three is optimal) for strength. Depending on the training goal, an aerobic routine for distance runners would require six days a week and some multiple daily sessions. For strength training, the typical routine is a three-day full-body or a four-day split routine, with some power lifters and body builders using a five- or six-day split. Frequency is directly related to consistency, which means that if you expect to improve, then you must work out on a regular schedule and avoid skipping too many sessions. To help maintain consistency, a logbook is essential. Those missed sessions can sneak up on you without your being aware of how many days you actually skipped. Often it's more than you think. One method of staying on track is to mark each page of your log by the week for every week that you completed all of your workouts. If you miss a training session during a given week, start your notation over again at Week 1. This will help you track your consistency. If you never get above Week 5 or 6, then you aren't very consistent. If your notations are in the Week 30 or 40 range, then you're very consistent, and the results will show.

Intensity: The most important factor, and the one considered essential to demonstrate improvement, is intensity, which is essentially a matter of how hard one exercises. Aerobic intensity is based on a percentage of heart rate or Vo_2max. Muscular training can be expressed in terms of percentages of repetitions maximum (percent of RM) or in the number of repetitions, which is directly related to percentage. Low repetitions would indicate a higher intensity or more weight. High reps indicate the opposite.

Prescribing aerobic intensity through percentage is an attempt to target the proper workload, based on the reported linear relationship between heart rate and Vo_2. The actual percentage is highly individualistic and should be adjusted by each participant based on his perception of the work bout. If the work is perceived as too difficult or too easy, then the participant must make an adjustment, regardless of the percentage figure. Percentages are nothing more than an attempt to quantify a *feel*. The percentage based on an estimate of one's maximum heart rate can be calculated using a simple age formula, as shown in Table 12-1.

TABLE 12-1

ESTIMATING TRAINING HEART RATE

A 40-year-old with a resting heart rate of 80 who chooses to exercise at an intensity of 70 percent could determine his training heart rate by several methods.

Low estimate

220 – 40 (age) =	180
	x .70
Training HR	126

Median estimate

226 – 40 (age) =	186
	x .70
Training HR	130

High estimate

210 – (40 × 0.5) =	190
	x .70
Training HR	133

Karvonen Formula

The Karvonen version is calculated by taking a percentage of heart-rate reserve (HRR), which is the difference between maximal heart rate and resting heart rate (RHR).

220 – 40 (age) =	180
Minus HRR	– 80
	100
Percentage	x .70
	70
Plus HRR	80
Training HR	150

The Karvonen method adjusts the heart rate to the equivalent percentage of Vo_2max. It is not the same as a percentage of HRmax. The Karvonen method results in a higher training heart rate, as noted in the example above. This is one reason the ACSM expresses two percentage ranges, one for a heart rate of 60–90 percent and one for a Vo_2 of 50–85 percent.

The formula may exceed or underestimate a person's true maximum. One standard deviation is plus or minus 12 bpm, meaning that any heart-rate calculation can only be viewed as a general guide. I once tested a fifty-five-year-old firefighter on a maximal stress test who recorded a maximum heart rate thirty beats higher than predicted by the formula. Usually training intensities are expressed as a range to compensate for this discrepancy. The most accurate method is to determine an individual's maximal heart rate using the maximal test. Otherwise, the participant can self-adjust the intensity of his training, based on his response.

The Borg RPE scale, as depicted in Table 12-2, is the most commonly used method of rating perceived exertion. It is at home in the clinical setting where treadmill tests are conducted. The RPE ratings grow linearly with exercise intensity, heart rate, and Vo_2.

Occasionally treadmill tests are interrupted because of symptoms of pain, such as angina or muscle aches. Consequently, there was a need to develop a scale that could also be used for pain, as well as other sensory perceptions. This led to the development of the CR-10 scale, shown adjacent to the RPE. The Category-Ratio (CR) scale is anchored at the number 10, which represents extreme intensity. It is a general-intensity scale that can be used to estimate pain intensity. The CR-10 is more complicated to use and is often misinterpreted. It is suggested that the reader consult with the Borg text for additional information on the proper use of both scales.

TABLE 12-2

BORG'S CR10 AND RPE SCALES

CR10			RPE	
0	Nothing at all	No "P"	6	No exertion at all
0.3			7	
0.5	Extremely weak	Just noticeable	8	Extremely light
1	Very weak		9	Very light
1.5			10	
2	Weak	Light	11	Light
2.5			12	
3	Moderate		13	Somewhat hard
4			14	
5	Strong	Heavy	15	Hard (heavy)
6			16	
7	Very strong		17	Very hard
8			18	
9			19	Extremely hard
10	**Extremely strong, max "P"**		20	Maximal exertion
11				
//				
●	Absolute maximum	Highest possible		

After Gunner Borg (1998), *Borg's Perceived Exertion and Pain Scales.* Champaign, IL: Human Kinetics. Reprinted by permission of the author.

One metabolic equivalent, or MET, is equal to 3.5 $mL \cdot kg^{-1} \cdot min^{-1}$, the amount of oxygen used by the body at rest. Thus, a workload of 10 METS would equal 35 $mL \cdot kg^{-1} \cdot min^{-1}$ of oxygen, or ten times the resting rate. The minimum recommendation for a firefighter or police officer is 12 METS, or a Vo_2 of 42 $mL \cdot kg^{-1} \cdot min^{-1}$. The MET is most commonly used in the clinical setting in conjunction with treadmill testing.

The use of percentages with strength training is also a highly individualistic estimate. Although most values are based on a percentage of one-repetition max (1RM), good estimations are achieved by using a percent of three- to five-rep maximum, or a repetition trial-and-error method. Following are several methods of estimating maximal lift and determining percentage or intensity.

The first method is the equivalent single (1RM) estimation. Multiply the corresponding rep factor by the weight lifted. For example, if an individual performs five repetitions with 170 pounds, multiply 1.18 x 170. The answer, 200.6 lbs., is the estimated 1RM. From the 1RM estimate, a percentage is taken to determine training weight (Goodman, R., *The Theory and Practice of and Training for Power in the Bench Press,* California: self-published, 1983).

The rep factors to use for this method are as listed in Table 12-3.

TABLE 12-3

ESTIMATING THE ONE-REP MAX

Two reps:	1.07
Three reps:	1.12
Four reps:	1.15
Five reps:	1.18
Six reps:	1.21
Seven reps:	1.24
Eight reps:	1.27
Nine reps:	1.30
Ten reps:	1.33

The Brzycki Formula (Brzycki, 1993) uses the same values as in the preceding method, but it uses a different single factor. The 170 lbs. x 5 reps would result in a 1RM estimate of 191 lbs. This estimate is then multiplied by the desired intensity, or 70–90 percent for most training. This method results in a maximal estimate nine pounds greater than the equivalent single estimation. It is a good method for estimating a max lift for testing purposes.

TABLE 12-4

ESTIMATING THE ONE-REP MAX USING THE BRZYCKI FORMULA

$$1RM = \frac{170 \text{ (weight lifted)}}{1.0278 - .0278 \times 5 \text{ (number of reps)}}$$

$$1RM = \frac{170}{.889}$$

1RM = 191 lbs.

A third method is the mid-intensity method, in which intensity is determined by a weight-per-rep value. To identify the weight-per-rep value, the total workload is divided by total repetitions for the workout (James, S., 1988). This method of determining intensity is best left to the experts who are most familiar with the procedure. Still, it may be worth investigating for that high-fit group consisting of serious weight trainers.

The fourth method, the repetition mode, is the easiest and probably the best choice for most. The average person involved in a weight training program should avoid maximal lifts for safety reasons, as well as for training effectiveness, since heavy training at the two-to-three rep range produces better results than performing singles. When conducting strength tests, either the Brzycki formula or the rep factors should be used to estimate the 1RM. Although the estimation method is frowned upon for research purposes, it will serve quite well for fitness program testing. Most individuals, particularly novices, won't identify their true maximum using single attempts, and therefore the rep procedure may actually produce a more accurate single effort. The best approach for the rep method is to start with a weight that one can perform for about ten repetitions, then add about 10 to 15 percent more weight and again perform as many reps as possible, adding weight each time until one reaches near-max, or two to three reps. After each set, record the weight and corresponding number of reps so that you'll know exactly how much weight to use for each rep routine. If your program specifies eight to ten reps, you'll know your starting weight for eight to ten reps and won't have to use a percentage or perform any calculations. You will also know approximately what percentage of max you're working. The percentages shown in Table 12-5 are ballpark estimates.

TABLE 12-5

ESTIMATING REPS AND PERCENTAGE OF 1RM

Reps	Percentage
10	75%
8	78%
6	82%
4	86%
2	93%
1	100%

Time: This refers to the duration of the exercise session, which in turn depends on the frequency. For example, a strength workout performed three days every week may last forty-five minutes, whereas the sessions for a four-day split routine may be only thirty minutes each. The recommendation is for thirty to forty-five minutes for strength training. Longer sessions haven't been proved to be more effective, since testosterone levels decline when a workout approaches one hour in length. The male hormone is necessary to stimulate the training response (Ballor et al., 1989).

Aerobic exercise is typically prescribed at a minimum of three twenty-minute sessions per week. Individuals who participate in off-duty running events will commit a significantly greater amount of time to aerobic training. Still, excessive aerobic training will have a negative effect on strength levels and should be reduced if strength is a problem (Hickson, 1984).

Type: The type or mode of training refers to the activity, both in general and specific terms. The general types are resistance training, aerobic training, and so on. Specific subcategories of aerobic training might include running, cycling, swimming, and walking, all of which would elicit a different response specific to the training mode. The mode can be used to adjust intensity. Track intervals, for example, are high intensity, whereas slow long-distance runs are low intensity.

TRAINING SESSION

The training session is typically divided into three parts: warm-up, conditioning, and cooldown. The warm-up phase should include some mild stretching or low-level aerobic activity, such as step-ups. The purpose is to elevate the body's core temperature by one or two degrees and increase the heart rate while providing a gradual transition to the workout. An activity-related warm-up is preferable to a general-body warm-up (Stamford, 1987, and deVries, 1974). If you're lifting weights, the warm-up should consist of some light repetitions for each specific exercise. The best approach for an aerobic warm-up is simply to start the activity at a low level of intensity, such as beginning a run as a slow, easy jog. A lengthy stretch routine prior to workout doesn't constitute a warm-up and is unnecessary in most cases.

The conditioning phase is the actual workout, in which the principles of training are applied and the training effect is produced. For aerobic activity, the desired heart-rate intensity is predetermined, and the exercise is performed at the desired level for the duration of the workout. Training intensities are selected based on physical condition and the desired results of the training session. The minimum training intensity should be at least 60 percent of the MPHR and expressed in a range; e.g., 60–70 percent. A starting guide for aerobic and strength training based on fitness levels is expressed as a percent of heart rate for aerobics, or a percentage of the one-repetition max for strength, as shown in Table 12-6.

TABLE 12-6

TRAINING INTENSITIES BY FITNESS LEVEL

	% HR	% 1RM
Fitness Level 1	80-90%	85-100%
Fitness Level 2	70-80%	75-90%
Fitness Level 3	60-70%	70-80%
Fitness Level 4/5	50-60%	60-70%

Note: The fitness levels refer to categories identified in the previous chapter using the Pompier table.

A basic strength-training program would include exercises for all major muscle groups, arranged in descending order: large group first, smallest group last. The muscle-group order is legs (L); chest (C) or back(B); shoulders (S); arms (A); and abdominals/low back (ABS) (Pauletto, 1988). These exercises can be performed according to a three-day, total body routine, or a four-day split routine. The most common splits are shown in Table 12-7.

TABLE 12-7

SPLIT WEIGHT ROUTINES

Three-Day, Total Body

Mon.	Wed.	Fri.
L, C, B, S, A, ABS	L, C, B, S, A, ABS	L, C, B, S, A, ABS

Four-Day Split Routine

Mon.	Tues.	Wed.
C, B, S	L, A, ABS	Off
Thurs.	**Fri.**	**Sat./Sun.**
C, B, S	L, A, ABS	Off

The aerobic workout would be performed on Tuesday, Thursday, and Saturday of the three-day routine, and on Wednesday, Saturday, and one other day of choice of the four-day split. If performed on the same day, it should be done after the weight training session to help in the removal of lactic acid.

Both aerobic and strength training should follow a four-phase cycle.

The strength workout remains the same whether using free weights, calisthenics, or variable resistance machines. Simply choose the mode of exercise and body part from the exercise menu, and include exercises from all groups. Perform the exercises in the order shown in Table 12-8.

Cooldown refers to the gradual reduction in work intensity following completion of the conditioning phase. Cooldown helps blood return to the heart through muscle pumping action to prevent pooling in the extremities. The prevention of blood pooling is most important and simply requires low-level activity similar to the warm-up until the heart rate is reduced to near normal. The cooldown period is where the stretching routine should be emphasized, particularly with weight training. This helps reduce the delayed onset of muscle soreness (DOMS) typically experienced within a day or two following the workout. The axiom "No pain, no gain" merits some attention. Delayed muscle soreness should not be confused with pain, which signifies injury or overexertion. Anyone who has performed heavy weight training or manual labor knows that some muscle soreness can be expected if you're working hard. Unfortunately, the nonphysical individual will often interpret soreness as a sign of injury and therefore reduce his work intensity to levels that aren't conducive to gain. Expect the delayed onset of muscle soreness, and practice at least a half-hour of stretching following a heavy routine to help minimize the effect.

Table 12-8
Exercise Menu

Muscle Group/Part	Body Weight	Free Weights	Machine
Group 1			
Quadriceps	Lunge Squat (one leg)	Squat	Leg press Leg extension
Hamstring	Squat	Squat Straight-leg deadlift	Leg curl
Gluteals	Front lying leg lift Hyperextensions	Deadlift Lunge	Low-back M
Gastrocnemius	Heel raise	Weighted heel raise	Calf raise
Hip abducters/adducters	Side-lying leg lift		Hip abduction
Anterior tibialis	Toe raise	Weighted toe raise	Seated toe raise
Group 2			
Latissimus dori/rhomboid	Chin-up (M)	Bentover row Pullover	Lat pull-down Cable row
Low back	Lying chest lift Hyperextensions	Deadlift	Low-back M
Group 3			
Pectoralis major	Push-up	Bench press Flyes	Chest press Pecdeck
Group 4			
Deltoids	Decline push-up	Upright row Military press	Lateral raise Shoulder press
Group 5			
Triceps	Parallel dips Bench dips	Tricep extension Kickbacks	Tricep press
Biceps	Chin-up	Bar curls	Bicep curl
Forearm	Fingertip push-up	Wrist curl	Wrist curl
Group 6			
Abdominals	Trunk curl	Weighted trunk curl	Ab crunch

PRINCIPLES

PROGRESSIVE OVERLOAD

The principle of progressive overload is the single most important training factor. The ability to produce physical adaptations that result in an improvement in any fitness dimension requires a system overload. The physiological response to overload is adaptation, by way of increased strength, endurance, and the like. Do not confuse overload with overwork, since the latter could result in injury or a temporary setback in gains. It is better to underestimate than overestimate. If the workout is too easy, adjustments can be made without the threat of injury. If you start a program that is too hard, you may become injured and not be able to adjust. If you start slowly, you can always increase the level of intensity. Most people who routinely visit the weight room overtrain. It's based on the false logic that if performing three sets works well, doing four will produce even better results. It may, and then again, it may not.

To increase strength, for example, it is necessary to apply greater demands on the muscular system by gradually increasing the workload using one or all of the four factors of intensity, frequency, duration, and mode. Using these factors, overload can be achieved by increasing the amount of weight lifted, performing slower repetitions, or reducing the rest period between sets (intensity). One can also lift more often (frequency), increase the number of sets or exercise, and thus the workout session (duration), or change to a more difficult exercise (mode), such as forced reps or eccentric contractions. Eccentric contractions (negatives) are high-intensity exercises, but they also produce the best gains (Spassov, 1988). Examples of some adjustments that can be made are depicted in Tables 12-9 and 12-10.

TABLE 12-9

STRENGTH-TRAINING SAMPLE OVERLOAD

Base model

Mode	Bench press (free weight)
Intensity	200 lbs. × ten reps, two-minute rest between sets
Frequency	Two days per week
Duration	Three sets

Possible adjustments

Adjustment to the mode: Bench press (variable resistance machine)

or...

Adjustment to the intensity: 220 lbs. X eight reps, two-minute rest between sets

or...

Adjustment to the frequency: Three days per week

or...

Adjustment to the duration: Four sets

Increasing one's fitness level depends on the application of this very basic and essential principle. Matching an individual's physical profile with the demands of the occupation is the principal goal of the fitness program. To accomplish that goal, overload is essential.

THE SAID PRINCIPLE

The Specific Adaptation to Imposed Demands (SAID) concept means that the body will respond to endurance training by increasing endurance, or to strength training

TABLE 12-10

AEROBIC TRAINING OVERLOAD EXAMPLE

Base model

Mode	Running (track)
Intensity	70% MPHR
Frequency	Three days per week
Duration	Thirty minutes

Possible adjustments

Adjustment to the mode: Running (hills)
or...
Adjustment to the intensity: 80% MPHR
or...
Adjustment to the frequency: Four days per week
or...
Adjustment to the duration: Forty-five minutes

by increasing strength, but not by increasing agility or speed. There is some overlap in training; however, the principle implies that if you choose to improve a specific dimension of fitness, then the training should be specific to that dimension (McCafferty and Horvath, 1977). Simply stated, you don't prepare for a marathon by running 40-yard sprints. If the goal is to prepare someone for public service work, then the training should correspond to the particular requirements of the job (Lemon, 1983).

The SAID principle is relevant to all forms of training and activity, as exemplified in our earlier discussion of test modes. An individual who chooses to perform his aerobic training on a cycle ergometer will typically demonstrate a higher Vo_2 score on a cycle ergometer test than on a treadmill. Although some overlap occurs, which is why any aerobic test is able to distinguish between the high and low performers, the best results are obtained by matching the test mode with the training mode.

When designing an individual fitness program, pay attention to the purpose of the training. If the training is for general conditioning, the plan will differ from one designed to enhance the performance of a specific job. Therefore, the specific demands of a given job assignment should be identified before a training program for it is developed. Pay attention also to the exercise recommendation as it relates to job demands, since an inappropriate plan may actually result in decreased performance, in spite of a general improvement in conditioning. For example, excessive endurance training can result in strength losses that may or may not be critical, depending on the individual. A public safety officer with marginal strength capabilities would benefit by focusing on strength training and limiting aerobic, endurance-type activities. An individual who performs desk duties may profit more from aerobic activity, where strength is not a factor, whereas someone assigned to field duty should focus on enhancing strength.

VARIATION

By definition, the term *variation* means *variety* or *change*. To maintain the stimulation of growth (muscle) or physiological enhancement through exercise, it is necessary to vary the type, quantity, and quality of training stimuli. As the body adapts to increasing workload demands, it ultimately reaches a peak, beyond which continued efforts no longer produce improvements. This plateau, or period of staleness as it is sometimes called, indicates a need for change. Forcing the body through this period may occasionally succeed, but most often the attempt will fail, and an injury or illness may result. Fitness improvements, the result of physiological adaptation, occur during the rest period between workouts, not during the exercise session itself. If the body isn't allowed to recuperate and adapt to the higher work demand, it won't improve. One way to provide a period of recuperation, besides exercising on alternate days, is to reduce the intensity level. For example, if you were to perform aerobic exercises on Monday, Wednesday, and Friday, you could exercise at 80 percent of the target heart rate on Monday, 70 percent on Wednesday, and 85 percent on Friday. One can also vary the exercises, such as by switching from a flat bench press to an inclined bench press, or by substituting hill runs for flat runs. Variation can be achieved by simple or complex means, but fundamentally it involves manipulation of the FITT factors.

One approach to variation that can be very specific is through the use of intervals, a form of training that intersperses work with periods of rest. The work/rest periods are expressed as a ratio, such as 1:2, 1:3, and so on. Intervals are designed to target a specific energy system, so the short-rest type would tax the anaerobic system, with longer rest periods aimed at the aerobic system. Intervals can be quite complex, but they are basically designed to allow an individual to work at a higher level of intensity than he would normally be capable of achieving. For example, a given participant might not be able to run a six-minute mile, but he may be able to accomplish the equivalent amount of work by running a mile in intervals of 440 yards at 90 seconds each, with a recovery period in between. This type of training allows one to prepare for higher workloads without excessive strain (Fox and Matthews, 1974).

Maintaining awareness of your body is fundamental. Knowing when you have reached your physical limit is important both in exercise and work. One way to develop an awareness is by monitoring your heart rate and comparing that with how you feel. Both the heart-rate monitor and RPE can be used for training; however, once you get more in tune with your own body, you won't need such devices. This mental check of your physical status is necessary if you are to prevent overexertion when performing difficult tasks. Dispense with the headphones and turn off the TV when exercising—you need to pay attention to your body if you're serious about training. Exercise is a form of work, and when you work hard, you need to remain conscious of the variation principle and its related concepts. If you worked too hard on a given day and feel exhausted, don't exercise. Go home and recoup.

OPPOSITION

Muscles can only pull, and therefore are arranged in opposing groups to facilitate movement. One physiology instructor of mine once commented, "If muscles didn't work in pairs, we would end up getting ourselves in positions that we couldn't get out of."

The maintenance of muscle balance necessitates training all muscle groups to prevent imbalances, which might lead to injury. Muscle balance doesn't imply equal strength, however, since some muscles are physiologically or biomechanically stronger than others. Usually the leg muscles are the principal concern. The relationship between quadricep and hamstring strength varies with sport activity, but in general, the hamstring should have approximately 60 percent of the strength of the quadriceps on the same leg. If one devotes equal time to both muscle groups, there is little concern for maintaining balance; however, the weaker group should receive emphasis (Thomas, 1986).

Developing and maintaining muscle balance entails working all major muscle groups. Postural symptoms of imbalance, such as slumped shoulders (weak traps), spinal curvature (weak abs), and forward lean (tight hip flexors) all require the application of this principle to correct. Common imbalances found in runners are depicted in Figure 12-2 (Dornan and Dunn, 1987).

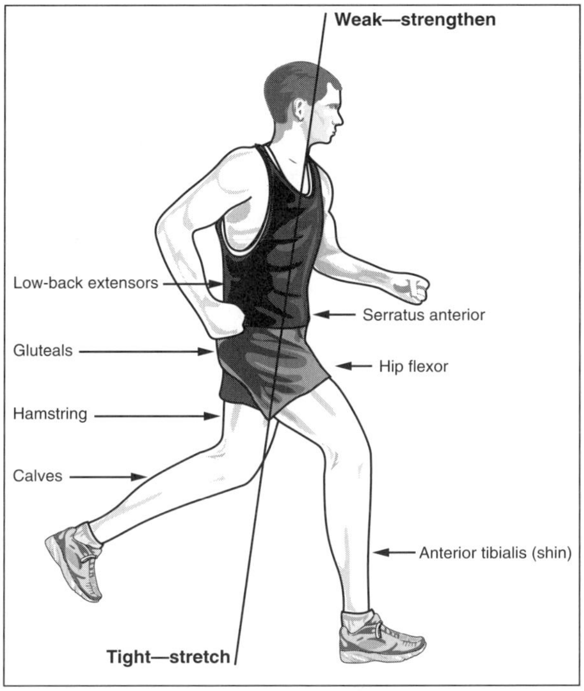

Fig. 12–2 Common Imbalances in Runners

The muscles of the anterior part of the body are typically weak yet flexible, whereas those on the posterior side are typically strong but inflexible. Therefore, correcting the imbalances by strengthening the anterior groups and stretching the posterior groups is desired.

In Chapter 8, I presented a variety of test options, including some that I favor: the bench press, leg press, sit-ups, the sit and reach, and the one-mile walk. My preference for these tests is based on their widespread use. Also, one may address injury prevention to some degree by motivating individuals to perform these particular exercises. The theory is that if you test for sit-ups, the subjects will perform sit-ups in their exercise routine. Since the ability to perform multiple sit-ups increases dramatically when one performs them on a regular basis, this exercise provides an easy way to improve one's fitness score while strengthening an important region of the body. For this reason, it is considered somewhat of a freebie in the test battery. Through sit-ups, we find an easy way to demonstrate improvement (as opposed to increasing aerobic capacity, which is far more difficult). Sit-ups are primarily used to address the low-back injury problem, although they have been shown to correlate well with some occupational tasks. The other tests recommended for the fitness battery have a similar relationship with injury prevention. The recommended test battery not only helps to assess the basic fitness components, it also serves to encourage exercises that address common occupational injuries.

The following general recommendations are relevant to these objectives.

1. The walk test strengthens the weak anterior tibialis muscles and gluteals.
2. The sit-and-reach exercise assesses low-back and hamstring range of movement, emphasizing the need to stretch this group.
3. The sit-up measures abdominal strength and endurance (Humphrey, 1988). It encourages strengthening of the midsection, the source of low-back injury.
4. The leg press emphasizes quadracep strength, needed for climbing and tasks such as hose advancement.
5. The bench press emphasizes strengthening the pressing motions of the upper body. Such actions are performed less frequently than pulling motions. Consequently, there is motivation to achieve

upper-body balance, particularly between the bicep and tricep muscles. In addition to strengthening the pectoral muscles, the bench press also strengthens the anterior deltoid, which is a common site of injury among firefighters. Such injuries are often associated with ladder raises, the use of pike poles, and other overhead work.

PERIODIZATION

Workouts should be planned in cycles that progress from a base level of training to a peak of intensity, followed by a rest period and a repeat of the cycle. The cycle can range from several weeks to several months and repeat throughout the year. If one cycle is three months long, then you can incorporate four cycles in a year. A four-phase cycle for strength training and aerobic training is provided in Table 12-11.

The final phase of task-specific training and active rest overlaps Phases III and IV. I refer to this period as being transitional. Fundamentally, it involves the use of weight circuits (Gettman et al., 1982) and job tasks performed while wearing exercise gear, then progressing to full turnouts and personal protective equipment. The last part of the phase then segues into a period of active rest; i.e., sports activity and perhaps a vacation. Otherwise, the participant might follow a maintenance-level routine. When returning to Phase I, the workload could be slightly

TABLE 12-11

TRAINING STAGES

Strength Training	Aerobic Training
Phase I: Hypertrophy Low intensity: 10–20 reps, 3–5 sets.	Base work: 60–80% training heart rate. Slow long-distance runs.
Phase II: Strength/Power High intensity: 2–8 reps, 3–5 sets. Add weight each set (pyramid, reps 8/6/4/2).	Anaerobic threshold: 80–90% THR.
Phase III: Circuit Training 10–12 stations; as many reps as possible in 40 seconds; 15–20 seconds rest between stations; 3 circuits, 3–5 days per week. Task-type events included at the end of this phase.	Anaerobic Training: 90–95% THR Interval Training Mode. Task-type events, including stair runs with equipment.
Phase IV: Task-Specific Training and Rest Period.	Task-Specific Training and Rest Period.

higher or lower than the first routine, depending on the length of the layoff. Usually it won't be at the same intensity as in the previous cycle (Figure 12-3).

The cyclic pattern of an exercise program can be made to coincide with the cycles of the occupation, as demonstrated in Figures 12-4, 12-5, and 12-6. For firefighting, the cyclic pattern can be overlaid on the annual fire pattern reported in the U.S. Fire Administration report *Fire in America*. The pattern formed by the fluctuation of structural and wildland fires parallels the recommended training cycles. This seasonal pattern is fairly constant nationwide, but of course, local variations can be made. Strength training should peak in Phase IV, just prior to the peaking of structure fires in the fall and winter, since these incidents impose the greatest strength demands. Alternately, aerobic training should peak in Phase IV during the peak of the outdoor fire season. Peaking aerobically during the summer months also helps to combat heat stress. Each of the four-phase training periods is called a mesocycle. The number of such cycles per year can vary according to the individual training agenda. The combination of mesocycles over a one-year period is called a macrocycle. The purpose of cycling is to remain fresh physically and to prepare more specifically for recurring annual work demands.

REGRESSION

Simply stated, if you quit training, your hard-earned gains will gradually disappear. Both aerobic fitness and flexibility decline rapidly with a layoff, while strength tends to drop slowly. Peak strength declines 10 to 15 percent with a two-week layoff, 23 percent in two months, 40 percent in six months, and approximately 80 percent by the end of one year (Spassov, 1988a). Muscle endurance will drop after a two-week layoff, while speed and agility remain minimally affected (Wilmore and Costill, 1994). The reason speed and agility remain relatively stable is that they are fundamentally nonmodifiable and are therefore less responsive to training or the lack thereof.

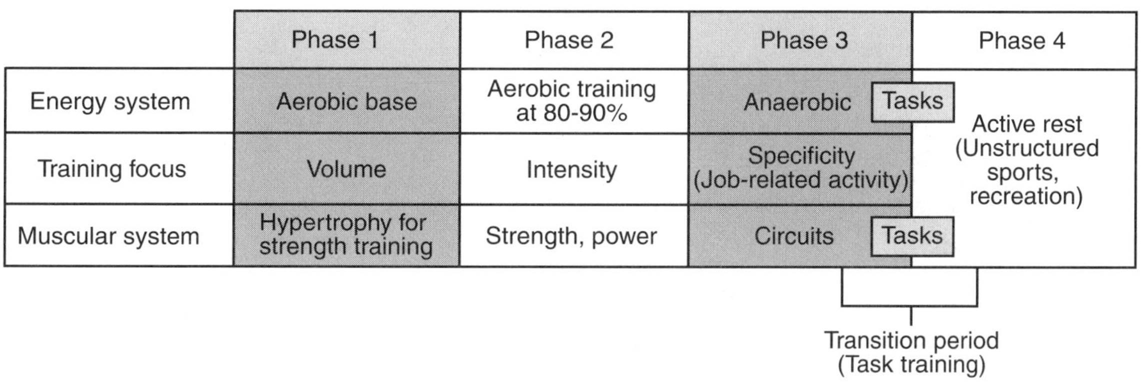

Fig. 12–3 Workout Training Phases

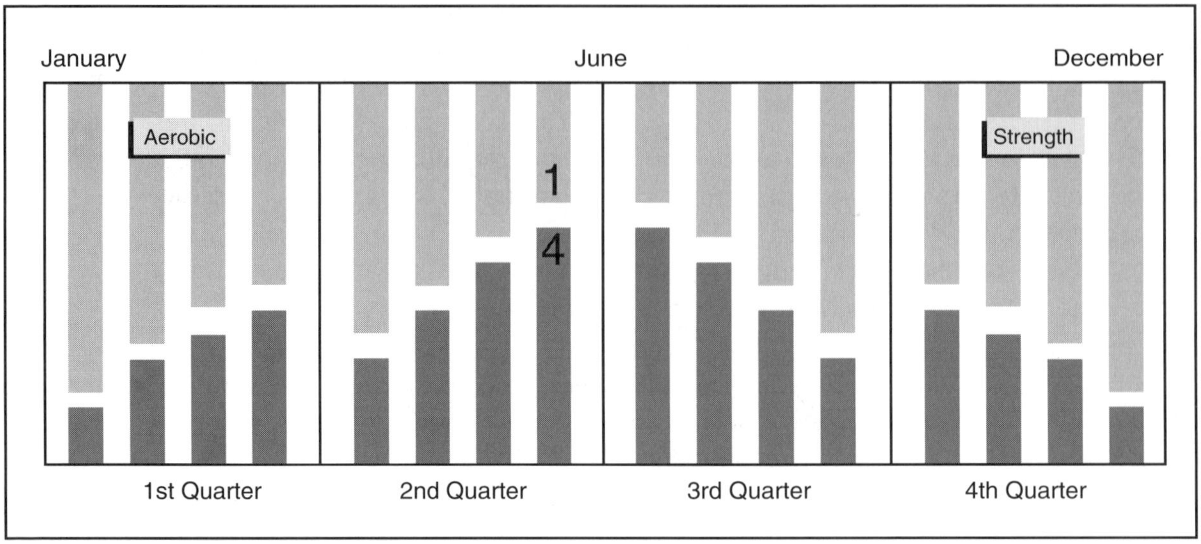

Fig. 12–4 Mesocycle

The macrocycle is divided into four mesocycles, each three months' in duration. As shown, when the aerobic cycle is in Phase 1, the strength cycle is in Phase 4. By this manner, the aerobic cycle peaks during wildland firefighting season, and the strength cycle peaks during structure firefighting season. This format produces four cycles per year, which can be further reduced to eight cycles and so on.

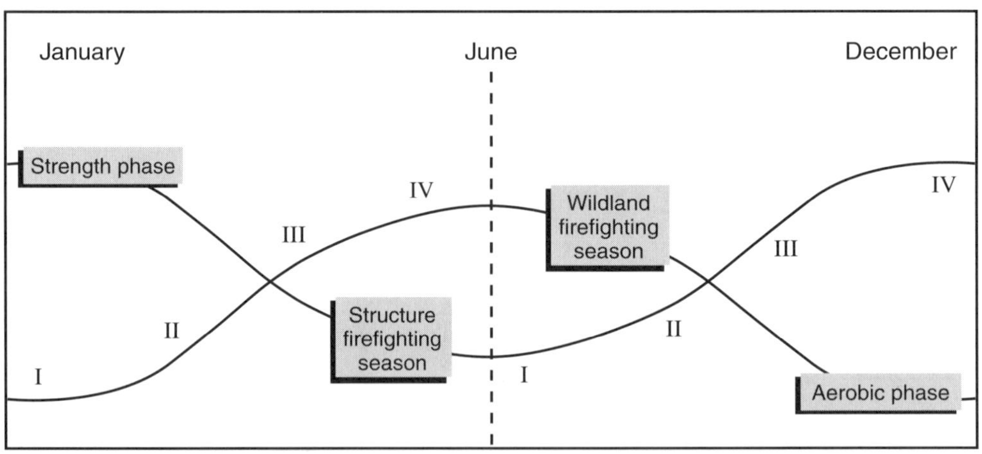

Fig. 12–5 Macrocycle

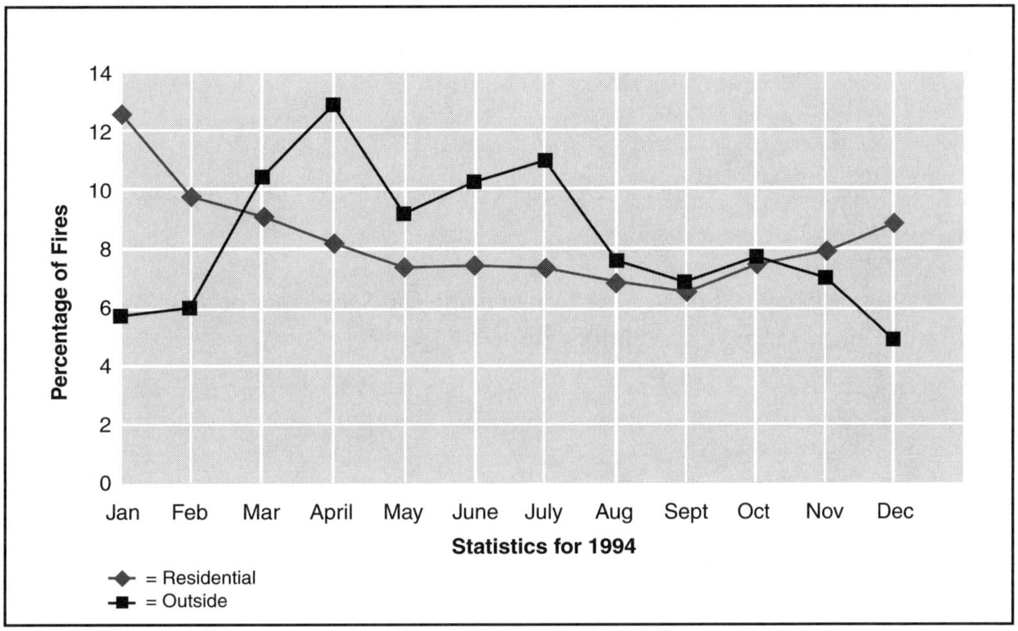

Fig. 12–6 Annual Fire Pattern
Data source: *Fire in the United States, 1985-94, Ninth Edition.* FEMA/USFA.

RECOMMENDED READING

Fleck, S.; and Kraemer, W. (1987), *Designing Resistance-Training Programs*. Champaign, IL: Human Kinetics.

Pauletto, B. (1988) *Strength Training and Conditioning for Speed: A Coaches Guide*. The National Strength and Conditioning Association, pp. 61-77.

CHAPTER 13

EVALUATING THE PROGRAM

An effective evaluation program includes two forms of evaluation: summative and formative. By using these two types, the program manager can be sure that the evaluation process is ongoing and comprehensive. A summative evaluation targets the product—the program itself and its results. A formative evaluation focuses on the process, or how the program is being conducted.

Three components of a program need to be addressed during an evaluation: product, impact, and process. The first two are summative qualities. The third is formative.

The pilot program is useful in that it serves as a product evaluation prior to full implementation. Each aspect of the program can be inspected during the pilot. To evaluate the impact of a program is to assess its overall effect on the department. In terms of the formative component, the program manager can determine through process evaluation whether the plan produces the desired results. Monitoring such aspects as adherence to the program, member participation, and test administration will provide a benchmark of effectiveness. How the program is conducted may indicate a need to revisit the level-of-participation choices made during the development phase.

Exercise prescription demands process evaluation. When prescribing an exercise program to an individual, you continually evaluate the process, testing to gauge whether the recommendations were effective. If the recommendations don't produce the desired results, then you must investigate the reasons for failure.

CRITERIA AND METHODS

Each step of this book involves an evaluation of alternatives, culminating in a preliminary plan. The pilot turns that plan into action, and the plan again becomes the subject of evaluation. In fact, there may be more than one pilot involved. The program is evaluated at all levels and on a day-to-day basis by the fitness coordinator, by the members, and in routine discussion among the staff members and delegates on the safety committee. Still, the program manager or safety director will most likely perform formal evaluations internally. There are a variety of instruments and procedures to be used.

SURVEY

As always, surveys or questionnaires can elicit feedback from the members. Following are some of the standard questions to be asked.

1. Is there agreement with the need for a program?
2. Were the goals and objectives of the program clear?
3 Did everyone understand the purpose statement of the program?
4 Was the relationship between the fitness tests and the job requirements apparent?

5. Were the tests appropriate and the instructions clear?
6. Did everyone understand the training requirements, principles, and practices?
7. Was the relationship between standards and job demands clear?
8. Were the handouts useful and effective?
9. Can you provide any feedback on the fitness coordinator's administration of the fitness tests?
10. Do you have any recommendations for improving the program?

The fitness coordinators who administer the program should also provide feedback. Were the selected tests reliable? Were the tests easy to administer? Was it easy to follow the protocols?

AUDIT

After the program has been in place for about three years, consider conducting an audit. This is best done by an outside source; however, since an audit can be expensive, it can otherwise be conducted internally by the program director or the city's risk management office. NFPA 1500 requires an audit of the health and safety program on a three-year cycle, and it includes a compliance checklist in the appendix. The checklist is an excellent guide for use by outside auditors. The fitness program is an integral part of the overall health and safety program, and an audit is a good tool for evaluating its relationship to other programs.

DATA COLLECTION AND RECORDS

The preferred methodology of many managers is to rely on statistics. Tracking scores, program costs, fitness levels, and injury data can all serve as measures of a program's effectiveness. Listed below are some of the factors worthy of assessment.

1. Injury statistics, including the total number, types, and causes, as well as the overall rate of injury. Review annual reports compiled by NFPA, IAFF, and OSHA for the type of information required.
2. Physical fitness scores.
3. Program costs.
4. Workers' compensation costs associated with injury.

In terms of record keeping, confidentiality is an important consideration. Medical records should remain strictly confidential, accessible only through the written consent of the member. Records such as fitness scores and injury data should be considered personal and not provided at random to other members. However, scores should be available to the program manager and department administrators if the tests are to be used to assess fit-for-duty status. The specific details of a personal profile need not be provided. A status report will suffice (Frost and Dufore, 1986).

COST-BENEFIT VERSUS COST-EFFECTIVENESS

Since the concern of many departments is how to justify funding of an exercise program, it would be beneficial to conduct a cost-benefit or cost-effectiveness analysis.

Some experts contend that there is no difference between cost-benefit and cost-effectiveness. I am inclined to differentiate between the two in terms of the ability to put a number on such an analysis. I interpret cost-benefit as being a comparison of two columns: Column A containing the direct and indirect costs of the program, and Column B itemizing workers' compensation costs, sick leave costs, and the like. I would say that cost-effectiveness is a comparison of the Column A costs to increased fitness levels, reduced injury rates, and other positive changes that can't be gauged in dollars and cents. There is an element of faith associated with this method. One might argue that it's difficult to convince anyone to fund a program measured by faith. Still, there is sufficient research to support implementing a fitness program with or without financial assessment of the results. In the words of corporate executive Leo Beebe, CEO of K-Tron International, "A study to measure the cost-effectiveness of our fitness program would cost $100,000. We did a lot of research before we went into it. We believe that the program is financially viable and that we don't need to measure it." (Pritchard and Potter, 1990.)

In contrast, an automobile manufacturer used a cost-benefit approach in deciding not to fix a fuel tank problem in one of its vehicles. The company assessed the cost of replacing the defective part—about eleven dollars per vehicle—and decided that it would be cheaper to pay the medical expenses of those who suffered burns. Sadly, I must admit hearing a similar response from the fire service.

Several years ago, I attended a lecture presented by a ranking chief officer of a large metropolitan fire department that had an unusually high incidence of heart attacks among its members. When an attendee asked whether the department had a fitness program or provided medical exams, the response was no. The presenter stated that since the members were in such poor physical condition, the department couldn't afford such programs. Instead, the department had opted to follow the practice of one large corporation that emphasized treating the victim's family graciously so as to avoid a lawsuit. The audience response was a resounding boo, and the post-lecture surveys affirmed their disapproval. How anyone charged with the responsibility of protecting life and property could adopt such an idea is incomprehensible. The public safety industry, although nonprofit, apparently isn't immune to focusing on cost above all else.

Because of this, I will state that faith is a logical and ethical option. We know that elevated cholesterol levels and hypertension are risk factors for coronary artery disease, and if we can reduce these risk factors, we can faithfully assume that we have reduced the potential for myocardial infarction. If one's fitness level, strength, and aerobic capacity are increased, we can faithfully assume that we have reduced the risk of injury attributable to muscular overexertion and cardiovascular stress. The literature clearly supports that assumption (AHA, 1998).

By documenting both medical and physiological changes, we must acknowledge that improvements will benefit the individual and the organization, making the program cost-effective. True, we can't measure something that hasn't happened, but when we introduce a rescue or survival technique to a firefighter recruit, we do so on the assumption that the procedure might save his life. The recruit in turn accepts the procedure on a condition of faith that it will work. The U.S. Marines taught an ambush survival technique that required soldiers to stand and walk into gunfire, contrary to the natural impulse to hit the deck. Since the survivor of a well-constructed ambush is a rare individual, the effectiveness of the method cannot be evaluated to the extent desired. There are some things that one cannot afford to learn from experience. One can only assume and have faith that it does work. The alternative would be to teach nothing and assume the consequences, and we know that doing nothing simply does not work.

Another example, more relevant to the program, involves an ergonomic measure that I introduced to our department. The recommendation was that toolbox weights be reduced by splitting the load between two separate, smaller toolboxes. The idea was based on a principle known to physical therapists, that asymmetrical loading of the spine contributes to degeneration of the disks. Knowing that such degeneration is slow and long-term, the decision to eliminate the potential for such an occurrence is based on the assumption that it will, or would, occur. The recommendation was based on faith, but in this case, it does have some scientific validity.

The same is true of health risk and fitness interventions. If you implement a program with the belief that it will work and produce the desired results, positive responses will eventually surface and support continuation of the program and, hopefully, the necessary funding.

COST-BENEFIT

Departments can determine a cost value associated with preventing injuries using a number of different methods. In terms of workers' compensation costs, the total workers' comp costs can be compared year to year as an indicator. Since costs fluctuate rapidly, particularly with open cases that span several years, it's difficult to acquire an accurate picture using this method. A long-range view, however, can give you a general idea as to how well or how poorly you're doing, particularly when you compare it to departments that have a profile similar to your own. Your injury claim history will affect your workers' comp premium, so it makes sense to track and control the costs.

In the private sector, one large corporation estimated and compared the cost of providing in-house rehabilitation to what they typically spent for services in the past from outside providers. The corporation determined its in-house program to be cost-beneficial and so added an on-site fitness program as a supplement in the belief that it would further reduce costs. Comparing total compensation costs with program expenditures indicated a five-dollar return for each dollar invested in the rehab program.

A second way to assess the cost-benefit of an injury prevention program is by means of the lost-workday case rate. David Bell developed this formula while employed with OSHA. Bell stated, "Lost workday cases are considered to be the best primary measure of safety performance because they represent the major losses to workers and their employers...." (Bell, 1987.) He developed a formula that allows a company to compare its lost-workday case rate with the industry average. Bell believes that this is helpful in setting reasonable goals and program budgets. At the same time, it illustrates the value of investing in a safety and health program. The formula uses the industry rate derived from the Bureau of Labor Statistics' (BLS) annual report, available from the U.S. Government Printing Office or from the bureau itself. The rate factor is multiplied by the employees on site and then divided by 100. The result is the number of lost-workday cases that can be expected for the company. The expected cases are then compared to the actual number of cases recorded for the organization. To determine a cost savings or expense above the industry rate, an average cost figure is selected and multiplied by the number of cases above or below the expected rate for the industry.

$$\frac{\text{LWCR} \times \text{Employment}}{100} = \text{Expected LWCRs}$$

$$\begin{array}{r} \text{Expected LWCRs} \\ -\ \text{Actual cases} \\ \hline = \text{Injuries avoided} \end{array}$$

Mr. Bell prefers using the lost-workday cases rather than the total cases, stating that the latter is more susceptible to judgmental and record-keeping errors. Since the Bureau of Labor Statistics data is based on a forty-hour week, fifty weeks per annum, and since fire departments work a variety of schedules, I propose using the total case figure for departments. The total case rate is perhaps as reliable as the lost workdays, due to the difficulty in calculating the lost-workday case rate for fire departments, given the variances in schedules and based on available data. Bureau of Labor Statistics data isn't available for many public safety occupations. The IAFF report provides both injury rate and some data on lost work hours, and it compares injury rates with those of other industries, as reported by the BLS. Therefore, I recommend referencing the IAFF's Death and Injury Survey and the NFPA's annual Injury Report, the latter of which provides similar data rates, including fireground statistics, for comparative analysis.

An average cost figure is recommended by Bell, based on the average loss time in the private sector. The cost-per-injury value is then multiplied by four to include indirect costs such as productivity, disrupted schedules, replacement personnel, and accident investigations. Once again, differences between the private sector and public safety organizations (work hours and nonprofit status, among others) prompt me to substitute Mr. Bell's high estimate derived from the private sector with a figure that may better represent the public safety industry. An average cost figure can be computed using Bell's method or simply using the average cost of an injury from your department. The resultant figure in either case will be less than that suggested by Bell, and should therefore provide an estimate that is at worst *realistic*. This more conservative approach is recommended to avoid overestimating the benefits.

A third way to determine the cost-benefit of a program is to measure the severity of the injuries. The NFPA's 1997 injury report defines severity based on the number of cases requiring hospitalization. Others may use lost workdays as Bell suggests, or an itemized cost-per-injury with a focus on special problems, such as back injuries. I recommend looking at all three. Hospitalization does infer that a serious injury has occurred; however, hospitalization costs can often total less than one back injury that didn't require hospitalization but, rather, a lengthy recovery period. Therefore, give some consideration to the cost-per-injury value. A somewhat arbitrary figure can be used as the index for severity. A department that normally incurs costs in the six-figure bracket will view a \$10,000 injury differently from a department with a \$30,000 annual injury cost. One could use the department's average injury cost multiplied four times (to cover indirect costs) as a baseline, then subtract the number of injuries that exceed that value. For instance, if the cutoff figure selected was \$10,000, simply record the number of injuries that cost more than \$10,000 as a measure of severity. The number can then be compared on an annual basis. Since the research supports the claim that higher-fit members tend to incur injuries that are less severe, the figure may provide some indication of the effectiveness of the program. Identifying individual injuries classified as severe, then comparing them to other data, such as individual fitness scores, can also provide useful information for program planning. If the number of severe injuries is reduced, one can safety conclude that the total cost of injuries will also decline.

COST-EFFECTIVENESS

Severity is one of the two aspects of injury that are analyzed; the other factor is frequency. Fundamentally, measuring frequency means counting the number of cases. The case number is then reported as the total incidence, or tabulated and expressed as a rate per 100 workers, per the number of alarms, or by some other criteria. The NFPA, IAFF, and BLS provide injury data in this manner.

Analyzing your injury data and comparing it to the annual reports published by national organizations is one way to estimate the effectiveness of your department's health and safety program. The NFPA report also includes data on populations served, allowing additional comparison by department size.

FITNESS STANDARDS

The most obvious evaluative tool of a fitness program is the fitness test. Recording scores and comparing results with pre- and post-testing and departmental or national standards is an excellent method. Measuring the program's effectiveness may also include comparison with organizations of similar size. This can be done by reading the literature, conducting surveys, or through direct contact with neighboring departments.

The results of fitness tests are basically used to categorize individuals by fitness level. The resultant profile is then referenced as a basis for exercise prescription, followed by a retest to determine any change in status and the effectiveness of the program. A secondary use of the fitness profile is to provide comparison of fitness levels for strength, aerobic capacity, and other performance measures with specific injuries or rate of injury. Fitness test data can also be translated into financial terms. One study estimated that the utility of using a strength test to select employees had a projected worth in terms of injury cost savings and productivity of $3,500, per the standard deviation (Arnold, 1982). In other words, an employee who scores two standard deviations above another is worth $7,000 more.

The analysis of fitness group scores with reference to injury data can also help determine the goals of the program. Some department fitness profiles have reported high aerobic capacity levels, while other departments are reported by researchers as being well below the norm. Only through the analysis of your fitness test data, with comparison to your injury data, will you be able to identify your problem and develop objectives to address the issue.

OTHER METHODS

Another method that can help you evaluate the program and organize your effort is done by using a work sheet, as shown in Figure 13-1. The work sheet lists goals and objectives, with specific data to be collected and the method of collection. This approach may prove beneficial for all aspects of the program, since it identifies what information you need to gather and how. If you intend to collect injury data, for example, you should know what specific data you want and how to acquire it. Decide these matters early in the program, and avoid waiting five years to discover that you should have been recording something else. A chart will help you focus on what specific information you need and the method or source by which to obtain it.

Another option for evaluating the program is to contract with a private consultant. Companies that offer such services may offer customized health assessments, questionnaires, individual reports, and a management report on member status that includes health care costs, risks, and benefits (Anderson and Jose, 1987). If you have more funds than expertise, a consultant may be the desirable option.

LONGITUDINAL STUDY

The Aurora Fire Department is located in central Colorado, bordering the City of Denver to the east. The department employs 295 career personnel—265 uniform and 30 civilian—and it provides fire and emergency services to 252,000 residents, covering a geographic area of approximately 150 square miles.

FIGURE 13-1

GOALS AND OBJECTIVES WORK SHEET

Goal	Data	Specific Measures	Procedure
Fitness levels	Fitness scores	Strength/aerobic/flexion	Annual fitness test
Health status	Risk appraisal	Risk factors, lipids, etc.	Medical exam
Injury rates	Annual reports	Rate per 100 workers	Compute/compare

Source: National Fire Academy course *Training Program Management*, student manual.

In the late 1970s, Aurora responded to a rising interest in public safety fitness programs, publicized in a number of research reports (Cady et al., 1979; Lemon and Hermiston, 1977; Barnard and Duncan, 1975; Davis et al., 1979; and Raven et al. 1977). The department's initial response was to purchase some exercise equipment, primarily sport-type equipment, which the members were allowed to use while on duty. No formal direction or specific exercise activity was provided. Participation was voluntary, unstructured, and without an evaluation process or standard. Occupational injuries for the department throughout this period were at or above the national average, and fitness injuries, principally related to sport activities, represented approximately 40 percent of the department's total claims. At the time, statistics such as these were often the reason many departments decided to eliminate their so-called fitness programs.

In 1978, the EEOC published its *Guidelines for Employee Selection Procedures.* Within a year, a more structured program was introduced in the recruit training academy. This program consisted principally of aerobic activities, and it later included an assessment process with scores compared to normative data. No minimum standard or fitness requirement was established or enforced. The program did not extend to line personnel, and fitness-related injuries remained high, averaging 29 percent of total department injuries incurred.

By 1983, a new physical performance assessment (task-type test) was introduced for the purpose of selecting entry-level candidates. This test was administered to all incumbents as part of the validation process, but no schedule or requirement was made to repeat the test with incumbents.

A physical fitness committee was formed to focus primarily on entry-level testing. The task-type test was restricted for use as a selection tool for recruits. Sport activities continued to be the primary fitness mode for incumbents. Some thought and effort was directed toward structuring on-line incumbent activities; however, no test or standards were introduced.

By 1985, the academy's program, formerly based on aerobics, took a more balanced approach, with an emphasis on strength development. Standard tests were introduced, and general-population normative data was used for the evaluation. Increased fitness levels, particularly the strength component, were recorded, with a concomitant decline in fitness-related injuries. Sport activities continued among incumbents, but an increased interest in a more structured exercise program began to develop.

In 1987, the first edition of NFPA 1500, *Standard on Fire Department Occupational Safety and Program,* was issued. The standard included a mandate that fire departments establish a physical fitness program. That same year, the Aurora Fire Department adopted the NFPA 1500 standard and formulated a plan to comply that included the development of a structured fitness program. There was a decision to extend the recruit program to incumbent personnel, requiring some substitution and modification of the tests. The changes were contrary to the accepted methods of the time and would not come into vogue for another ten years. The department designed a pilot study and selected a group of volunteers. An introductory class concerning the fundamentals of exercise was developed, and the modified fitness test was administered to the pilot group following instruction.

Criterion standards were also developed during this time to replace normative data as a means of scoring the fitness test. Although developed and adopted by Aurora in 1987, the use of criterion standards wouldn't be accepted in the mainstream for another ten years. The standards were based on research studies and the identification of physical demands specific to job functions.

A composite scoring system was also developed during this time. Following testing and the tabulation of results, members of the pilot group were instructed to perform structured routines and return for a retest within twelve weeks. The pre- and post-test results, which overlapped 1988-89, appear in Figure 13-2.

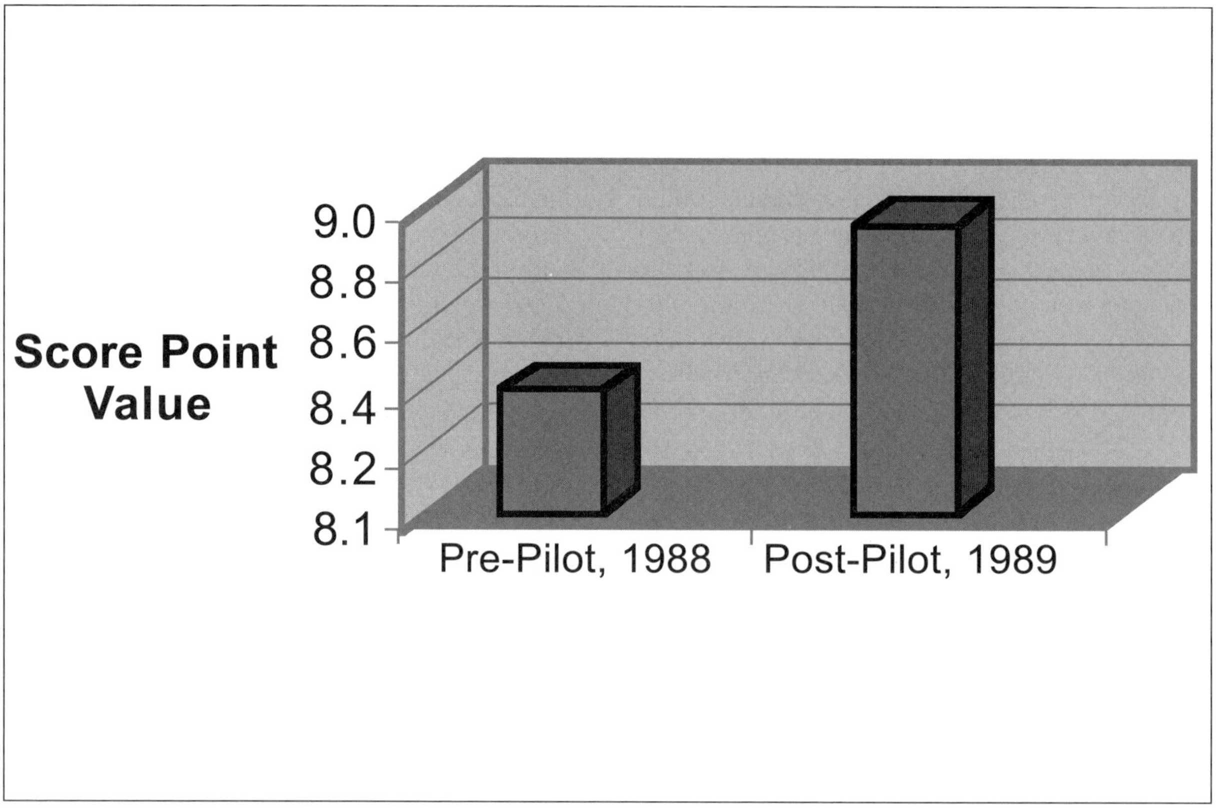

Fig. 13-2 Pilot Group Fitness Scores, Pre- and Post-Test

Around this time, fitness coordinators were selected and trained in-house. The principal responsibility of the coordinator was to assist the program developer in conducting fitness tests. Besides NFPA 1500, NFPA 1501, *Fire Department Safety Officer,* was also introduced.

Beginning in 1989–90, on completion of the pilot tests, the remaining department members were administered the fitness test. The department fitness program standards and testing were formally adopted, and it became mandatory to comply with NFPA 1500. Conditioning programs were provided to all members who didn't meet the minimum standards. Exercise logs were kept by assigned members and reviewed during the conditioning period and prior to the retest. The conditioning period was typically eight to twelve weeks, depending on individual response and need. A 100-percent compliance rate was achieved through exercise programming. The members weren't restricted from duty during the initial prescription period, but they were required to perform specific, monitored activities on-line. The department also purchased additional exercise equipment, and the members contributed to the purchase of selected weight-training equipment by way of a station recreation fund. There began also a noticeable shift from sport activities to structured exercise routines.

In accordance with NFPA 1582, *Medical Requirements,* the department selected a physician and entered into a formal agreement outlining his duties and responsibilities. The department also provided an orientation program for the physician, including ride-alongs and participation in training exercises.

Another innovation at this time was an annual medical exam, including a submaximal graded exercise test. The test was adopted and provided by a local health facility. Each exam was funded in part by the member's insurance and performed by the department's physician. Incumbent medical exams were voluntary unless a pretest health screen indicated that medical clearance was required prior to participation. Both physical and occupational therapists were enlisted to assist in the development of functional job descriptions, and a liaison was established between the providers and the department's fitness coordinator.

In 1990, the Americans With Disabilities Act (ADA) was introduced. NFPA 1582, *Standard on Medical Requirements for Fire Fighters,* was developed by a subcommittee of NFPA 1500 to replace medical requirements previously contained in NFPA 1001, *Standard for Fire Fighter Professional Qualifications.*

By 1991, all department members were evaluated for the second time on the fitness test. A pattern of decline in overall injuries emerged that was lower than it had been for the previous ten years. The concept of job-matching was emphasized in rehabilitation, and a host of studies were conducted in cooperation with providers and a local university. Also, a nutritional analysis program, including the use of hydrostatic weighing techniques to determine body composition, was adopted. At the time, the program was restricted to selected individuals.

Beginning in 1992, all department members were evaluated using the fitness test. Entry-level medical requirements and back evaluations were also modified in response to then-current practices.

Also in 1992, to address both fitness and physical performance requirements, the NFPA assembled a task group to begin work on drafting NFPA 1583. The second edition of NFPA 1500 was also issued. It included a section on physical performance requirements. The first edition of NFPA 1582 was issued, and 1521, *Fire Department Safety Officer,* replaced the 1501 first edition.

In 1993, the Aurora Fire Department fitness program began to target the lower-fit groups. Members of fitness levels 3, 4, and 5 were identified and tested. Those classified in the highest categories, 1 and 2, were not tested in 1993. Following a conditioning period, the target members were reevaluated. Significant improvement was documented, with some members achieving Level 1, a category attainable by only 10 percent of the department.

In response to ADA and the work of the NFPA 1583 task group, a proposal was made within the Aurora Fire Department to develop and validate a physical performance test to replace the 1983 version. The test was developed by an outside contractor and validated on-site. The department also assigned a full-time health and safety specialist to manage the health and safety program. Individual fitness files were created, and a data collection procedure was developed.

The following year, the new physical performance test, a criterion-task test, was introduced to all department incumbents. The test was conducted without a time-required mandate. A 94-percent pass rate was achieved during the introductory phase. In deference to the task test, the fitness test wasn't conducted this year so as to reduce out-of-service time.

In 1994 also, a phase-in program and policy were developed to ensure the compliance of all members with physical performance requirements by April of 1996. Members who didn't meet the performance standard were provided assistance and reevaluated periodically throughout the period. A conditioning program specific to the performance test was developed and integrated with the fitness program.

To familiarize the therapists and physician with the job requirements, the department strongly recommended that they participate in the performance test wearing full PPE. Curiously enough, all of the medical providers participated and *passed,* though passing was not a requirement.

The department also developed an injury rehabilitation program to reflect job requirements, and an injury investigation procedure that targeted causal factors was developed in cooperation with medical providers (LeCuyer, 1997).

Ergonomic interventions were developed and introduced in response to investigations and observations of performance tests. The emphasis was on the physiological stress associated with firefighting demands as represented by the performance test. Specifications for PPE were revisited in response to performance test observations. Meanwhile, the procurement of exercise equipment continued, with contributions from both members and the administration of the department.

In 1995, all members were evaluated on the fitness test. Four of the ten fitness coordinators were certified by an outside agency. A work-hardening program was developed in conjunction with a modified return-to-work procedure. Medical exams and procedures were updated. In response to requests by the members, the department purchased additional exercise equipment. And the draft of NFPA 1583 was made available and open for comment.

The following year, all members of the department were again evaluated on the fitness test. Performance testing was completed, and compliance with the performance standard was achieved by all incumbents. The average fitness score peaks were at two standard deviations above the criterion standard. The remaining fitness coordinators, plus eight additional members, were certified by the outside agency. The health and safety specialist assumed the role of program manager, a role later identified in the 1997 edition of 1521.

Also in 1996, ADEA exemption for public safety officers was granted, with the stipulation that both fitness and performance evaluation methods be determined by NIOSH within a designated time frame; i.e., September 30, 1998, to September 30, 1999. The NFPA 1583 document was returned to committee by the Standards Council following appeal.

In 1997, all of the department's members were again evaluated on the fitness test. Changes to PPE and other equipment were completed in response to ergonomic investigations and recommendations.

The 1997 edition of NFPA 1500, 1582, and 1521 were issued. The 1521 document, *Fire Department Safety Officer,* identified two specific safety officer roles: health and safety officer (HSO) and incident safety officer (ISO).

In 1998, the role of the fitness test was expanded, following an interdepartmental research study to validate, the test served as a tool to certify members for SCBA (LeCuyer, 1998). A maximal graded exercise test was added to the annual medical exam, replacing the submaximal version on a triennial basis, regardless of age. Performance tests methods were reevaluated, and a research project was set in motion to provide one method of resolving the performance test issue.

The information provided in the above time line is general, highlighting key events only. Numerous research projects were conducted in response to legislative directives, standard updates, developments in exercise science, and advances in safety. Throughout this period, projects were designed and conducted in cooperation with local universities and other health professionals, and these were instrumental in the design and development of the program.

The period described above encompasses sixteen years, which can be divided into two segments, referred to as the Pre-8 (1982–89) and the Post-8 (1990–97). Dividing it this way allows us to examine the program from a perspective of eight years before and eight years after implementation. The general characteristics of the two periods are profiled in Table 13-1.

TABLE 13-1

PRE-8, POST-8 PERIOD COMPARISON

Pre-8 (1982-89)	Post-8 (1990-1997)
Voluntary	Mandatory
Sport-type activity	Fitness training activity
Unstructured	Structured
No standards	Job-based standards adopted
No evaluation test	Annual fitness test
No program supervisor	Program manager assigned
No medical exam	Medical exams available

The program can be evaluated by inspection of the three principle categories: the fitness/performance profile, analysis of the injury data, and analysis of the cost-effectiveness and benefits. The fitness components are tested and scored independently, then averaged to produce a composite score. The composite scores are graded on a 1-10 scale, with 10 points indicating the highest possible score and equating to the top fitness category, Level 1. The fitness categories with corresponding scores are depicted in Table 13-2.

TABLE 13-2

FITNESS LEVEL POINT VALUES

Fitness Level	Score Range
Level 1	10.0>
Level 2	8.0–9.9
Level 3	6.0–7.9
Level 4	5.0–5.9
Level 5	<5.0

For the pilot, a group of 67 volunteers were evaluated in the fall of 1987, and they were retested in the spring of 1988. The average score was 8.35 following the pretest, indicating an above-average group. The results were as expected from voluntary participants, who typically comprise the higher-fit group. The post-test average increased to 9.00 following a self-monitored conditioning program. The fitness scores of the pre- and post-test are shown in Figure 13-2. Following completion of the pilot, an assessment was scheduled for all department members, to be conducted on an annual basis beginning in 1990.

Fitness assessments were administered departmentwide from 1990 through 1997, except for 1993-94. The results are presented as a composite score average by year, in Figure 13-3. A steady improvement is documented during the seven-year period, reaching a plateau in 1996-97. The peak period represents an average score that is almost two standard deviations above the minimum standard, a highly desirable yet unexpected achievement. Since the average score is quite high, a continued rise in the average group score isn't expected, although individual improvement is anticipated. The maintenance of the present fitness-level average is the current program goal.

The years 1993 and 1994 are excluded, since no departmentwide fitness test was conducted. The performance test (task-type) was administered in 1994, in place of the fitness test, which was waived for logistical reasons. In 1993, only the target group, members classified as Level 3 or lower, were tested and reevaluated. Although fitness Level 3 encompasses an acceptable score range from minimum to average, the desired goal of the program for all members was to be above average. The initial test and follow-up results are shown as an average in Figure 13-4, and by fitness level in Figure 13-5. A significant improvement was documented, with 11 percent of the group surprisingly reaching the highest rank, Level 1. This remains relatively constant at about 10 percent throughout the period, as expected. The most noticeable change in score pattern appears at Level 2 and 3, with a shift from the lower-fit Level 3 (average) to the higher, above-average Level 2.

In terms of job performance, higher fitness levels are associated with increased productivity and improved job performance. This is evident in the comparison made between fitness levels and performance times on a criterion (task-type) test. Table 13-3 shows the relationship using a group of recruit firefighters, where a 26-percent improvement in fitness scores translated into a 25-percent improvement in performance following a twelve-week training program.

The relationship between individual fitness levels and injuries is documented in the research, with some investigators reporting significantly higher rates of injury for lower-fit workers compared to the most-fit workers (Cady et al., 1979; and Arnold, 1987). Evidence of the relationship is exhibited in Figure 13-6, where fitness scores are compared to total

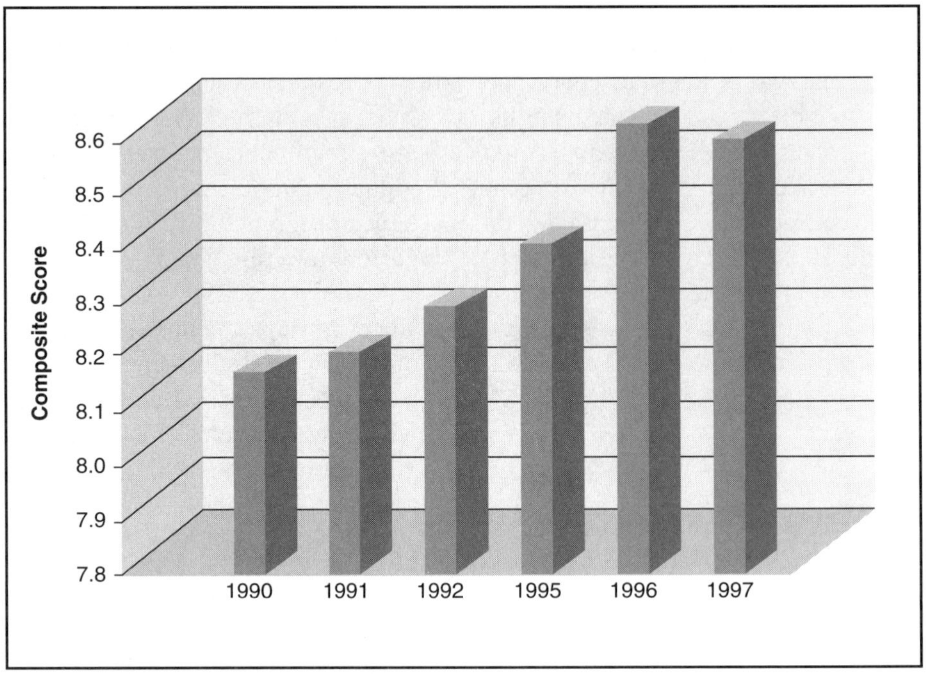

Fig. 13–3 Physical Fitness Composite Score Average

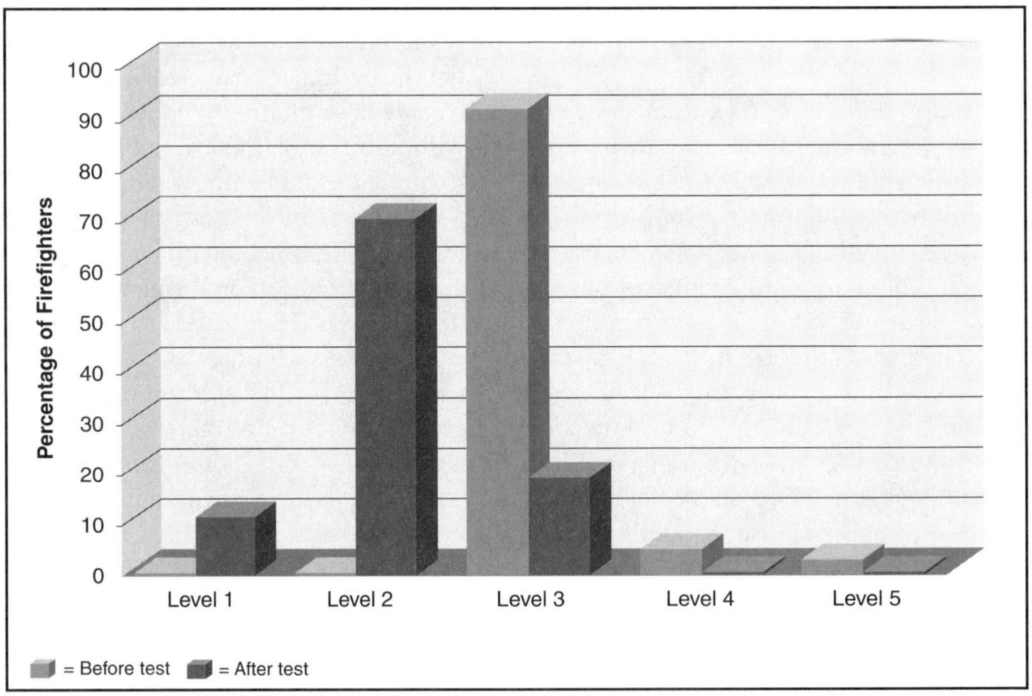

Fig. 13–4 1993 Retest Scores, Pre- and Post-Test

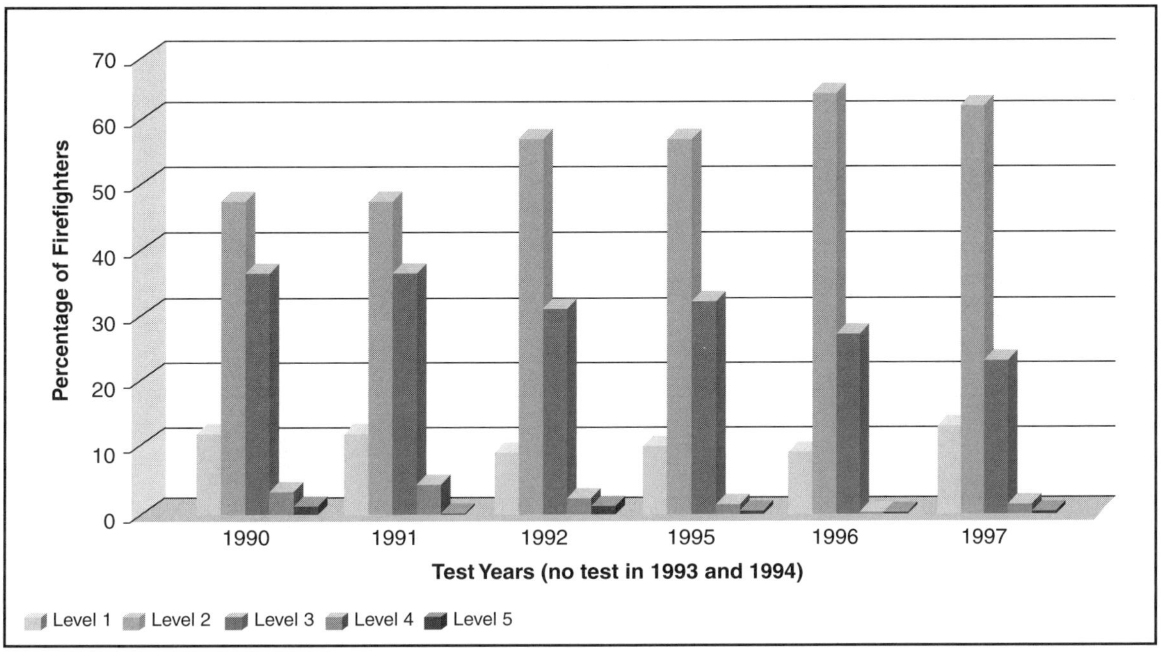

Fig. 13–5 Fitness Score Levels

injuries. The correlation between increased fitness levels and the decline in total injury claims for the Aurora Fire Department was significant (r = -0.94), lending added support for continuation of the program.

ANALYZING INJURY DATA

Two characteristics of injury data typically inspected are the rate of injury and the severity. In terms of fitness-related injuries, a comparison between the Pre-8 and Post-8 periods, expressed as a total figure and annual rate, appear in Figure 13-7. When comparing all injuries for the two periods, the Post-8 period shows a 40-percent reduction in the average annual rate per 100 firefighters, as per Figure 13-8. The average injury rate per 100 alarms for the Pre-8 period (data available for 1984-89) was double the average rate for the Post 8 period, as shown in Figure 13-9.

The NFPA provides data that allows departments to compare statistics with other departments of comparable size, based on the population of the community served. The NFPA fireground injury rate per 100 firefighters shows a rate three times that of Aurora in the latest 1996 report (Figure 13-10), with a rate per 1,000 fires for the Post-8 year of 1996 at approximately five times the Aurora rate. This represents a significant reduction from the Pre-8 period, when

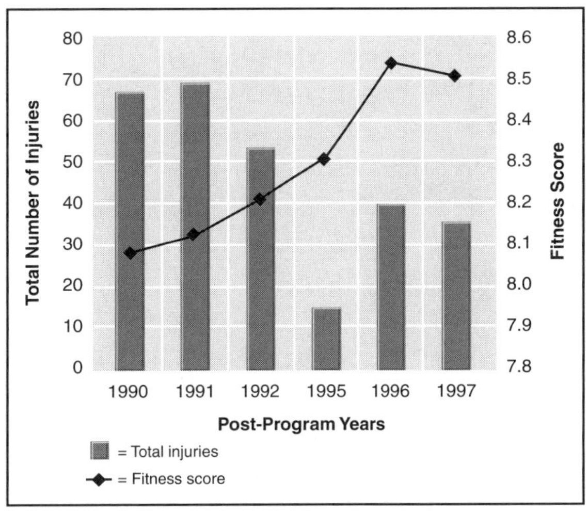

Fig. 13–6 Relationship Between Total Injuries and Fitness Scores

TABLE 13-3

RELATIONSHIP OF FITNESS TO PERFORMANCE USING A CRITERION TEST

Fitness Score		Performance Test		
Before training	After training	Before training	After training	Improvement (sec.)
6.86	9.1	4:16	3:22	:54
6.16	8.9	4:48	3:31	:77
3.96	8	5:04	3:34	:90
6.32	9.5	4:12	3:09	:63
8.52	10	4:32	3:37	:55
7.7	9.3	4:09	2:51	:78
6	6.8	5:56	4:36	:84
4.56	9.1	4:40	3:25	:75
9.16	10	4:20	3:10	:70
7.66	9.5	3:43	3:02	:41
5.46	9.2	5:12	4:11	:61
8.52	9.3	4:14	3:22	:52
8.52	9.6	3:49	3:06	:43
4.56	8.6	4:02	3:12	:50
8.16	9.6	2:44	2:15	:29
6	8.6	3:33	3:00	:33
3.6	8.8	7:52	5:12	:160
8.06	9.7	4:42	4:02	:40
8.06	10	4:33	3:24	:69
7.82	9.8	4:52	3:57	:55
Avg. score				
6.78	9.17	4:53	3:41	
Fitness level				
3	2	Avg. improvement: 63 seconds		

A 26 percent improvement in fitness scores resulted in a 25 percent improvement in performance test time. The fitness scores are based on a 1–10 scale, and the performance scores are recorded as time. Task-specific training wasn't provided; therefore, the improvement in performance can be attributed to fitness training.

Aurora rates were comparable to the national data, as shown in Figure 13-11. A similar comparison was made with the IAFF report, since Aurora is a local affiliate of that organization. In 1987, when the NFPA introduced the 1500 standard, requiring that departments establish a fitness program, the Aurora Fire Department showed a comparable rate of injury, though slightly lower, as compared with other affiliates. Nine years later, following development of the program, the Aurora rate had decreased significantly. The injury rate per 100 firefighters in the IAFF 1996 Death and Injury Survey was 32.6 per 100, and the rate for Aurora for the same

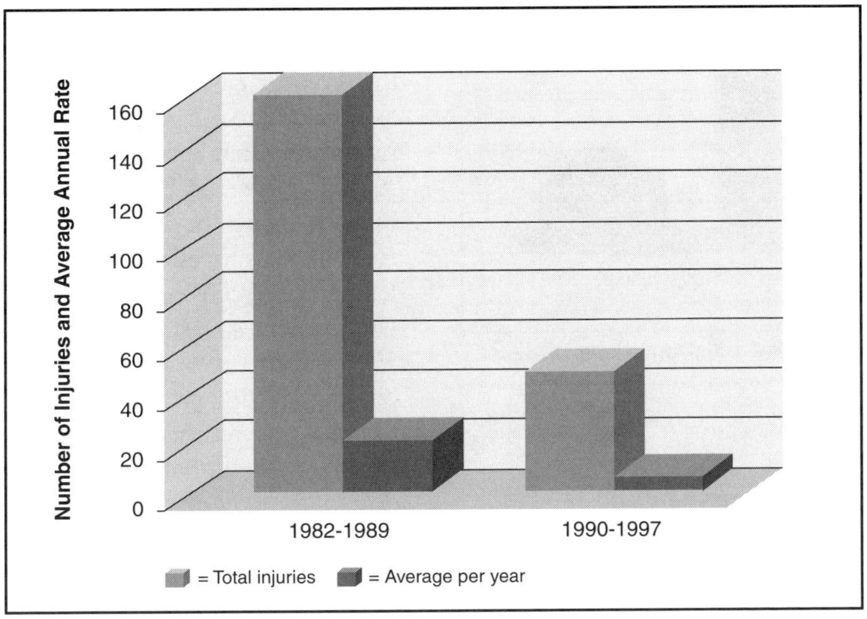

Fig. 13–7 Physical Fitness Injuries, Pre- and Post-Program

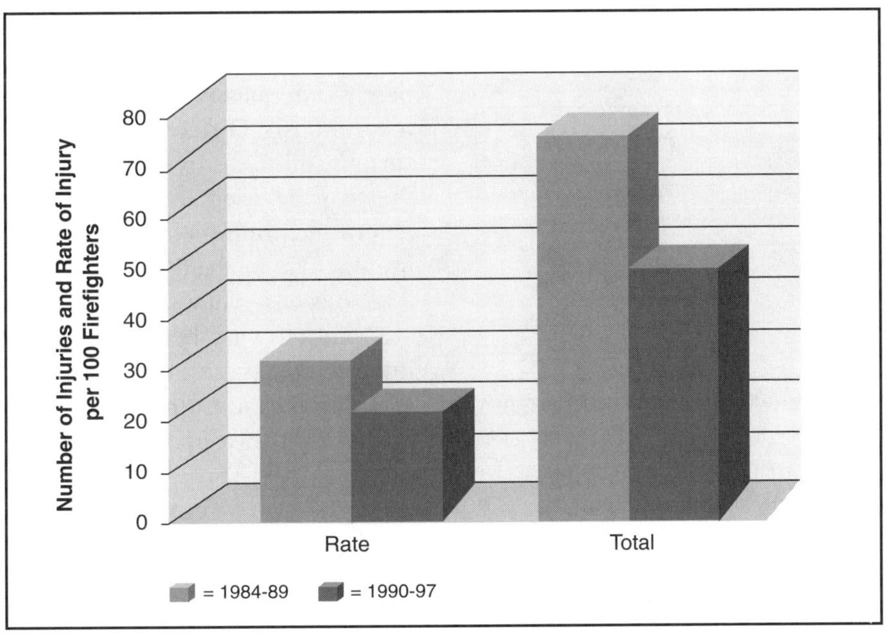

Fig. 13–8 Pre- and Post-Program Rate of All Injuries, 1984-97

year was 17 per 100, approximately half the average of reporting affiliates (Figure 13-12).

Injury rates, when compared with national averages as reported by IAFF and the NFPA, can provide a fair benchmark of program effectiveness. If the desired goals are achieved, then the cost of producing that result may be deemed cost-effective and beneficial.

Based on the fitness and injury data detailed above, the Aurora program is decidedly cost-effective, since the desired outcome—a reduction in the injury rate with a concomitant

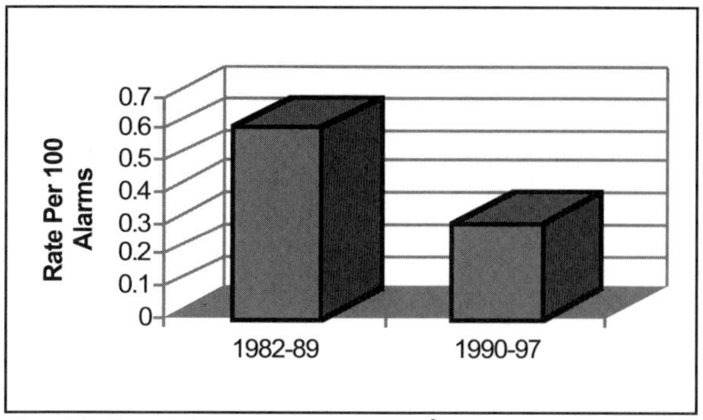

Fig. 13–9 Injury Rate Per 100 Alarms From 1982-97

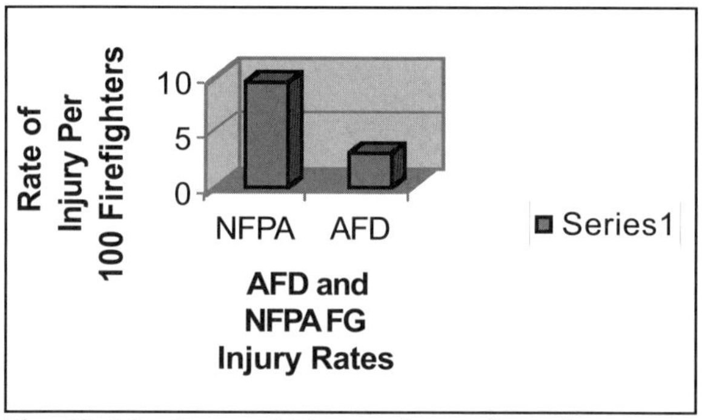

Fig. 13–10 Fireground Injury Rate by Population of Community

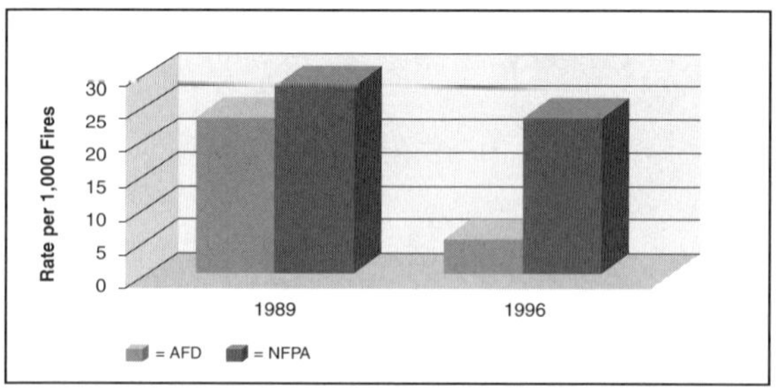

Fig. 13–11 Injury Rate Per 1,000 Fires: Pre- and Post Program

Aurora Fire Department rates compared to national average per NFPA reports.

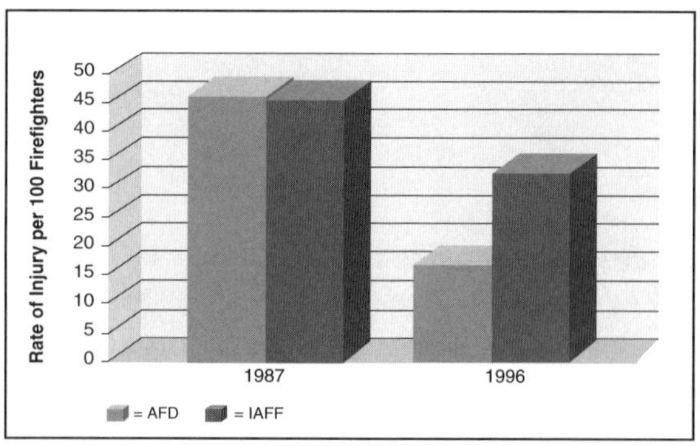

Fig. 13–12 AFD and IAFF Injury Rate Comparison

increase in fitness levels—was accomplished. Although difficult to prove, the cost-benefit of the program can be reasonably estimated as long as conservative figures are used so as to avoid overstating that benefit (Kipp and Loflin, 1996). A cost figure can be determined, based on frequency, using the lost-work case rate (LWCR) formula developed by David Bell (Bell, 1987). Using the LWCR as a measure, the Aurora Fire Department's estimated cost benefit was $73,800 in 1996, with workers' comp costs compared with program costs, both direct and indirect. The return on investment was three dollars for every dollar invested, and it rose to five dollars for every dollar in 1998-99.

The severity of injuries can be assessed by the number of lost workdays, the cost of the injury, or as defined by the NFPA in its 1996 annual report on injuries that require hospitalization.

Using an injury cost approach as an index of severity, one can track the number of injuries that exceed a specific dollar amount and compare the frequency of their occurrence on a year-to-year basis. Differentiating between minor and serious injury provides important information that can be used to develop interventions and determine program goals. There is no preset figure that can be used to identify what constitutes a serious case versus a minor one, and estimates are likely to vary between departments. Serious injuries constitute approximately 25 percent of all injuries, and estimates of $25,000 have been proposed as the upper-dollar indication of severity. Based on Aurora's injury data and associated costs, a very conservative figure of $4,000 was selected and recommended by the city's risk management office as the criteria.

Since higher fitness levels are associated with a reduction in the severity of injury, and since fitness levels had increased during the Post-8 period, it was assumed that the total number of claims that exceeded the $4,000 figure would decline. The total number of claims were identified each year and used as an estimate of program effectiveness. The tally from 1991 through 1997 provides some indication that such a relationship does exist (Figure 13-13).

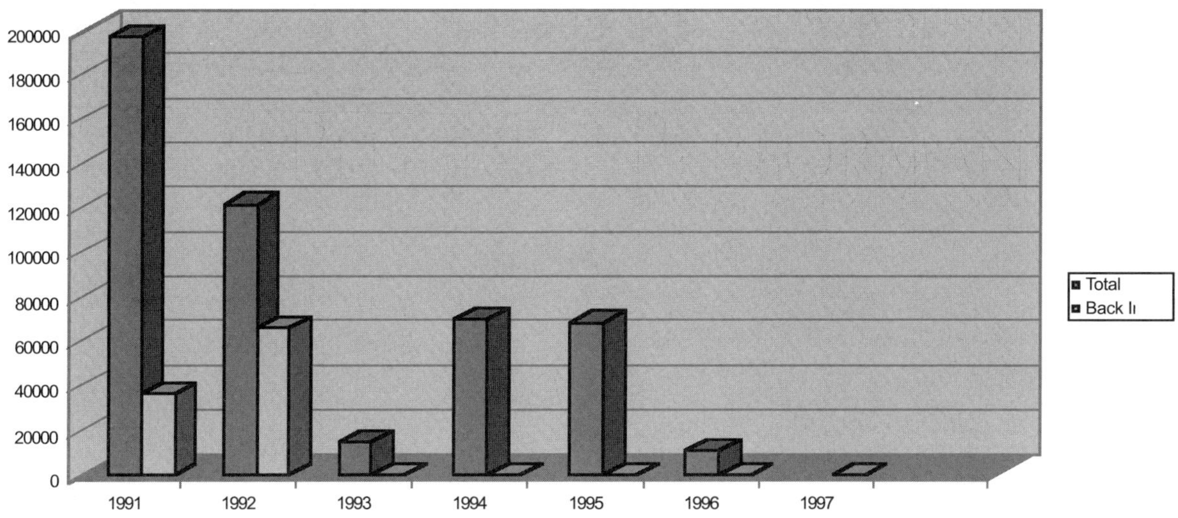

Fig. 13–13 Claim Cost of Injury

During the final edit of this text, the injury data for 1997 through 1999 became available for the Aurora case study. The injury rate continued to fall further, extending the trend described above. The IAFF injury data for 1997 reported that 40 percent of all firefighters were injured in the line of duty. That same year, the Aurora Fire Department reported 10 percent, one-quarter that of the IAFF affiliates.

The program can be described as effective, based on the achievement of the desired goals of reduced injury rates and increased fitness levels. When the program is viewed in the presence of an increasing workload (doubled alarms), an increase in the average age of firefighters (from 26 to 41), and a modest decline in staffing (eight members), one can see that its effectiveness is actually greater than reported.

Clearly, this case study supports the need to address fitness throughout the fire service through the implementation of a structured, mandatory fitness program. The data collected during the eight years following implementation, as compared to the preprogram years, exhibits a positive change and indicates a successful, effective, and beneficial fitness program.

RECOMMENDED READING

American College of Sports Medicine (1998). Position stand: the recommended quantity and quality of exercise for developing and maintaining cardiorespiratory and muscular fitness and flexibility in healthy adults. *Medicine and Science in Sports and Exercise*. 30 (6): 975–985.

Fleck, S.; and Kraemer, W. (1987) *Designing Resistance-Training Programs*. Champaign, IL: Human Kinetics.

Gledhill, N.; and Jamnik, V.K. (1992) Characterization of the physical demands of firefighting. *Canadian Journal of Sport Science*. 17 (3) 207–213.

Pauletto, B. (1986) *Strength Training and Conditioning for Speed Development: A Coaches Guide*. National Strength and Conditioning Association, pp. 61–75.

PART V REFERENCES

American College of Sports Medicine (1983). Position stand: proper and improper weight-loss programs. *Medicine and Science in Sports and Exercise.* (15): ix–xiii.

American College of Sports Medicine (1998). Position stand: the recommended quantity and quality of exercise for developing and maintaining cardiorespiratory and muscular fitness and flexibility in healthy adults. *Medicine and Science in Sports and Exercise.* 30 (6): 975–985.

American College of Sports Medicine (1993). *ACSM Resource Manual for Exercise Testing and Prescription, Second Edition.* Philadelphia: Lea and Febiger, p. 231.

Anderson, D.; and Jose, W.S. (1987) Employee lifestyle and the bottom line: results of the StayWell Evaluation. *Fitness in Business.* 3: 86.

Arnold, J.; Rauschenberger, J.M.; Soubel, W.G.; and Guion, R.M. (1987) Validation and utility of a strength test for selecting steelworkers. *Journal of Applied Psychology*, 67 (5): 588–604.

Astrand, P.O.; and Rodahl, K. (1970) *Textbook of Work Physiology*. New York: McGraw-Hill.

Ballor, D.L.; Becque, D.M.; Marks, C.R.; Nau, K.L.; and Katch, V.L. (1989) Physiological response to nine different exercise/rest protocols. *Medicine and Science in Sports and Exercise.* 21 (1): 90–94.

Bell, David (1987). Gauging safety outlays and objectives. *Occupational Hazards,* pp. 57–59.

Brzycki, M. (1993) Strength testing: predicting a one-rep max from reps to fatigue. *Journal of Physical Education, Recreation, and Dance.* 64 (1): 88–90.

Cady, L.D.; Bischoff, D.P.; O'Connel, E.R.; Thomas, P.C.; and Allen, J.H. (1979) Strength and fitness and subsequent back injuries in firefighters. *Journal of Occupational Medicine.* 24: 269–272.

Caillet, R. (1988) *Low-Back Pain Syndrome, Fourth Edition.* Philadelphia: F.A. Davis.

Davis, P.O.; Biersner, R.J.; Barnard, R.J.; and Schamadan, J.S. (1982) Medical evaluation of fire fighters: how fit are they for duty? *Postgraduate Medicine.* 72 (2): 241–248.

Davis, P.O.; and Dotson, C.O. (1987) Physiological aspects of firefighting. *Fire Technology.* 23: 280–291.

Davis, P.O.; Dotson, C.O.; and Santa Maria, D.L. (1982) Relationship between simulated fire fighting tasks and physical performance measures. *Medicine and Science in Sports and Exercise.* 14 (1): 65–71.

Davis, P.O.; and Dotson, C.O. (1987) Physiological aspects of fire fighting. *Fire Technology.* 23 (3): 280–291.

Davis, P.O.; and LeCuyer, J.E. (1995) Performance and fitness. *Fire Chief.* August, p. 20.

deVries, Herbert (1974). *Physiology of Exercise: For Physical Education and Athletics.* Dubuque, IA: Wm. C. Brown Co.

Dickinson, A.; LeCuyer, J.E.; and Work, W. (1997) *Validation of an Entry-Level Fire Fighter Test.* Lakewood, CO: West Metro Fire District.

Doolittle, T.L. (1979, March) *Validation of Physical Requirements for Fire Fighters.* Seattle, WA: Seattle Fire Department.

Dornan, Peter; and Dunn, Richard (1987). *Sporting Injuries.* New York: University of Queensland Press, p. 247.

Duda, M. (1988) Plyometrics: a legitimate form of power training. *Physician and Sports Medicine.* 16 (3): 213-215.

Fox, E.L.; and Mathews, D.K. (1974) *Interval Training: Conditioning for Sports and General Fitness.* Philadelphia: W.B. Saunders Co.

Frost, S.; and Durfor, W. (1986) Confidential fitness reports. *Fire Command,* November. pp. 36–37, 42-44.

Gettman, L.R.; Ward, P.; and Hagan, R.D. (1982) A comparison of combined running and weight training with circuit training. *Medicine and Science in Sports and Exercise.* 14 (3): 229-234.

Gledhill, N.; and Jamnik, V.K. (1992a) Characterization of the physical demands of firefighting. *Canadian Journal of Sport Science.* 17 (3): 207–213.

Gledhill, N.; and Jamnik, V.K. (1992b) Development and validation of a fitness screening protocol for firefighter applicants. *Canadian Journal of Sport Science.* 17 (3): 199–206.

Heyward, V. (1984) *Designs for Fitness: A Guide to Physical Fitness Appraisal and Exercise Prescription.* New York: Macmillan Publishing. p. 77.

Heyward, V. (1998) *Designs for Fitness: A Guide to Physical Fitness Appraisal and Exercise Prescription, Third Edition.* New York: Macmillan Publishing Co.

Hickson, R.C. (1980) Interference of strength development by simultaneously training for strength and endurance. *European Journal of Applied Physiology*. 215: 255–263.

Humphrey, D. (1988) Abdominal muscles strength and endurance. *Occupational Medicine: State of the Art Review*. 3 (2): 219–237.

IAFF/IAFC (1997). *Fire Service Joint Labor Management Wellness-Fitness Initiative.* Washington, DC: International Association of Fire Fighters.

Jackson, A.; Watkins, M.; and Patton, R. (1980) A factor analysis of twelve selected maximal isotonic strength performances on the Universal Gym. *Medicine and Science in Sports and Exercise.* 12: 274.

James, Steve (1988). Methodology of monitoring weight training. *NSCA Journal,* 10 (6): 20.

Jacobs, D.T. (1976) *Physical Fitness and the Fire Services.* Quincy, MA: National Fire Protection Association.

Johnson, B.L.; and Nelson, J.K. (1986) *Practical Measurements for Evaluation in Physical Education, Fourth Edition.* Minneapolis: Burgess Publishing Co.

Khalil, T.M.; Asfour, S.S.; Martinez, L.M.; Waly, S.M.; Rosomoff, R.S.; and Rosomoff, H.L. (1992) Stretching in rehabilitation of low-back-pain patients. *Spine.* 17 (3): 311–317.

Kilbom, Asa (1980) Physical work capacity of firemen. *Scandinavian Journal of Work-Environment Health.* 6: 48–57.

Kipp, J.D.; and Loflin, M.E. (1996) *Emergency Incident Risk Management: A Safety and Health Perspective.* New York: Van Nostrand Reinhold.

Kraemer, W.K.; and Fleck, S.J. (1988) Resistance training: basic principles (part 1 of 4). *Physician and Sports Medicine.* 16 (3): 160–171.

Kraemer, W.K.; and Fleck, S.J. (1988) Resistance training: physiological responses and adaptations (part 2 of 4). *Physician and Sports Medicine.* 16 (4): 108, 110–112, 117–118.

Kraemer, W.K.; and Fleck, S.J. (1988) Resistance training: physiological responses and adaptations (part 3 of 4). *Physician and Sports Medicine.* 16 (5): 63–66, 69, 72–74.

Kraemer, W.K.; and Fleck, S.J. (1988) Resistance training: exercise prescription (part 4 of 4). *Physician and Sports Medicine.* 16 (6): 69-72, 77, 80–81.

LeCuyer, John E. (1989) *The Relationship Between a Fitness Profile and Job Performance Test.* Unpublished report. Denver, CO: The Rocky Mountain Fire Academy.

LeCuyer, J.E.; and Kakos, Scott. (1992) *The Margaria-Kalamen Test Versus a One-Minute Leg Press as a Predictor of Performance on a Hose-Drag Task.* Unpublished report. Aurora, CO: Aurora Fire Department.

LeCuyer, John E. (1994) *A Comparison Between and Absolute and Relative Measures of Aerobic Capacity and the Ability to Predict Fire Fighter Job Performance.* Unpublished report. Aurora, CO: Aurora Fire Department.

LeCuyer, J.E. (1998) *A Comparison of Aerobic Scores With Job-Performance Test Times.* Unpublished report. Aurora, CO: Aurora Fire Department.

Lemon, P.W. (1983) Conditioning: determining what's important for your sport. *NSCA Journal.* February-March, pp. 52–53.

Lemon, P.W.R.; and Hermiston, R.T. (1977) The human-energy cost of fire fighting. *Journal of Occupational Medicine.* 19 (8): 558–562.

Lusa, S.; Louhevaara, V.; and Kinnunen, K. (1994) Are the job demands on physical work capacity equal for young and aging firefighters? *Journal of Occupational Medicine.* 36 (1): 70–74.

Mathews, D.K. (1973) *Measurement in Physical Education, Fourth Edition.* Philadelphia: W.B. Saunders Co., p. 47.

May, K.; and Nicewater, W.A. (1994) Reliability and information functions for percentile ranks. *Journal of Educational Measurement.* Winter, 31 (4): 313–325.

McCafferty, W.B.; and Horvath, S.M. (1977) Specificity of exercise and specificity of training: a subcellular review. *Research Quarterly.* 48 (2): 358–370.

McCardle, W.D.; Katch, F.I.; and Katch, V.L. (1986) *Exercise Physiology: Energy, Nutrition, and Human Performance, Second Edition.* Philadelphia: Lea and Febiger, p. 179.

NFA student manual: *Training Program Management* (1997). Emmitsburg, MD: National Fire Academy.

Pauletto, B. (1986) *Strength Training and Conditioning for Speed Development: A Coaches Guide.* National Strength and Conditioning Association, pp. 61–75.

Pritchard, R.E.; and Potter, G.C. (1990) *Fitness Inc: A Guide to Corporate Health and Wellness Programs.* Homewood, IL: Dow-Jones Irwin, p. 11.

Runner's World Magazine: Runner's Training Guide (1977). California: World Publications.

Schoenfield, B.R.; Doerr, D.F.; and Convertino, V.A. (1990) An occupational performance test validates program for fire fighters at the Kennedy Space Center. *Journal of Occupational Medicine.* 22 (7): 638–643.

Sharkey, B.J. (1986) *Coaches Guide to Sport Physiology.* Champaign, IL: Human Kinetics.

Sharkey, B.J. (1978) Fit for work? *JOPER*, September, 18: 21.

Sharkey, B.J. (1990) *Physiology of Fitness, Third Edition.* Champaign, IL: Human Kinetics, p. 399.

Sharkey, B.J. (1997) *Fitness and Work Capacity, Second Edition.* USDA Forest Service. Missoula, MT.

Shepard, R.J. (2000) Exercise and training in women, part I: influence of gender on exercise and training responses. *Canadian Journal of Applied Physiology.* 25 (1): 19–34.

Sothman, M.S.; Saupe, K.; Jasenof, D.J.; and Blaney, J.(1992) Heart rate response of fire fighters to actual emergencies: implications for cardiorespiratory fitness. *Journal of Occupational Medicine.* 34 (8): 797–800.

Sparks, P.J. (1987) *Minimum Medical Standards for Firefighters.* Olympia, WA: Association of Washington Cities.

Spassov, A. (1988a) *Program Design: Special Considerations When Programming for Strength and Power Athletes, Part I.* National Strength and Conditioning Association. 10 (4): 58–61.

Spassov, A. (1988b) *Program Design: Constructing Training Programs, Part II.* National Strength and Conditioning Association. 10 (5): 65–70.

Spitler, D.L.; Jones, G.R.; Wade, C.D.; and Williams, D.L. (1987) A wellness menu for law enforcement agencies: one police department's solution to wellness assessment. *Fitness in Business,* pp. 48–54.

Stamford, B. (1987) Exercise advisor: warming up. *The Physician and Sports Medicine.* 20 (2): 168.

Stamford, B. (1989) Exercise advisor: what is interval training? *The Physician and Sports Medicine.* 17 (9): 193.

Stamford, B. (1992) Exercise advisor: exerting yourself listening to your body's signals. *The Physician and Sports Medicine.* 20 (2): 187–188.

Thomas, T. (1986) *Muscular Fitness Through Resistance Training.* Dubuque: Eddie Bowers Publishing Co., p. 20.

Vanderburg, P.M.; and Considine, W.J. (editors) (1995) *The Fitness Handbook, Second Edition.* St. Louis, MO: Wellness Bookshelf, pp. 131–156.

Williams, T.; and Evenson, S. (1988a) Physically fit for duty? by whose standards? part I: mandatory physical fitness programs. *Fire Chief.* March, pp. 43–46.

Williams, T.; and Evenson, S. (1988b): Physically fit for duty? by whose standards? part II: sliding-scale fitness standards. *Fire Chief,* April, pp. 58–61.

Wilmore, Jack; and Costill, Daniel (1994). *Physiology of Sport and Exercise.* Champaign, IL: Human Kinetics, pp. 309–314.

Wollack and Associates (1986). *Denver Fire Department Physical Skills Test Validation Report.* Denver, CO.

Yessis, M.; and Trubo, R. (1988) *Secrets of Soviet Sports: Fitness and Training.* New York: Arbor House.

PART VI

SUPPLEMENTAL PROGRAMS

CHAPTER 14

WORK/INJURY MANAGEMENT

Legislative directives introduced during the past decade will pose a challenge to all department administrators on issues of employment decisions. The greatest impact will come from the Americans With Disabilities Act of 1990 and its effect on preemployment testing, medical screens, and return-to-work determinations. Departments will have to develop tests and methods for determining fit-for-duty status that don't violate federal employment law or expose members to unnecessary risks or needless assessments. Deciding how best to evaluate employees who have suffered an injury or illness prior to reassignment to full duty is the central problem.

PHYSICAL WORK CAPACITY

The concept of physical work capacity (PWC) has been applied by cardiologists, physicians, physical therapists, occupational therapists, and other medical providers. It has been defined by physiologist Herbert deVries as the maximal level of metabolic work that an individual can perform (deVries, 1974). Physical work capacity as defined is measured through assessment of maximal aerobic capacity, or Vo_2max. The term *physical work capacity* can be somewhat misleading, however, and it is often inappropriately applied.

In the early 1960s, Kenneth Cooper, M.D., was the principle promoter of aerobics, claiming that aerobic capacity was the ultimate measure of fitness. Serving as the exercise guru for the layman, Cooper was responsible for getting thousands off the sofa and into running shoes. In this respect, he was very successful, and he contributed substantially to the improved health of many. The downside was an inappropriate interpretation that continues to confuse the issue surrounding the fitness/work relationship. This definition has led to the assumption that aerobic capacity is the *single* measure of work capacity. Aerobic capacity certainly is very important, but it isn't the only factor, nor is it always the most important factor. The extent to which aerobic capacity or any other physical factor is deemed most important depends on the nature of the work or task.

Physical work capacity isn't unidimensional, as some have been led to believe. Work capacity also involves the muscular system and the characteristics of the job in relation to the physical demands. A physical or occupational therapist, therefore, wouldn't define PWC in the same manner, but rather, through several factors, including the frequency and time of effort; environmental and mental stressors; and other characteristics as they relate specifically to the job requirements (Rogers, 1988).

I'll provide a documented example to illustrate the difference. Firefighter A had an aerobic capacity of 55 mL.kg-1.min-1, and Firefighter B had an aerobic capacity of 36 $mL \cdot kg^{-1} \cdot min^{-1.}$ Using the deVries/Cooper definition, Firefighter

A had the greater work capacity. Following an aerobic test, I administered a job-task performance test. Firefighter A actually failed to complete the test. Firefighter B, on the other hand, performed well and easily finished within the time standard. Although the classic definition categorized Firefighter A as having the greater capacity for work, he wasn't able to perform the job. In terms of work evaluations and job decisions, the preferred method of determining work capacity is through demonstration of a job task or simulation as opposed to conducting an aerobic assessment and determining aerobic capacity. Public safety candidates have actually challenged job-simulation tests on the basis that they don't directly measure aerobic capacity, which they proclaim a la Cooper to be the most important measure. Aerobic capacity is important, but it won't tell you whether you can drag a victim. Only dragging a victim will provide that answer.

Conducting a job-task analysis, which includes a physical demands analysis, is most appropriate and necessary toward developing a functional job description and matching test format for return-to-work and employment decisions. A functional job description is one in which the physical demands of the essential tasks are delineated. An example appears in Figure 14-1.

The functional description is one part of a complete job description. It provides key information needed for a physical therapist to prepare an individual for his job and for a physician to make a proper determination. When weights, measures, and time constraints are included in the job description, medical providers are given the information that they need. The physical demands analysis is usually performed by a physical therapist and generally includes the muscle groups involved, as well as the frequency, intensity, and duration of the task. The job-task analysis should include a physical demands assessment, though some test developers don't include this component.

The job-task analysis provides the basis for developing a performance-type test, and the physical demands analysis provides the specific descriptors that are essential to a rehabilitation program.

FIGURE 14-1

TASK DESCRIPTION

Simple: Drag a mock victim to simulate a rescue.

Descriptive: The subject shall grasp the mannequin with both arms and drag it while walking backward to a designated location in simulation of a rescue attempt.

Functional: The subject shall move to a squat position at the head end of a 175-pound mannequin, grasp the mannequin's shoulders, and push it into an upright position. The subject will then wrap his arms around the mannequin's chest and lift it to an upright position. Walking backward, he will then drag the mannequin for a distance of 100 feet. The event will be performed within sixty seconds, and the time will be stopped when the mannequin's feet cross the 100-foot line.

RETURN TO WORK

A health and fitness program includes a broad range of disciplines and activities that are typically beyond the capability of the average certified fitness coordinator or program manager. Thus, it is imperative that the program manager establish a working relationship with medical providers.

The physician is the key player in the return-to-work process. Historically, the physician treats the patient, then makes a determination as to whether he's once again fit for duty. The procedure often follows a very simple and subjective pattern, with the physician merely asking the patient whether he feels ready. The natural impulse for a breadwinner in such a situation is to answer in the affirmative. If a physician isn't provided a functional job description, he has no specific information available to make an informed decision, other than the patient's opinion. Although common, this practice isn't considered acceptable or in anyone's best interest. The problem isn't unique to public safety occupations, but when a physician returns a firefighter or police officer to full duty with limitations such as "cannot lift more than twenty pounds," the patient and his coworkers are put at risk and the department is exposed to increased liability.

This problem was identified by the workers' compensation system, and the demand for more specific and objective information from physicians was required. Physicians began to solicit help from other specialists—occupational and physical therapists—who were more in tune with the patient's capability and limitations. A kinesiophysical approach used by therapists and based on objective patient observations was developed, replacing the psychophysical approach in which the patient himself essentially makes the decision (Isernhagen, 1992a). The more objective approach is favored, though both are still in use today. This objective approach ultimately led to the development of a functional capacity evaluation, or FCE. The FCE is popular with the workers' compensation system since it allows comparison between the employee's tested ability and essential job functions. In order to make such a comparison, the physical requirements must be identified, and that is accomplished through the job-task analysis or the physical demands assessment.

JOB-TASK ANALYSIS

The job-task analysis starts with an inspection of the department's job descriptions, operational procedures, and policies. A visit to the job site is also required to collect data and information on equipment, facilities, and tools connected with the job. Incumbents are often interviewed at this time and are asked to complete a questionnaire or survey to identify specific job requirements. The physical demands assessment, if performed in conjunction with the job-task analysis, would include measurement of tools and equipment, noting weights, distances, duration of work time, and other relevant data as to how a specific task is performed. Once all of the data has been gathered, the result is compared to current job descriptions for discrepancies, or it is used to develop a new description if none exists. The description must reflect the findings of the physical demands assessment.

The job-task analysis and physical demands assessment provide the basis for employment decisions, such as when an employee may return to work. The end product of the job-task analysis may be referenced in legal matters associated with employment decisions. The first issue would be whether a job-task analysis was completed and, if so, whether it was adequate, appropriate, and consistent with professional procedures. Therefore, it would be beneficial if the job-task analysis and physical demands assessment were conducted by a professional contractor.

JOB DESCRIPTION

The ADA doesn't require functional job descriptions, but it does highly recommend that such descriptions be developed. When any employment decision is made with regard to physical requirements, the job description will be the first choice of reference. Therefore, all relevant data must be documented in this format.

Job descriptions should be written by professionals. However, the most common practice is to assign the responsibility for it to a ranking officer. In this situation, the *Dictionary of Occupational Titles* may be the reference of choice. A companion book, entitled *Selected Characteristics of Occupations Defined in the*

Revised DOT (1993), provides more specific information on physical requirements. This text is often referenced by physical therapists. Although the text is being replaced by the Department of Labor's new O*NET system, the data is derived from the DOT, and the text will still be available in 2000. One of the reasons for the change is that many new jobs aren't included in the unrevised DOT. This, however, is irrelevant for our purposes because the physical requirements of public safety work have remained unchanged for eons. The DOT will still be a good source. Perhaps one of the most important aspects of the DOT is the delineation of physical work requirements based on job assignments. For example, a battalion chief is assigned a strength rating of L, or light, versus a firefighter's rating of VH, or very heavy. The difference is eighty pounds in weight-lifting requirements. Environmental requirements also vary considerably. This brings up the issue of matching physical requirements with the individual's job assignment. With this in mind, the return-to-work evaluation would also be different. Since many departments use the same test for everyone, employment decisions could easily have an unlawful, adverse effect on some individuals. Therefore, identifying the physical demands by job description is critical. Another reference for writing job descriptions for firefighters is NFPA 1001, *Standard for Fire Fighter Professional Qualifications, 1997 Edition.* This standard provides job performance requirements written in a format that is easily transferable to a code book, as mentioned earlier.

RETURN-TO-WORK EVALUATIONS

A flow chart for making return-to-work decisions is presented in Figure 14-2. The process will work for most injuries. In the figure, the return-to-work decision can be made at any time in the process, depending on the nature and severity of the injury. The choice of procedures should be left to professionals; i.e., physicians and physical therapists.

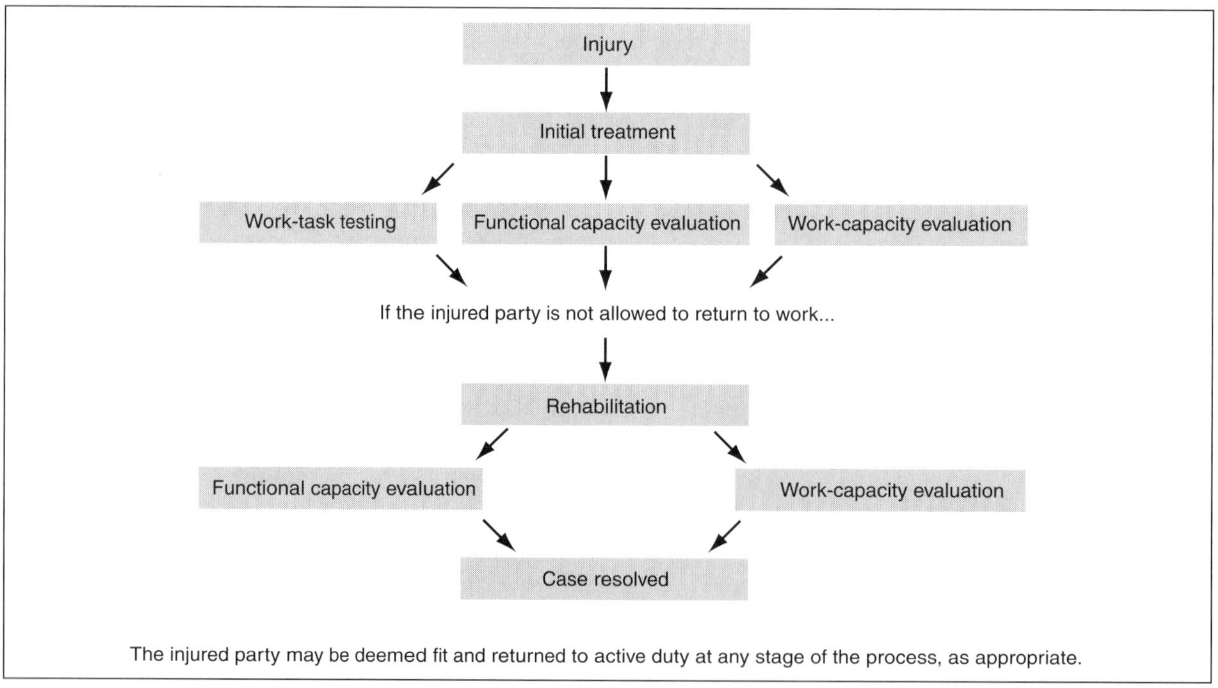

Fig. 14–2 Course of Return-to-Work Decisions

The chart makes use of three tests: the functional capacity evaluation, the work capacity evaluation, and the work task test. All have different applications and meanings within the physical therapy profession. The functional capacity evaluation is a full-body evaluation. It is a general test applicable to a class of occupations. Candidates for manual-labor jobs would all receive the same general test, and the lifting requirements would be matched to suit the particular occupation. In this sense, it is standardized, which makes it appealing to workers' compensation groups (Isernhagen, 1992b).

The work capacity evaluation differs in that it is more occupation-specific. Technically, a physical performance test or CTTest qualifies as a work capacity evaluation, since both employ specific job tasks. The distinct advantage that public safety jobs have over most occupations is that most departments have a validated CTTest in use. This makes the physical therapist's job easy, in that he won't have to develop a test, which is time-consuming and costly. The department need only supply the test protocol to therapy groups.

The work-task test is even more specific, since it usually involves a specific body part and task. Figure 14-3 shows a firefighter simulating a ladder raise using a work simulator. Such a simulator is capable of duplicating hundreds of tasks at varying workloads, and it is available in most rehab settings.

REHABILITATION

The purpose of a rehabilitation program is to restore an individual's functional ability to preinjury status, if and when possible. Following the basic medical treatment for the injury, the patient is evaluated and his functional status is used as a baseline to develop a rehab program. The nature of the rehab program and the potential for restoration depends on the type and severity of the injury. The most common public safety injuries, strains and sprains, can easily be handled with a work-conditioning program. The more serious and complex injuries may require a work-hardening program to make the transition back to work. The specific goals of work conditioning aren't unlike those of a fitness program. It is fundamentally an exercise program that emphasizes general fitness. It is often combined with an educational component to help prevent further injuries. Work hardening differs in that it uses simulated work tasks. This requires access to department facilities and equipment that will enable an occupational therapist to construct a work circuit (Altug et al., 1990). A physical conditioning program and an educational component are also provided when appropriate.

Typically, public safety personnel who are injured are assigned to light-duty status until cleared for full duty. The term *light duty* and the assignment of personnel to "created" positions has produced a great deal of controversy over the past few years. I have personally known firefighters who have spent twenty years or more on light duty. The practice can be traced back to a time when public safety officers "took care of their own," so to speak—a time when disabled members were left unemployed and uncompensated. Since the chances of suffering a disabling injury are high, the practice was relatively common in these occupations. Today, with disability pensions, there is less need for informal, unofficial forms of protection. Furthermore, few departments can afford to create assignments that they neither need or want, nor does the ADA require them to do so. This is why job descriptions are important. An injured public safety officer will have the opportunity to apply for another position, but that position won't be regarded as light duty. The preferred term, therefore, is *transitional work program* (Shrey and Mital, 1994). This term implies temporary assignment, a transitional phase between rehabilitation and full-duty assignment.

A transitional work program keeps the member involved in the day-to-day operations of the department without exposing him to

Fig. 14–3 Ladder Raise on a Work Simulator

unnecessary risks associated with full duty. The training division of a department is an excellent assignment for such personnel. The program should be designed in consultation with a physical therapist, who can determine the types and limitations of selected activities.

ERGONOMICS

Ergonomics is the science that addresses human performance and wellness in relation to the job, equipment, tools, and environment. The basic principles of this discipline were applied following the Second World War. In 1960, England established its Ergonomics Society and applied the term to consumer products. The idea would eventually catch the attention of industry, where the potential benefit to be accrued from improving the work station would be an increase in productivity. Later, the potential for reducing work-related illnesses and injuries became the primary motivator. In both the private sector and public safety occupations, musculoskeletal disorders and related injuries are the leading cause of work-related disability. Reducing injuries means minimizing the risk factors, and one approach with that principal goal is ergonomics.

Ergonomic knowledge has traditionally been used to improve the design and use of consumer products. It advances today with an added focus on the worker and the reduction of job-related injuries. Applying ergonomic principles requires an understanding of human anatomy, physiology, kinesiology, behavioral science, and psychology. Experts are well aware that prevention is the key to workplace safety, and the application of ergonomic controls can help a department reach its goals in this regard.

Ergonomic controls are also used to reduce worker stress, which may manifest itself as the fatigue of a specific muscle or muscle group due to repetitive motion or overexertion. There are two types of ergonomic controls, administrative and engineering. Administrative controls applicable to public safety are as follows.

1. Incident management system.
2. Fitness/wellness programs.
3. Skill-training programs.
4. Standard operating procedures.
5. Personnel assignment, including the safety officer.
6. Accountability systems.

Engineering controls focus on the hardware associated with the occupation, such as tools, equipment, and vehicles. This is addressed in NFPA Standard 1500, *Occupational Safety and Health Program, Section 4-5.1 (1997 Edition)*. This standard states that *"... The fire department shall consider safety and health as primary concerns in the specification, design, construction, acquisition, operation, maintenance, inspection, and repair of all tools and equipment."*

Unfortunately, some manufacturers and administrators have lost sight of this and specify apparatus for style and comfort, forgetting that the vehicle is primarily a tool. The utility of tools and equipment and their relationship to the job is paramount if injuries are to be reduced.

New legislation and the removal of age, gender, height, and weight requirements has changed the demographics of public safety. Large anthropometric variations exist in the workforce today, and the design and specification of equipment and apparatus require attention to ensure that people can reach and use all the tools and equipment needed to perform the job. Ignoring this can create problems that currently don't exist, and you can literally spec some of your members out of a job. An example of where design specifications have increased worker stress is shown in Figure 14-4.

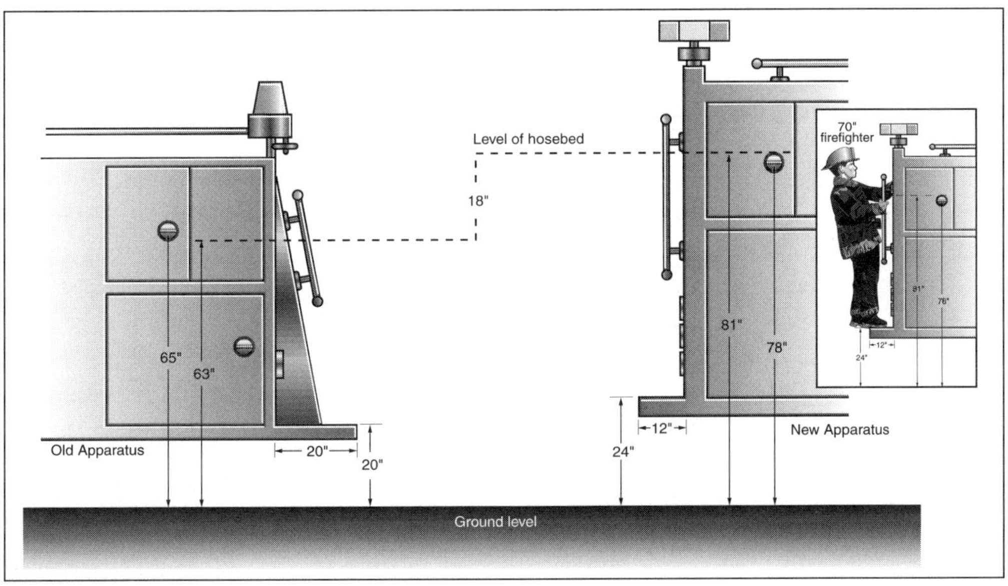

Fig. 14–4 Comparison of Old and New Apparatus

The older apparatus was more in consonance with personnel (or "user-friendly," in today's parlance) than the newer-style apparatus prevalent today. The compartment latch is thirteen inches higher on the new rig. Even if the handle were lowered to accommodate more firefighters, there would still be the problem of having to reach in and retrieve tools safely once the door is opened. Another obvious problem is the height of the hosebed, which was sixty-three inches on the old vehicle and is eighty-one inches on the new one. To reach the hosebed, you have to negotiate a step that is now four inches higher (pull up those baggy bunker pants) and stand on a tailboard of only twelve inches' accommodation, eight inches less than the old one. Another problem is the location of ladders. The difference between ladder mounts on these two rigs is twelve inches. That's significant. Specifying a ladder lift (additional cost) to lower the ladder was necessary for safe removal and stowing. Since this is the principal workstation of a firefighter, the job has suddenly become more difficult and less safe. The transition from old rig to new rig in this example initially resulted in one back injury and two knee injuries for one department, and a plethora of related injuries soon followed.

Obviously, the public safety officer's principal workplace (the emergency scene) cannot be controlled completely. There are inherent hazards associated with the occupation. Still, the risks can be minimized and should never be created as a result of the equipment selected. The work methods (administrative control) and the equipment used at the scene (engineering controls) can and should be modified to reduce the dangers present at all emergency incidents.

The first step toward implementing an ergonomic program is to conduct a workplace assessment. This includes a review of workplace injury records to help identify problem areas. The best approach is to correct as many potential ergonomic problems as possible before they develop. Engineering controls should be considered initially because they have the greatest impact on a worker's behavior and performance.

To prevent injuries, practical recommendations must be sought and implemented. This requires the support of management and an understanding of the cause-and-effect relationship between the workplace and injuries. Placing a rehabilitated firefighter back into the same environment that induced the injury without correcting the problem is poor injury management. Assessment and consultation services are available from physical therapy groups that provide rehab services to your members. These therapists should be included in the investigation of accidents as part of the treatment process. I recommend the use of a field investigation report, complete with a simple graphic and a statement that describes the incident. This will provide the therapist with some very useful information that could help prevent further related accidents. The therapist will visit the site if requested and consult with the members so as to develop some viable intervention. This approach should also be used internally. Your members will most likely have the best ideas for improving safety, since they study their jobs every day as they work and are in tune with the inherent problems. For example, a firefighter may discover from work experience that it is easier to push a heavy weight than pull it, or slide a weight rather than carry it. He may not know why (the force required to lift a weight is approximately thirty-two times greater than is required to slide it), but it doesn't really matter. If a modification works, take his advice and make the change. If it doesn't work, revisit the problem. Doing nothing is a decision in itself, and it is one that doesn't work. Anyone who has worked at heavy manual labor quickly discovers an easier way, and firefighters are no exception. Supervisors must remain receptive to worker feedback and provide alternatives when possible. Human behavior plays a major role in the accident chain, and it's important to involve everyone in the cure.

A thorough injury investigation procedure combined with employee feedback will help identify some key stressors in your department. A simple and inexpensive example of an ergonomic intervention that can be introduced in any department pertains to the common toolbox. More than one firefighter has experienced having to carry a fifty-pound toolbox through an apartment complex only to discover that he only needs a screwdriver. Change requires management's support in allowing a firefighter to decide what tools are most often used and to provide a smaller kit for that purpose. Reducing the weight of heavy boxes by splitting the load between two smaller units will remove the chance of injury associated with handling asymmetrical loads. Carrying a heavy weight in one hand creates a spinal curvature that can accelerate the process of disk degeneration over time. If the load cannot be split, such as with an extrication tool, then it should be carried with both hands or balanced on the shoulder, if possible, to reduce stress on the back. The preferred approach is to assign two people to the task. Simply because an injury doesn't occur directly in association with a specific task doesn't mean that it's nonproblematic. Physical therapists know that this practice will lead to back problems; therefore, it makes sense to remove potential risk factors before those problems accrue.

When implementing worker or consultant recommendations, take care to avoid a misdiagnosis and possibly create a problem that currently doesn't exist. In an effort to reduce cab noise, manufacturers installed rear-mount engines, which necessitated a new design, resulting in an increase in workstation height and shortened tailboards. Steps are often added to the apparatus to offset the increased height. These I have found to be of little use and more hazardous than not having them. They are usually small and hard to find with your foot, particularly at the emergency scene. This is a feeble attempt to overcome poor design. The very act of stepping up is a major contributor to back injury because it forces an arch in the lower back, particularly if one has tight hamstring muscles. This modern design created more problems than it solved, and it resulted in an increase in the injury rate. The problem also extended outside the department, since it created access and work problems for mechanics. Give me the old rig. It was practical, and I'll wear ear protection to deal with the cab noise. Another example found everywhere today is the use of back-support belts. According to one physical therapist, the routine use of back-support belts may lead to a weakening of the torso musculature and actually create back problems. There are situations in which the use of this device is appropriate, but generally speaking, it is unnecessary, and one should consult with a therapist before using a belt. A back belt doesn't increase the ability to lift, and it is no compensation for muscle inadequacy, to which ninety percent of back pain may be attributed. Instead, the development and implementation of a personal fitness program is encouraged to improve strength, endurance, and flexibility of the muscles and joints that support the spinal column. Yet another problem that evolved in an effort to solve one was the installation of a nonskid material in compartments to keep equipment from sliding around. It works, but it also created a retrieval problem. Since the equipment cannot slide or be pulled closer for removal, it has to be lifted with outstretched arms. This places tremendous stress on the lower back. Biomechanical studies have shown that it isn't unusual to expect 800 to 1,000 pounds of compressive force to develop at the lumbosacral disk while lifting a load of only 35 pounds. If the load is bulky or heavier than 35 pounds, or if it cannot be held close to the body, the forces on the spine may exceed 2,000 pounds. The solution here is to provide slide-out shelves to eliminate lifting with outstretched arms. That way, the advantage of using nonslip material is retained, and body mechanics aren't compromised. Below are listed some other equipment modifications or specifications that will enhance productivity and reduce potential injury.

FIRE

1. Work gloves for nonfire emergencies. Fire-resistive gloves reduce grip strength by 40 to 60 percent and are unnecessary for many tasks.

2. Wearing the appropriate clothing (not bunker gear) for rescue and assistance work.
3 Lightweight leather boots that fit better than the traditional rubber boots and reduce fatigue.
4. Lightweight hose.
5. Use rescue straps to drag victims rather than carry them.
6. Lightweight SCBA.

POLICE

1. Wear light-colored uniforms to reduce heat stress during hot weather and lightweight (nonpolyester) trousers with cargo pockets.
2. Wear military-style suspenders to carry the load of the duty belt.
3. Spec vehicles to meet the needs of the police officer. Controls should be readily accessible and standardized in terms of function. Standardization is predicated on the consistency principle in human factors.
4. Car seats should be customized to allow for seating comfort while wearing the duty belt.
5. Use rescue straps to drag victims rather than carry them.

There is no ergonomic cure-all that will fit every situation. The best approach is to keep it simple and address the obvious based on personal experience. Public safety occupations are physically demanding, and it is unrealistic to think that all task loads can be reduced or simplified; however, it makes even less sense to contribute knowingly to their difficulty.

Administrative controls in our diverse work environment will require rethinking as to how and what we do. The assignment of personnel is a key instrument. Although personnel costs are significant for all departments, the alternative is to face increasing workers' compensation costs. Assigning personnel to positions that they can perform well and safely requires an assessment of your workforce and your standard operating procedures.

The National Institute of Occupational Safety and Health recommends that jobs that create compressive forces on the lumbosacral disk of greater than 770 pounds be monitored, and that jobs involving forces in excess of 1,430 pounds be redesigned. The institute recognized the growing problem of work-related back injuries and published the *Work Practices Guide for Manual Lifting,* or WPG. The WPG was based on the theory that "overexertion injury is the result of job demands that exceed a worker's capacity." The guide employs a somewhat complex formula to identify acceptable load weights for select populations and tasks. It is available on the Internet, but it requires a professional to interpret and use. It was intended for use in the industrial setting, but it does have some application in the realm of public safety. For example, using specialized equipment found in the therapy setting, a firefighter could be evaluated to determine whether a ladder raise is within his safe-lift limits. If found to be excessive, ergonomic intervention would be required (Figure 14-5). To determine the physical requirements of a ladder raise, the weight and dimension of the ladder, combined with the height of the firefighter, must be recorded. The force required is a resultant of two components: the vertical force (Fv) required to lift the ladder, and the horizontal force (Fh) required to push it forward. Calculations are based on the impulse/momentum formula *Moment equals force times lever arm,* or M = F x L. Initially, the moment (or torque) of the 28-foot ladder (108 pounds) at rest is 108 pounds x 8 feet, or 864 foot pounds (8 feet indicates the center of gravity). To lift the ladder from the top rung, the lever arm becomes 16.5 feet, so the force required to initiate the movement to head height is 864 ÷ 16.5, or 52.4 pounds. The remaining angles and forces are similarly calculated, with height measures in increments of two inches. As shown in Table 14-1, a 60-inch person raising a 108-pound, 28-foot extension ladder must apply a force of 86 pounds at the heaviest point in the lift (rung 8 at 44.2 degrees), where he is bearing about 80 percent of the ladder's total weight. In comparison, a 72-inch person raising the same ladder must apply a maximum force of 72 pounds (rung 7 at 46 degrees), where he is lifting 67 percent of the ladder's weight. The shorter person, therefore, works harder, lifting an additional 14 pounds. This is a relatively light ladder, and a larger ladder would introduce a greater difference combined with problems of technique, since the raise requires a hand-to-hand walking of the rungs. Thus, the 60-inch firefighter is actually performing a one-hand overhead press of 86 pounds. Depending on individual strength, the task may require reassignment or an increase in staffing. Since the weight of modern ladders has increased somewhat, many departments do opt for changes in ladder assignments as a defense against overexertion injuries.

A similar problem in terms of staffing can be found in police departments. The New York State Police had height and weight requirements until the early Seventies, and you would have been hard-pressed to find any trooper shorter than 6 foot 3 or 4. Due to physical stature alone, troopers were less likely to meet resistance, particularly when they worked in pairs. When the requirement was dropped, the assault rate on officers jumped dramatically. If and when changes to standards aren't desired, attention should be given to particular assignments. Fire departments have historically assigned larger members to perform the heavier work. Police departments handling tough districts where physical encounters are common should consider selective assignments. Although the introduction of martial arts techniques has helped many officers, the visual impact of size alone is often an effective deterrent.

From a safety standpoint, departments must be ready for change and take appropriate action when necessary. The study of ergonomics and safety is now sophisticated enough to have a significant effect on the prevention and cure of the most prevalent worker injuries, and the best approach is the implementation of an effective prevention program. Proper workplace ergonomics, supplemented by a sound fitness program, can greatly benefit the cause of injury prevention.

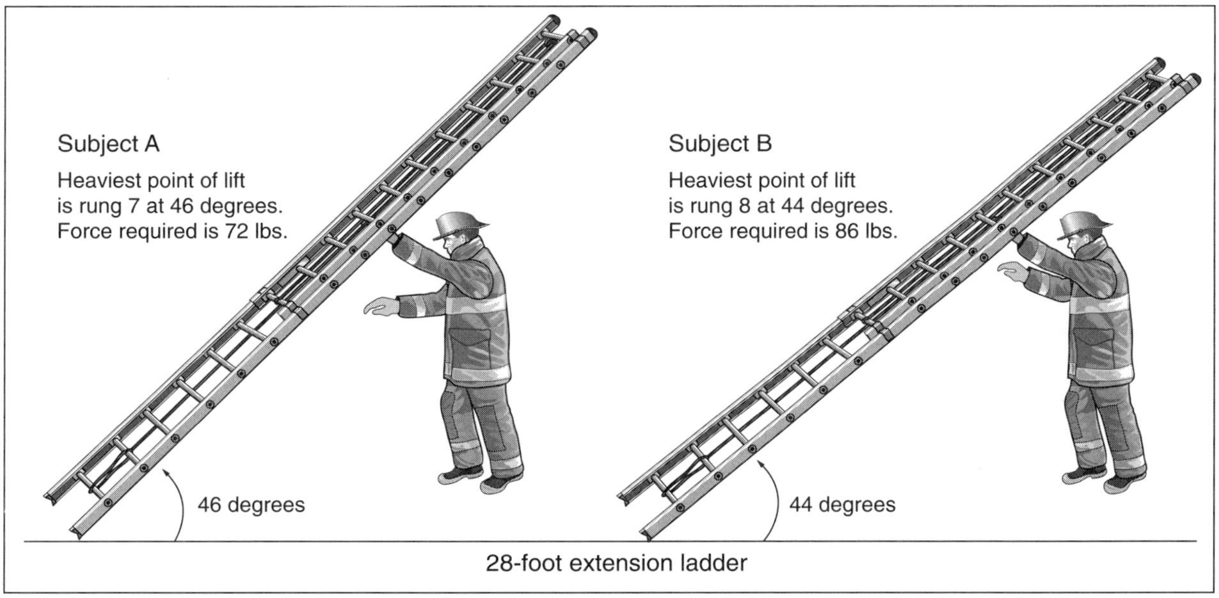

Fig. 14–5 Anatomy of a Ladder Raise

TABLE 14-1

FORCE REQUIRED TO LIFT LADDER BY RUNGS

Height of f.f.	Lift to Waist	Lift to Head	2	3	4	5	6	7	8	9
60″	56	54	57 (20.7°)	61(22.6°)	66 (25.0°)	72 (28.0°)	77 (31.8°)	83 (36.9°)	86 (44.2°)	80 (56.4°)
62″	56	54	57 (21.4°)	61 (23.4°)	66 (25.9°)	71 (29.0°)	76 (32.9°)	81 (38.3°)	84 (46.1°)	73 (59.4°)
64″	56	53	57 (22.1°)	61 (24.2°)	65 (26.8°)	70 (30.0°)	75 (34.2°)	80 (39.8°)	81 (48.1°)	66 (62.7°)
66″	56	53	56 (22.8°)	60 (25.0°)	65 (27.7°)	69 (31.0°)	74 (35.4°)	78 (41.3°)	77 (50.1°)	58 (66.4°)
68″	56	53	56 (23.6°)	60 (25.8°)	64 (28.6°)	69 (32.1°)	73 (36.6°)	76 (42.8°)	74 (52.3°)	47 (70.8°)
70″	55	53	56 (24.3°)	59 (26.7°)	64 (29.5°)	68 (33.2°)	72 (37.9°)	74 (44.4°)	70 (54.5°)	34 (76.5°)
72″	55	52	55 (25.1°)	59 (27.5°)	63 (30.5°)	67 (34.2°)	71 (39.2°)	72 (46.2°)	66 (56.8°)	0 (90.0°)

Based on a 28-foot, 108-lb. ladder with 16.5-foot bed length. The balance point is at 96 inches, and the rungs are 14 inches on center.

RECOMMENDED READING

Chaffin, D.B., and Andersson, G.B. (1991) *Occupational Biomechanics.* John Wiley and Sons Inc., NY.

Rasch, Philip, Ph.D., and Burke, Roger K., Ph.D. (1960) *Kinesiology and Applied Anatomy.* Lea and Febiger, Philadelphia, PA.

Salvendy, Gauriel (1987). *Handbook of Human Factors.* John Wiley and Sons Inc., NY.

CHAPTER 15

WELLNESS

An eight-year study conducted through the Cooper Institute for Aerobic Research in Dallas investigated the relationship between physical fitness and mortality in 25,341 men and 7,080 women. The study found low cardiorespiratory fitness to be a strong predictor of mortality, and participants who recorded higher values—even if they possessed risk factors such as smoking, hypertension, or elevated cholesterol—had lower death rates. The study supports the current philosophy that activity is perhaps the more important factor (Blair, 1989). Supported in the research and consistent with the subject of this text, personal fitness is essential if we are to address the health/injury problem, and the emphasis should be on fitness for duty. Truly, this implies a higher level of effort typically not emphasized in a private-sector wellness program. The benefit of this approach is revealed in the positive influence that general fitness has on various health factors. The relationship between the exercise dose and one's health profile is depicted in Figure 15-1.

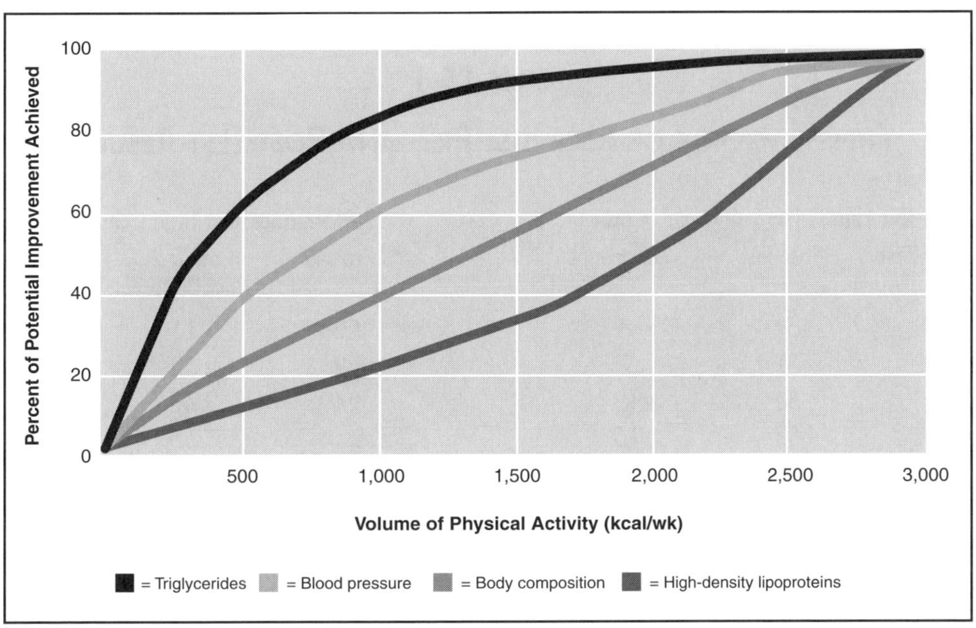

Fig. 15–1 Relationship of Exercise Dose to Body Response
Note: To achieve improvements in many health indicators, the amount of physical activity can be attained through any combination of frequency, intensity, and duration. Still, improvements in aerobic fitness and strength can only be achieved by working above a threshold intensity for an appropriate frequency and duration. Schematic developed by N. Gledhill and V. Jamnik, York University. Source: *The Canadian Physical Activity, Fitness, & Lifestyle Appraisal: CSEP's Plan for Healthy Active Living*, 1996. Reproduced by permission of the Canadian Society for Exercise Physiology.

In addition to the health benefit, the relationship between improved fitness levels and job performance is also documented. On reviewing the test results of twenty firefighter recruits following an eleven-week program, it can be seen that across-the-board improvement in all fitness scores translated into improved performance times for all recruits, as measured by a job-task simulation test battery (Table 15-1).

Although physical fitness as described in this text is considered to be of paramount importance, it should also be regarded as the first step in a complete program and not be left to stand alone. In other words, there are factors beyond fitness that should be addressed once the basic program has been established. In the words of one corporate executive who is an avid supporter of wellness programs, "It is necessary to develop people in three dimensions: body, mind, and spirit" (Pritchard and Potter, 1990). Thus, the wellness concept is next on the agenda.

According to Dr. K. Cooper, "To function properly, the human body should be in a state of balance; neither too much or too little—food, work, exercise, stress, or rest—a state of well-being." (Cooper, 1982) One author defines well-being, or wellness, as "a conscious and deliberate approach to an advanced state of physical and psychological/spiritual health." (Ardell, 1984) A model expounded by Elisabeth Kübler-Ross includes four components: physical, mental, emotional, and spiritual well-being, all superimposed on one another and contained in the spiritual, the last phase of the human growth cycle (Figure 15-2) (Kübler-Ross, 1981).

The components developed through the various stages of life are as follows.

1. Physical well-being: the optimal functioning of the physiological system.
2. Mental (intellectual) well-being: the optimal mental processing of information.
3. Emotional well-being: the ability to feel, express, and control the full range of human emotions.

TABLE 15-1

FITNESS PROFILE OF RECRUITS: PRE- AND POST-TEST RESULTS

Aerobic Fitness (Vo_2 mL • kg^{-1} • min^{-1})			Performance Test Time (Minutes/Seconds)		
Pre-Test	Post-Test	Difference	Pre-Test	Post-Test	Difference
48	54	6	4:28	3:11	:43
52	59	7	6:26	4:52	:94
43	56	13	4:40	3:42	:58
42	57	15	5:44	4:55	:49
43	55	12	4:15	3:42	:33
44	51	7	7:45	6:09	:69
49	59	10	5:13	4:19	:54
55	58	3	3:42	2:58	:44
54	56	2	3:39	3:19	:20

An improvement in aerobic scores (average 8.3 mL • kg^{-1} • min^{-1}) translated into improved performance test times (average 55 seconds faster) for all recruits. Source: Rocky Mountain Fire Academy. Denver, 1996.

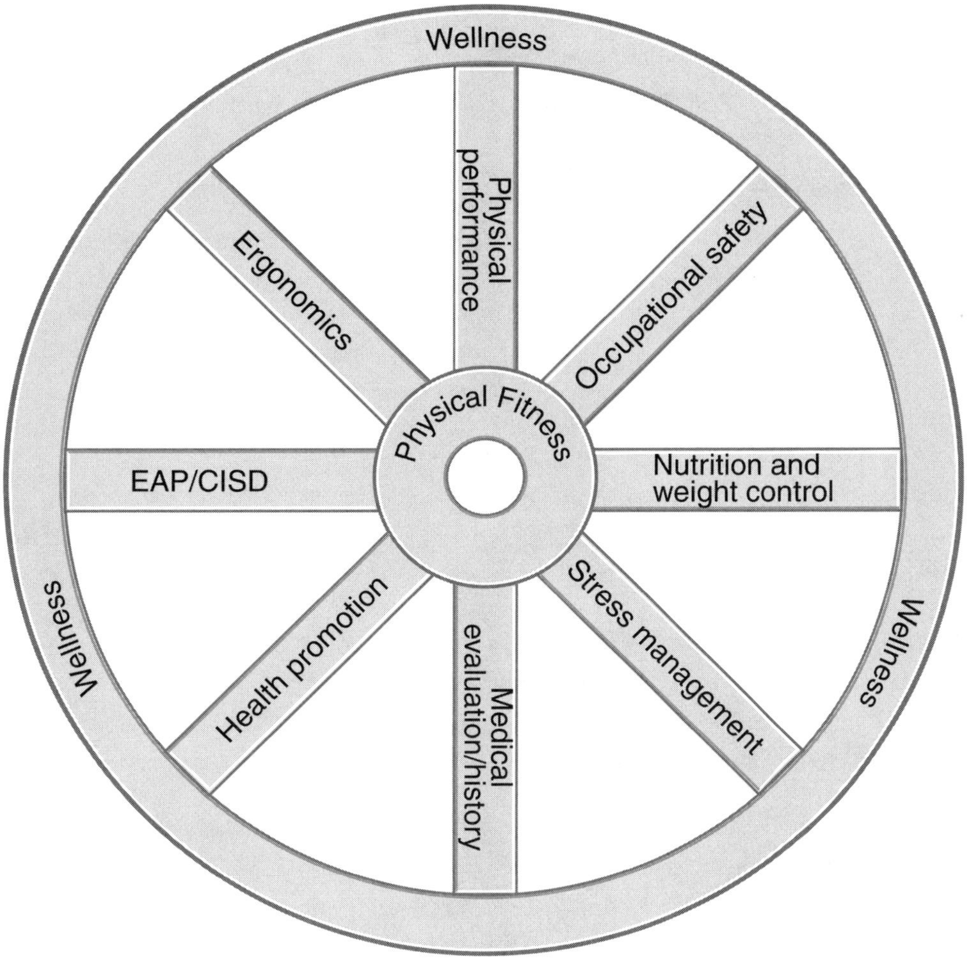

Fig. 15–2 The Wellness Wheel

4. Spiritual well-being: the maturation of a higher consciousness; the development of personal values and a meaningful purpose in life.

Wellness is the total balance and integration of these components, described by Seaward. From a holistic perspective, all areas must be addressed (Seaward, 1994).

The wellness wheel is but one model that shows the relationship of the fitness/wellness paradigm as it relates to our program design. Fitness may only be one aspect of a complete program, but in the context of a public-safety enterprise, it plays a critical role.

To the wheelwright, the hub is the most essential component, the starting point from which the outer rim evolves. Fitness rightly belongs at the hub. The other components that together comprise the wellness wheel are directly connected to fitness, and subsequently develop from the center.

When evaluating department members through fitness testing, test administrators can expect to find a proportionality in the various levels of fitness. This proportionality is represented in Table 15-2. Depending on department demographics and which fitness standards have been selected, approximately 5 to 15 percent of the membership will place in the highest fitness category. The majority of the members, about 70 to 85 percent, will score in the middle categories. The remaining 10 to 15 percent will fall into the low and borderline categories.

TABLE 15-2

THE TYPICAL FITNESS GROUP PROFILE

Percent of Members	Fitness Category	General Description
5–15%	High fitness	These are the most productive members. The average work output of a high-fit member is twice that of a low-fit member. They sustain fewer injuries and at lower overall cost (Cady et al., 1985; Arnold et al., 1982).
70–85%	Average to Above-Average Group	The injury rates and associated costs vary and tend to reflect the profile of the nearest group, high or low.
10–15%	Borderline and Below-Average Group	The injury costs of this group approach six times those of the high-fit group (Cady et al., 1979; Arnold et al., 1982).

WELLNESS PROGRAMS

According to *Ready Facts* (Bureau of Labor Statistics, 1995), of the medium and large private establishments reporting, 58 percent offered employee assistance programs that included services for drug and alcohol abuse, as well as counseling for emotional and financial problems. Additionally, 34 percent of the employers offered wellness programs that included programs for physical fitness, stress management, smoking cessation, and weight loss.

As shown in Table 15-3, a survey conducted by the Fairfax County Fire and Rescue Department reported the following from seventy-five fire departments.

In comparison to the private sector, the Fairfax survey reports a similar percentage of wellness program participation: 38.7 percent to 34.0 percent private, with substantially greater

TABLE 15-3

FAIRFAX WELLNESS PROGRAM SURVEY

	Yes %	No %
Wellness program	38.7	61.3
Stress management	74.7	25.3
Critical incident stress debriefing	89.3	10.7
Family assistance	57.3	42.7
Substance abuse	82.7	17.3
EAP	94.7	05.3
Physical Fitness	65.3	34.7
Annual exams	48.0	52.0
Mandatory program	55.1	44.9
Fitness awards	17.3	82.7

support for employee assistance programs among fire departments.

In a survey of public safety organizations, comparing police departments, fire departments, and corrections facilities, fire departments were shown to have the least participation at 36 percent (Table 15-4). Still, this is consistent with the Fairfax report of 38.7 percent. Given the variation in survey types and departments, one conclusion is that participation appears to be somewhat less than desirable, yet consistent in terms of percentages (Landy, 1992).

TABLE 15-4

PUBLIC SAFETY ORGANIZATIONS HEALTH PROMOTION SURVEY

Does your department have a health promotion program?

	Fire (N = 164)	Police (N = 177)	Corrections (N = 100)
Yes	36%	44%	42%
No	64%	56%	58%

The trend, however, appears to be moving in a positive direction, as indicated by the IAFF/IAFC joint labor management initiative, as well as the positive response from various corporate sources (Pritchard and Potter, 1990).

One approach to developing a wellness program is to follow the design method presented in this text. Start with an assessment of needs, and proceed methodically, through to the evaluation. Some corporations in the private sector offer as many as thirty different programs under the umbrella of wellness. Attempts to match such a level of service by public safety organizations are unnecessary and, in most cases, cost-prohibitive. Therefore, I would advise that you start with a survey of incumbents to identify your department needs. Once the initial assessment has been completed, proceed as you did with the fitness program and develop a plan that will address specific concerns.

Regardless of the approach and response to the survey, there are some topics that you should include in all public safety wellness programs.

1. Physical fitness.
2. Stress management.
3. Nutrition, weight control, and health promotion.
4. Employee assistance program.
5. Occupation-specific programs.

PHYSICAL FITNESS

Fitness is central to the wellness program. Therefore, it must be the first priority. The sooner the program is implemented, the better, particularly in public safety occupations, in which members are actively involved in emergency operations. The common denominator in the occupational programs presented here is individual fitness. This is the principle preventive measure; the one variable over which we as individuals have control.

STRESS MANAGEMENT

In both the private sector and public safety occupations, the management of stress will always surface as a primary issue (Ornish, 1982; Healy, 1988; Yeung, 1991; and Pagani et al., 1991). The most-referenced expert on the subject, Hans Selye, MD, defines stress as a "nonspecific response of the body to a demand." (Selye, 1956) Stress is a necessary part of everyday life, but when it is prolonged or leads to frustration, it can become *distress* and harmful, throwing one out of balance. The psychotherapist Gerald Fishkin, Ph.D., coined the term *burnout* to describe the "total inability to maintain balance." (Fishkin, 1989.) Burnout connotes a state beyond the point of stress. The burnout concept is an extension of Hans Selye's general adaptation syndrome (GAS) theory. The three stages of this progression are alarm, resistance, and exhaustion. In the alarm stage, a person reacts to a stressor with a rush of adrenaline. During the second stage, the individual's learned physical and psychological defense mechanisms are called on to help the person cope with the situation. If the adrenal response cannot be deactivated, the person enters the third stage, typified by an inability to cope, or burnout.

The acknowledged association between stress and illness has some experts estimating that 70 to 80 percent of all health-related problems are either created or aggravated by stress. Twenty years ago, psychological stress wasn't identified on an injury report, and forty years ago, it was virtually nonexistent in the literature. The incidence of stress-related disability retirements was negligible in the early '80s, but they now represent 10 percent of firefighter disability retirements, ranking fourth by cause, according to the 1998 IAFF report (Figure 15-3). Stress is a serious problem with public safety officers, and a stress-management program, combined with professional services, should be provided.

A program model for stress management, offered by the United States Fire Administration, provides information on firefighter stress and the application of stress-management principles (USFA, 1991). The publication includes the results from a survey given to the participants of a pilot program, requesting them to provide feedback on areas of interest and issues from a checklist of fifty items. In Table 15-5, stress management is rated third of the top five areas, as reported by over two-thirds of the participants. Recognize, too, that the first two items are also stress-related in nature.

NUTRITION, WEIGHT CONTROL, AND HEALTH PROMOTION

To be properly addressed, the subject of nutrition would require at least another full volume. It is a topic that commands a great deal of attention, and it is often cited as a major area of interest in wellness surveys.

TABLE 15-5

USFA PROGRAM SURVEY, 1991

86%	Family communication
74%	Relaxation skills
69%	Stress management
66%	Diet for reducing the risk of heart disease
66%	Exercise for self and spouse

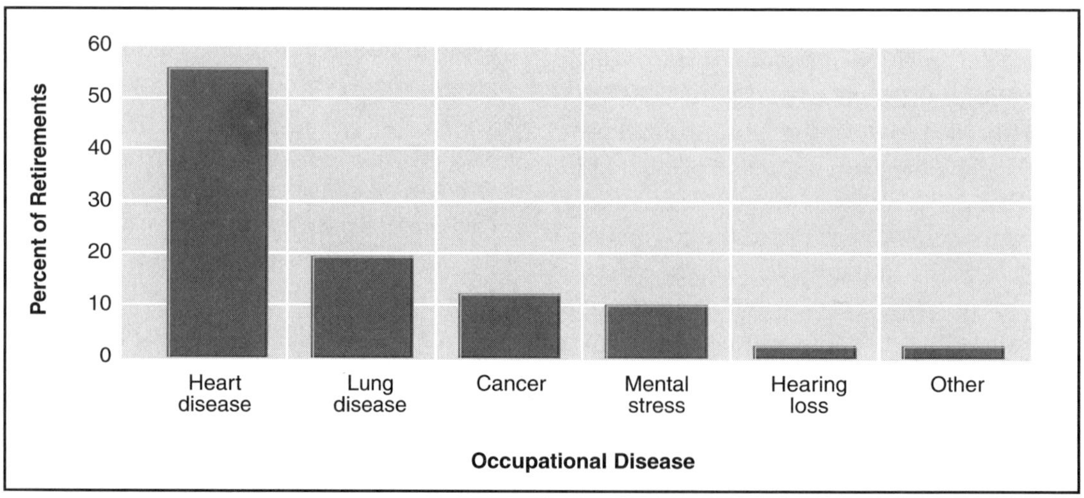

Fig. 15–3 Line-of-Duty Disability Retirements by Cause
Source: International Association of Fire Fighters, 1998 Death and Injury Survey

Virtually every book on the subject will advise you to eat a balanced diet. This statement alone would immediately rule out the low-carbohydrate diet, the high-protein diet, the no-fat diet, or any other unbalanced approach to eating. That 98 percent of diets fail, often due to exclusion of the exercise component, doesn't appear to deter people from trying the latest fad. A science-based society will religiously await the magic pill. Sadly, no magic pill exists, and it never will, in my opinion. The greatest breakthroughs are in marketing, and the conflict and confusion that abounds only serves to bewilder a public eager for accurate information.

In a 1995 survey conducted by the American Dietetics Association, the public made it clear that it wants simple and positive advice on healthful eating and not the latest research findings until there is general agreement among nutrition experts (Morreale and Schwartz, 1995). The contradictory reports on some food products create havoc with those who desire accurate information and not debate. The survey reported an increase in the use of two new education tools that help provide some general guidelines; namely, the Food Guide Pyramid and the Nutrition Facts Label. Selecting foods based on these tools is a simple approach that doesn't require a complete overhaul of your eating habits. Members would do well to start with these basic tools and stay clear of the quick-fix proclamations.

An important part of all wellness programs is in providing health-promotional information and services to members. Health promotion can be as simple as distributing a single-item handout, such as a nutritional label, with an advisory to compute the percentage of fat.

Most people are aware of some nutrition basics, such as limiting saturated fats and eating more fruits and vegetables. You don't have to be an expert to eat right. One recommendation is to eat 55 to 60 percent carbohydrates, 25 to 30 percent fat, and 10 to 15 percent protein. You'll see variations in the percentages, but fundamentally, all will echo the food pyramid. The food pyramid provides guidance in planning a well-balanced diet, and food labels provide a way to check on nutrient percentages.

There is a plethora of information available, and it can all be distributed in the form of memos, handouts, or through a department newsletter if you have no funds. If your department does have a budget for such matters, you may opt to provide professional counseling services through registered dieticians. A variety of consulting firms provide such educational programs for public safety organizations (Lafferty and Dillard, 1985). Arranging personal one-on-one counseling may be the preferred method, particularly with individuals who have serious weight problems. Another option is to educate a fitness coordinator on nutrition basics and let him, in turn, instruct the members. Nutrition courses that target public safety officers are available through several vendors (Appendix G).

Additionally, a number of software programs are available that can provide a variety of options, including a complete dietary analysis, diet plans, exercise plans, meal plans, and other useful information. You need not be a dietician to interpret the results, but I recommend viewing some sample printouts before you decide which program you want, since some can be labor intensive and provide only marginal information. Always get a sample printout or a trial period to test the program before you purchase it.

Numerous resources are available that will allow you to offer health education within even the smallest budget. Finding them simply requires a little ingenuity and effort. If you can't write a newsletter, subscribe to one. The American Heart Association is an excellent resource for literature on many subjects. Of course, health information is also available on the Internet. Since accuracy is a problem, I recommend the following sources reviewed by experts.

1. Journal of the American Medical Association: http://www.org/public/journals/jama/
2. American Cancer Society: http://www.cancer.org

3. On-line mental health: http://www.coil.com/-grohol/
4. Healthfinder: http://www.healthfinder.gov
5. Mediconsult: http://www.mediconsult.com
6. USDA's food and nutrition research briefs: http://www.nal.usda.gove.fnic.usda.fnrb
7. New England Journal of Medicine: http://www.nejm.org
8. American Heart Association: www.americanheart.org

Nutrition has a great effect on one's general health and physical fitness. Poor eating habits can cause nutritional deficiencies, contribute to the development of disease, cause a gain or loss of weight, and dramatically impair work performance.

Although nutrition and weight control play an integral part in maintaining one's health and well-being, nutrition is but one factor in the nutrition-fitness-performance triumvirate. The strenuous nature of firefighting, for example, combined with exposure to toxic substances, is perhaps a greater threat than poor nutrition. Therefore, view nutrition as a supplement toward achieving your personal fitness goal and not the sole solution.

EMPLOYEE ASSISTANCE PROGRAM

The employee assistance program (EAP), or member assistance program (MAP) as it is sometimes called, can easily include all aspects of the wellness wheel. Generally, however, it is intended to address behavioral issues. The EAP should be voluntary, unless mandatory referral is strongly indicated, and it should always be confidential. The EAP program typically includes stress management, substance abuse counseling, and family counseling. Substance abuse includes alcohol, drugs, and tobacco products. A departmental policy should be established that takes a no-nonsense approach to addressing such problems while providing every reasonable opportunity to assist a member (Wilkinson and Stelnicki, 1988).

OCCUPATION-SPECIFIC PROGRAMS

Programs that are specific to the occupation, although typically not considered part of a wellness program, should be viewed as a necessary component of wellness. The current trend in the private sector is to include occupation-related programs. In the realm of public safety, components such as infection control might be included in the wellness program (Cross and West, 1998).

A critical incident stress debriefing (CISD) involves an open discussion of feelings related to a specific incident. Information related to the event is provided, and ideas are exchanged through the interaction of incident members, professional counselors, and the debriefing team. Preferably, the CISD takes place immediately following the incident, although an on-site debriefing may be necessary, depending on the event. Establishing a CISD team is essential for emergency workers. Setting up such a team entails contacting mental health professionals for assistance, establishing a peer support group, and developing written policies and procedures.

Studies have shown that firefighter accidents and injuries can be reduced through a properly implemented emergency scene rehabilitation program. Prolonged exposure to either heat or cold can pose serious health risks to workers and reduce productivity. The Phoenix Fire Department is reputed to have the model emergency rehab unit.. Still, it is a misconception of some agencies that a city such as Phoenix, with temperatures that often reach over 100°F, is more in need of such extensive rehab than the average department. This is not true. Moderate temperatures combined with high humidity are equally dangerous. A 75-degree temperature at 90 percent humidity is as taxing as a 95-degree temperature with 10 percent humidity. In point of fact, all agencies need a rehab unit, and its function shouldn't be limited to handling heat injury or rehydration, but all types of thermal injuries. Firefighters do suffer cold injuries, and many have collapsed from dehydration in subzero weather. A good rehab program can be introduced at minimal cost and with little operational impact in any organization. The U.S. Fire Administration provides a sample rehab policy that outlines the role and responsibilities of the members, procedures for establishing rehab at the scene, and specific guidelines for monitoring members.

The responsibility for hazardous materials varies worldwide. In the United States, fire and police departments are typically assigned the task. Guidelines for the development of a medical surveillance program as it relates to hazardous materials are outlined in OSHA 29 CFR 1910.120. Some of the features include medical exam requirements, physical fitness certification for duty, and the maintenance of health and exposure records. Although exposure to toxic chemicals is the primary concern, the physiological stress associated with wearing a fully encapsulated suit must not be overlooked. The fitness levels of haz mat workers must be above average if they are to function safely inside this secondary, unnatural environment.

All public safety organizations should develop an infection control program that outlines procedures for control of infectious and communicable disease. Infection control connotes a comprehensive approach to managing risks, and it should be integrated into a department's overall risk management plan (Kipp and Loflin, 1997). The program policy should be aimed toward maximizing the protection of department members, patients, and the general public. A sample policy is included in the appendix of NFPA 1581, *Standard on Fire Department Infection Control Program, 1995 Edition.* Training guidelines for infectious disease are also available through the U.S. Fire Administration and the U.S. Department of Health and Human Services.

The OSHA requirements of a respiratory protection program (29 CFR 1910.134-40) include medical certification to wear and use SCBA. This requirement is supported by NFPA standards 1500, 1404, and 1582. If the use of SCBA reduces work capacity by 30 percent, as reported by some investigators, the relevance of the physical fitness program becomes obvious. A proper respiratory program cannot exist without due attention to fitness.

Health data on firefighters is limited, although some information is available that suggests that those in the profession are at increased risk of pulmonary disease, cardiovascular disease, and cancer. The link between heart disease and a high-fat diet is well established. The U.S. Department of Health and Human Services estimates that 80 percent of human cancer is related to lifestyle, diet, stress, environment, and inactivity (Levenstein, 1992). The percentage related to diet alone has varied among researchers, but it exhibits an increase of from 40 percent in early studies to 60 percent in some later reports (Dreher, 1988). Agreement on the relationship between diet and cancer, however, is strong among investigators (ACS, 1991; and Block et al., 1992).

Occupational disease retirements for firefighters in 1997 were at 11.8 percent, a value that has remained relatively constant over the years, with heart disease rated number one, accounting for 47 percent of the retirements. Heart disease is also rated number one for police. In the IAFF's Death and Injury Surveys, deaths in addition to those occurring in the line of duty are reported as they relate to occupational disease. In 1998, 39 firefighter deaths were attributable to occupational disease—44 percent by heart disease and 46 percent by cancer (IAFF, 1998). Table 15-6 depicts the percentages for the past ten years. As shown, cancer displays the higher percentage, with some earlier reports declaring the firefighter rate to be 100 to 300 times greater than that of any other occupation (Healy, 1988). It should be noted that the American Heart Association's *1998 Heart and Stroke Statistical Update,* p. 3, AHA, Dallas, TX, reported that cancer deaths exceeded those of coronary heart disease. This and the IAFF data clearly show that cancer may well be the number one issue for firefighters, replacing heart disease, which has so often been claimed to be the principal problem.

TABLE 15-6

IAFF ANNUAL DEATH AND INJURY SURVEY
PERCENT OF FIREFIGHTER DEATHS ATTRIBUTED TO OCCUPATIONAL DISEASE

	1989	1990	1991	1992	1993	1994	1995	1996	1997	1998
Cancer	34%	51%	55%	55%	53%	61%	67%	38%	50%	46%
Heart Disease	38%	27%	39%	39%	38%	38%	25%	43%	32%	44%

Source: Data from the International Association of Fire Fighters Annual Death and Injury Surveys, 1989-1998

RECOMMENDED READING

The American Heart Association annual reports.

Isernhagen, S.J. (guest editor) (1992). Iglarsh, Z.A.; and Richardson, J.K. (consulting editors). *Orthopaedic Physical Therapy Clinics of North America: Industrial Physical Therapy.* Philadelphia: W.B. Saunders Co., pp. 83–90.

The International Association of Fire Fighters *Annual Death and Injury Surveys.*

The National Fire Protection Association *Annual Fire Fighter Death and Injury Reports.*

Seaward, B.L. (1994) *Managing Stress.* Boston: Jone and Bartlett Publishers Inc.

CHAPTER 16

PHYSICAL PERFORMANCE

At a meeting held in Phoenix in 1992, the NFPA assembled a task group charged with developing a physical performance document. The general consensus of the task group was that the test format should be of the task type and responsive to legislative directives. The document was tentatively entitled NFPA 1583, *Recommended Practice for Fire Fighter Physical Performance Assessment.* The 1583 draft spurred considerable debate over fitness versus performance, and this ultimately led to the demise of the 1583 document four years later at a meeting of the Standards Council. The document defined both fitness and performance, and it provided a guide for the development of a physical performance assessment.

FITNESS VERSUS PERFORMANCE

A fitness test is health based and a performance test is job based. A person may be classified as somewhat unfit because he is overweight, hypertensive, or because he exhibits some other health factors that negatively affect performance. Still, he may be capable of passing a performance test. Conversely, a person may be highly fit and fail a performance test. The reason is that a performance test is designed to determine whether one can perform the required essential functions of a particular job, not to determine one's level of fitness.

Using the example of a typist, you would administer a typing test to determine whether one can perform; i.e., type. Whether the applicant is a cardiac patient, obese, or otherwise unhealthy is irrelevant to the intent or purpose of the performance test. The performance test answers the question "Can you type?" If the concern is predicated on health or fitness, then you would administer the appropriate fitness test.

Fitness, of course, is relevant, but it isn't a lone determinant. There are seventy-year-old men running marathons who would be considered fit but who certainly aren't suited for work as a firefighter or police officer. If such an individual were to apply for a public safety position that relied on a fitness test, he could possibly meet the requirements, particularly if age- and gender-based normative data were used as a standard. If a performance test designed to reflect the actual physical demands of the job were administered, it is unlikely that such an individual would qualify. Perhaps the term *suited for work* should replace the term *fit for duty.* The fitness test is the test instrument of the fitness program, which is categorically an assistance program designed to determine a level of preparedness. The performance test is the instrument for making job decisions (Davis and LeCuyer, 1995).

Historically, job performance tests were conducted on the job. A common practice of a major auto manufacturer was to throw a probationary employee into the most physically demanding position available. If he survived thirty days, he would be allowed to join the union and considered a permanent employee. If not, he would be terminated.

Title VII of the Civil Rights Act (CRA, 1964) and the entrance of women into traditional male-dominated occupations placed employee selection procedures under scrutiny. Employers, particularly those in public safety, where task-type tests were frequently used for selection, were forced to provide validation for their test procedures. The Americans With Disabilities Act (1990) prompted a rise in the use of task-type testing in private industry as an employment screen. Technological changes in tool design and work stations, combined with ergonomic interventions, reduced the physical demands of some occupations, thereby opening the door to individuals who had formerly been excluded. The first performance test that I administered involved a stair climb. The subject was outfitted with a total protective ensemble and hose pack weighing 165 pounds. That same task performed today involves a total weight of 85 pounds. A 50-percent reduction in equipment alone has had a significant effect on the number of candidates who can now qualify, using the same test event.

Criterion task tests are much easier to design in the private sector than in the realm of public safety. In private industry, employees are selected for specific jobs that have well-defined tasks and work stations. Public safety work, on the other hand, is characterized by hundreds of different tasks performed in a changing environment. The lift capability of a driver working for a courier service is easy to identify, since most carriers set a weight limit of about 75 pounds per package. The number of packages handled in a given workday is also fairly stable, so developing a test for such a position wouldn't be difficult. In a factory setting, work tasks are often even more defined, and the work stations are easily modified to incorporate lifting and handling aids. If a warehouse worker were required to lift a 100-pound box from the floor to a table height of thirty inches, and if that act would encompass all of the physical demands of that position, then the criterion test would be obvious: Lift a 100-pound box and place it on the table. That is the essence of a fair and objective test. If the myriad of fire and police tasks could somehow be represented in one simple task, our test process would be easy. Unfortunately, that is not the case.

The NFPA's Standard 1500 requires fire departments to develop physical performance requirements for candidates and members who engage in emergency operations. The resultant test is then used to evaluate and certify members as fit for duty. Members who are unable to meet the requirements aren't permitted to engage in emergency operations.

The Equal Employment Opportunity Commission (EEOC) developed a set of guidelines to assist employers in developing employment tests. The *Uniform Guidelines on Employee Selection Procedures (1978)* requires that tests used for employment decisions be job-related and validated if shown to discriminatc against a protected class. A physical performance assessment used for employment decisions should follow a validation process based on a thorough job-task analysis that identifies the essential job functions.

Since not all tasks can be incorporated in one test, the marginal functions are excluded, and only those deemed essential are selected. A job description should include all job tasks and serve as evidence of the essential tasks. Although physical performance assessments are simple in appearance, the development and validation process is complex. Therefore, it is recommended that the test be designed by a professional test developer, preferably someone with relevant occupational experience.

Remember that the purpose of the NFPA 1583 physical performance project was to develop a document that would provide general guidelines for the development of a physical performance assessment. Since it is virtually impossible and perhaps undesirable to develop a single test that would apply to all departments, the consensus of the task group was to provide categories in a menu format, from which tasks might be selected. The document was somewhat successful in accomplishing that objective, but it was more successful in producing controversy.

It is unlikely that police officers and firefighters could come to an agreement on a single test format. However, it is safe to assume that both agencies could arrive at some agreement as to the categories of essential tasks. For example, all fire departments use hose. Thus, a hose task would be appropriate for inclusion in all fire department test batteries. The specific task, however, may include hoses of different sizes. They may be charged or uncharged, or advanced up a stairway, over a hill, or down a street. The specifics would be left to the test developer, based on the individual job-task analysis. Consider the following task categories to be proposals.

FIREFIGHTER TASK CATEGORIES

Victim rescue: Rescue is the number one priority at any incident. There are few who would dispute that this task is essential and critical.

Hose advance: This is another obvious task to incorporate in a test battery. The hose is the basic tool of firefighting, used in both structure and wildland operations. The type and size, as well as whether the hose should be charged or uncharged, offer a wide variety of options. The exact format of the test is best left to the discretion of individual departments.

Forcible entry and ventilation: Techniques used to enter buildings, as well as to open roofs, walls, and ceilings, entail the use of lightweight hand tools, such as the ax, pike pole, halligan, pry bar, sledgehammer, and so on. A host of fundamental tasks are used to perform ventilation, overhaul, and forcible entry. No specific tool may be regarded as absolutely essential, since there are alternatives in many situations. Still, the evaluation of a member applying adequate force and performing multiple repetitions can represent a group of tasks deemed critical and essential in a given application.

Stairclimb/ladder climb: This is another common event that encompasses a variety of activities, whether carrying tools and equipment up or down a stairway, ladder, or grade. In most test batteries, the evolution is performed as a stairclimb event, representing a fire attack that involves connecting to a standpipe.

Equipment carry: This event can involve a wide variety of tools in scenarios that may include climbing or maneuvering over some sort of obstacle. Tools such as smoke ejectors, ground ladders, hose packs, and extinguishers are routinely used in the initial attack, and the ability to carry them is a basic requirement. If a tool cannot be carried to the site, it cannot be used. The difference between this set of tools and the forcible entry group is weight. These tools range from 25 to more than 100 pounds, and it is expected that a firefighter be capable of handling such equipment alone.

Rope pull: Hoisting tasks are implicit with the use of ladders, hose, equipment, and even victims. Rope is a basic tool, and the ability to grip and pull a rope is essential to many critical operations.

POLICE OFFICER TASK CATEGORIES

Victim rescue: A police officer can be expected to perform a rescue involving a citizen or fellow officer. The victim rescue is intrinsic to this line of work just as it is for a firefighter.

Obstacle course: This event lends itself to an infinite number of configurations. It should reflect the typical local environment in which an officer would pursue a suspect on foot.

Run event: This is a fitness-type test that can qualify as a task for police officers. Since police officers are frequently required to run in chase scenarios, a run-type test is very job-specific and justified. The distance and time of the event would have to be determined by the job-task analysis. The run event could be included in the obstacle course, or it could be administered separately on a standard track for safety reasons.

Arrest control: This is a very basic task that is difficult to simulate. Since handling a suspect requires gripping the individual, a grip test would be appropriate. As with the run event, this task is another case in which a traditional fitness-type test can qualify as a task test. The ability to control a subject is highly dependent on having adequate strength to grip and control the individual. Although a prop can be devised, the best prop, aside from a live subject, is a handgrip dynamometer. The minimum standard requirement would have to be identified in the job-task analysis.

Wall climb: Every jurisdiction, whether urban or rural, has any number of obstructions that might have to be traversed during a chase. For example, if the typical residential neighborhood has six-foot privacy fences surrounding the yards, it's reasonable to expect a police officer to be able to jump or climb such a fence. Again, the specifics would be based on the local environment and determined in the job-task analysis. This event could also be included in the obstacle course or evaluated separately.

Stair climb: This is another task with a reasonable expectation. This event could either be included in the obstacle course or evaluated separately; it could incorporate ascents or descents or both. The actual height of the stairs should be determined in the job-task analysis.

Using these events alone or in combination can allow you to develop an unlimited number of test variations, each to suit the individual needs of any organization while maintaining some degree of standardization among them. This will help to reduce conflict and contradiction between test developers and departments, and hopefully exclude those obscure, questionable tasks that might unfairly discriminate and often lead to litigation.

TIME STANDARDS

Validity is the single most important factor related to the development of tests, and time is perhaps the most controversial aspect, since successful performance is based on completing a test within a predetermined time frame. Time standards are difficult to identify, and developers vary in the choice of methods to establish them for tests. The opponents of timed tests claim that faster times do not necessarily indicate better job performance, and they qualify this assertion by stating that other factors, such as skill, intelligence, and so on, contribute to the makings of a good firefighter or police officer. Yes, other factors are important considerations, but the function of a physical performance assessment is only to quantify a subject's capacity for physical work—nothing more, nothing less.

Fitness tests such as endurance runs, which are used to predict aerobic capacity, are based on a framework of time. The faster that you complete the run, the greater your aerobic work capacity. Note how many Olympic competitions are scored by time. The individual who can complete the most work in a given time has the greater physical capacity to perform that work, and will therefore excel in it.

Consider whether the element of time is important to someone watching his house burn down, or to the victim of a burglary, waiting for the police to arrive. Is time important to a fellow firefighter who has run out of air, or to a fire victim in a third-story window, waiting for you to raise a ladder? There is an undeniable, compelling relationship between fitness, time, and performance. The potential for loss of life in an emergency situation often hangs on one's capacity for physical work, which will determine both the time and the outcome.

Time is a necessary part of all physical performance tests. It is directly related to performance. It should, in fact be *the* criterion by which the score is assessed. There are a variety of methods available for determining test standards. A few of the most common are presented here.

Rank order: You can avoid having to determine an appropriate completion time for a given test simply by not setting any time limit at all. Use rank order for all who complete the test, ranking the fastest number one and so on. The strength of a rank-order system is that you'll be able to identify those who possess the greatest work capacity. It also provides an easy way to identify and reject the unqualified. One disadvantage of a rank-order system is that it creates a competitive environment, which may lead to safety issues. A rank-order system also tends to be disadvantageous to female candidates and therefore mandates validation.

Pass-fail: You must be able to identify a specific time if you are to use a pass-fail system. Frequently a statistical procedure is used to identify the cut point, such as setting it one standard deviation below the mean. You can derive the mean score from an incumbent test sample. One problem with a pass-fail method is in selecting a sample that is representative of the population to be tested. If the sampling is poor, whether too fast or too slow, then the resultant test time may be unrealistic. The advantage of a pass-fail system is that it removes the competitive aspect inherent to the rank-order system. Pass-fail testing also provides greater opportunity for candidates who might not fare so well by rank order.

Banding: A band is a group of scores that indicate the same general category of capability. Essentially, bands are not unlike the letter grades that we received in school, where a score of 90 through 100 is an A, 80 through 89 is a B, and so on. The assumption is that there is little difference between individuals within a given band. When the assigned grade is based on the average of several different tests, it allows individuals who are weak in one area to compensate in another, and thus to compete. By this method, there is no need to establish an overall cut point, since everyone who completes the test receives some points. The disadvantage is that poor performers have little chance to compensate, although it is possible, leading to another problem: that of selecting the unqualified. There is also a problem in setting the cut points for each band.

Break point: This method is based on a naturally occurring phenomenon exhibited by all tests. There appears to be a point on the time line of any given test where few people record scores. This natural break point implies that if one doesn't finish the test by a specific time, then he won't finish it at all. In the example shown in Figure 16-1, the break point was very close, within thirty seconds, of the established time standard determined by experts. The subject column at the far right represents those who failed to complete the test. One problem with this method is that low performers may be able to qualify.

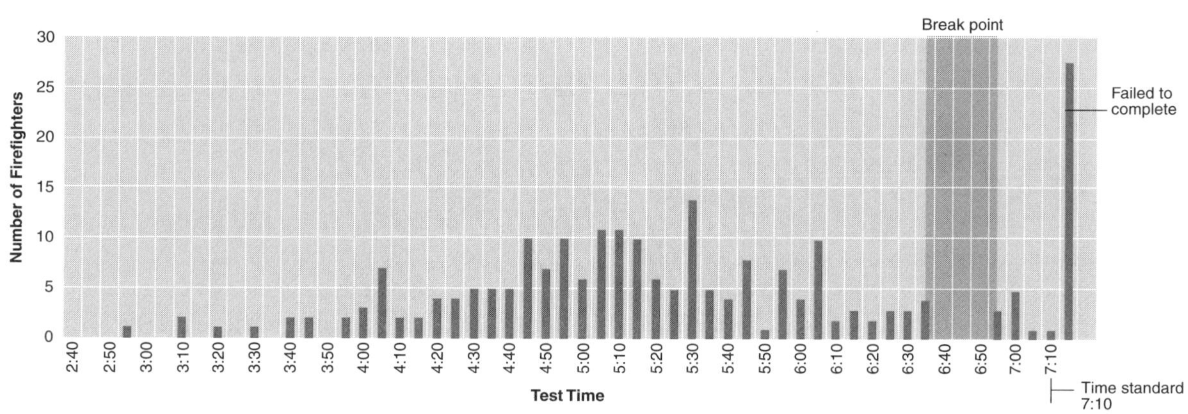

Fig. 16–1 The Natural Break Point on a CTTest

No time: By this method, there is no cut point, and all time requirements are removed. The criteria to qualify is to maintain forward progress and to complete the test. This method can be a reasonable approach if all candidates put forth an honest effort. The result would be similar to that of a natural break point. The effort problem becomes evident when a candidate attempts to conserve energy and consumes an undue length of time opening a hydrant—say, two minutes, which in the real world would be totally unacceptable. Another problem is that a slow-motion performance might actually prove detrimental, since some tasks, such as a hose drag, are best accomplished quickly to take advantage of the body's momentum. The credo of continuous forward progress also eliminates any rest periods for the applicant, another disadvantage. Since it is possible to assign completion times, there is no real reason to use this approach, which almost assuredly guarantees mediocrity.

Expert rating: Using qualified experts to establish a cut point by observing performance is a valid and reliable method that is gaining popularity. The method isn't new; rather, it has been practiced for many years in the realm of sports. Even in the Olympics, a panel of judges observe and grade performances in events such as diving, figure skating, boxing, and gymnastics. The scores are predicated on the expertise of the judges. The process may sound subjective, but it is quite effective. A correlation between a standard firefighter job performance rating and the elapsed time to complete a test battery was reported to be r = 0.949 (Dickinson, 1996).

TASKS

It is impractical to discuss all of the possible tasks that may be included in a test battery, so I have elected to present some key points related to task selection.

The list of tasks should be limited to those deemed essential or critical—critical in the sense that failure to perform it would compromise the overall operation or pose a direct threat to the health and safety of a member or citizen. Of relevant critical tasks, select the most arduous one. If a candidate can perform this task, you can safely assume that he can also perform those that require less effort.

The task that you select should require minimal skill to perform. Often tasks are challenged on the basis of the skills required. Unfortunately, the need for skill can only be reduced and never quite eliminated from any physical task. Even using a hammer requires knowledge of how to grip the handle and how to swing the head in an effective arc. If your intention is to put an ax, a saw, or a gun in an applicant's hand, no one who finds a simple hammer task to be complex could be an ideal candidate.

The task should be one that a single person can reasonably be expected to perform, even if it is something that is typically performed by a team. For example, two people are usually available to rescue a victim, since firefighters work in teams of two. Still, if one member of that team *is* the victim, then you would expect the other firefighter to be able to perform that task alone. Since the purpose of an entry-level test is to evaluate one person, teamwork cannot be permitted. Providing an assistant during the test to simulate a team will only corrupt the test.

Some test developers advocate linking all of the tasks into one continuous event, with a total time established for the test battery. Others prefer to separate the tasks and score them individually. Separating the tasks naturally introduces a rest period in between, and this presents administrative problems. Having conducted tests using this format, I have found control of the rest period difficult, since the candidates will have varying degrees of downtime. They tend to find ways to delay the proceedings, such as by complaining about problems with the equipment. Resting between the events also presents a problem of blood pooling in the extremities, which can lead to syncope, or fainting. Resting is contrary to the practices of exercise, in which a subject cools down appropriately after exertion. If a cumulative score is used, there is no reason to score the events separately, and the end result will be the same, based on total time.

The selection and arrangement of the tasks in a test battery may not reflect the actual job scenario. In firefighter tests, tasks that would typically be assigned to either an engine company or a truck company are combined into one event. Since a candidate may be assigned to either position, the essential tasks of both positions must be evaluated. If a department decides to hire for specific jobs, then a separate test could be developed for each position. The drawback to this is in the limitations of specialization. You may wind up with employees who cannot work in a variety of positions. Incumbents opposed to specific test arrangements state their preference for evolutions that would actually be performed on the job. Using standard operating procedures to select tasks is acceptable for an incumbent, but not for an entry-level candidate, who must display general suitability.

THE ENVIRONMENT

The environment in which a test is conducted has also been challenged, given that firefighters have to endure extremes of weather, smoke, flame, heat, reduced visibility, and other threats. Obviously, for health and safety reasons, realistic circumstances can't be imposed on candidates. A test performed in an ideal environment is to the candidate's advantage. Some elements of a real situation can, however, be incorporated into a test for incumbents, such as the use of live fires.

EQUIPMENT

Providing proper-size gloves, coats, helmets, and other apparel is a problem when conducting tests. Improper equipment and apparel can mean the difference between a pass and a fail. Still, using custom equipment or sizes to fit all potential candidates is impossible and unnecessary as long as a reasonable fit is available. You should always use standard equipment. Using outdated or nonissue equipment is unacceptable. Employing makeshift equipment is acceptable as long as it is compatible with a proper simulation of the task, both in terms of motion (biomechanics) and force (resistance). Makeshift equipment is used when the real item presents administrative or safety problems.

Personal protective equipment for firefighter candidates cannot be used in the same manner as for incumbents. An entry-level candidate may wear SCBA but not use it, since proper use requires training. The fitting of boots and PPE trousers is impractical. Instead, it is customary to add extra weight in the form of a vest or ankle weights. Personally, I do not recommend this practice of adding weight, since this introduces a variable that can negatively affect performance and safety. The preferred way to compensate for weight is to adjust the test time lower.

Both police and fire departments should incorporate a victim rescue in their test battery. The use of a rescue strap is often debated. If line personnel use a strap, it may be acceptable to use one in the test; however, it is probably better not to use it because of the element of skill that it requires. Without the strap, a partial lift of the mannequin is necessary, and this provides an opportunity to evaluate the candidate's back strength. Preemployment back evaluations are difficult, and this event serves well as an indirect evaluation tool.

ADMINISTERING TESTS

The most controversial issue surrounding the administration of tests centers on the decision to use a prescreen, or health risk appraisal. The Americans with Disabilities Act of 1990 (ADA) doesn't allow using a medical test prior to a job offer and therefore advises that the candidate present a medical release from a physician. Some legal advisors don't recommend this approach, even though it is contained within the law (AELE, 1997). The decision not to use a screen is also contrary to the standard of care as directed by the American College of Sports Medicine and the American Heart Association. The decision is a catch-22 situation, but I would recommend using a prescreen such as the PAR-Q, and if a physician referral is indicated, provide that individual with an exam prior to testing. The cost of one or two exams is the least expensive alternative and the most responsible approach.

A fair percentage of test challenges are the result of administrative problems. You should make every effort to ensure that all applicants are treated equally in every aspect. Provide specific instructions, and demonstrate or use an audiovisual aid to augment the instructions. The test area should be free of obstructions and distractions. Administer the test only in good weather. Use a standard wind-chill, heat, and humidity chart to make this determination. The question of precipitation should be obvious: None is the only acceptable amount.

RISK

The health risk of time-based testing is of concern to some. There will always be some health risk associated with physical testing, and this is one reason a health risk assessment is recommended. When testing potential public safety candidates, the test must be an accurate representation of the job demands. That means that it should include job tasks that reflect the physiological demands. If it doesn't, then you don't have a valid test instrument. Firefighting is a demanding occupation. It can elicit heart rates of 90 to 100 percent of one's estimated maximum heart rate. The test should induce the same response. If a candidate considers that to be too risky, then he is applying for the wrong job.

DISPARATE IMPACT

Develop and validate a quality test, and select those who qualify. These mandates sound simple, yet they present problems for small-statured applicants and raise the question of discrimination. Although, by definition, a test is designed to discriminate between those who can and cannot perform, the problem with physical tests is that women are at a distinct disadvantage when competing with men—a disadvantage that is principally related to size, though other gender factors contribute (McArdle et al., 1986; Shepard, 2000). The relationship between lean body weight and the performance times of a group (n = 6) of female firefighters on a criterion task test was r = 0.85, clearly revealing body size to be a valid predictor of completion times (LeCuyer, 1997).

Can women qualify and perform well in public safety occupations? Unequivocally, the answer is yes. The problem isn't one of qualification; rather, it is an issue of numbers, since women constitute a small percentage of the membership of police and fire departments.

An extensive test development and validation project conducted by the California Department of Forestry (CDF) in 1987 concluded that a grip strength of 39 kg was the minimum acceptable level for a CDF firefighter (Terra, 1988). A woman with a grip strength of 39 kg would be classified as excellent according to one resource, while the same score would classify a male in a very poor category (Heyward, 1984). The difference between male and female becomes obvious, and so does the dilemma. If a female candidate expects to compete and qualify for a position that demands a grip score of 39 kg, she will have to come from the top 1 percent of all women. This obviously limits the number of women who are able to qualify. Still, it does not eliminate women (Misner et al., 1987). In a grip test of incumbent female firefighters (n = 10), the average strength measure was 41 kg, and the range was 39 to 55 kg. Obviously, those who have been employed come from that top 1 percent of women.

The question of numbers is perceived as a problem, particularly within the U.S. Department of Justice, but it shouldn't even be a consideration. Firefighting and police work aren't occupations that typically appeal to women. An adequate number of women could certainly qualify, based on my experience in testing female phys ed students, but few women are interested in climbing onto a roof and chopping holes with an ax. A greater number of women may have an interest in other public-safety positions, however, such as training, public education, emergency medicine, and so on. The problem isn't one of numbers or standards, but rather, the opportunity to move directly into the desired position.

A cable company that seeks to employ a course developer for its training division isn't going to require the new hire to dig trenches and pull cable for five years before he gets to write lesson plans. Rather, that employee will quickly be assigned to the position that he was hired to fill. By the same token, the cable company isn't likely to begin its search by canvassing its legions of veteran cable pullers for someone who can write. Clearly, the avenue of entry is at the crux of the issue, and the career paths of public safety occupations should be investigated (Baltic, 1996). For those who desire the ax job on the roof, hire the qualified and let the percentages turn out as they may. Maintain standards commensurate with the job, and qualified women will step up to the challenge. For those who seek other career goals, provide options so that a motivated candidate won't be faced with an lengthy apprenticeship in some unsuitable or unwanted position.

APPLICATION

A physical performance assessment is appropriate for the following.

1. As a tool for preemployment testing in the selection of firefighter candidates, per NFPA 1500, Section 8-2.3.
2. To serve as a work capacity test in an injury rehabilitation program so as to determine the return-to-work status of a member, per NFPA 1500, Section 8-7.2.
3. To assess the physical capability of incumbents with respect to NFPA 1500 standards, per NFPA 1500, Section 8-2.4.
4. To make employment decisions that comply with disability law, per (in the U.S.) the Americans With Disabilities Act (1990).

The physical performance test may be developed and implemented for entry-level selection at any time. However, it should not be administered to incumbents until the fitness program has been established. The fitness program will provide a method of addressing physical fitness issues that may become evident during the assessment. Since the physical performance assessment serves a purpose different from that of the physical fitness assessment, departments must pay close attention to the legal aspects separating the two.

COMPANY STANDARDS ALTERNATIVE

The following is adapted from LeCuyer, J.E. (1999). Company standards: an alternative to physical performance testing. *Fire Engineering,* February, pp. 89–90, 92–94.

Is the physical performance assessment (PPA) appropriate for incumbents? Are there alternatives that could better serve in place of the PPA? The IAFF/IAFC developed a physical ability test (CPAT) for candidates and advised that it not be used for incumbents (1999). They did not provide an alternative, however.

I would propose the following alternative. Use company standards in place of the PPA (the more appropriate term for PPA is criterion-task test, or CTTest). The concept of substituting company performance standard evolutions for a CTT is based on the need to evaluate members in a manner that more accurately reflects the job, is commensurate with individual assigned positions, and includes the CTTest essential tasks.

The CTTest is designed to evaluate firefighter candidates with regard to the essential job tasks. Since restrictions with preemployment tests necessarily exclude skill-dependent tasks, don't require full use of SCBA, and incorporate a sampling of tasks that are out of sequence or that overlap standard company assignments, the value of using this test protocol for incumbents is questionable. The task restrictions are necessary for entry-level testing but irrelevant for incumbent evaluation. Incumbents are expected to possess the skills required, and through the use of several company evolutions, all essential tasks can be assessed in a format that reflects the actual job. Therefore, substituting a standard company evolution that includes the essential tasks is logical and most appropriate.

The CTTest is developed by extracting essential tasks from standard evolutions. When using company standards, we are basically putting the essentials back into the evolutions to evaluate incumbents. You're basically reversing the process while still evaluating the essential tasks. The advantages of this concept are as follows.

1. Company standards match actual job tasks. This helps to resolve the complaint that "We don't do all those tasks at one time or in that sequence."
2. It serves a dual purpose. Since it is a physical performance assessment and a training exercise that meets training hour requirements, it eliminates the need for two separate training sessions.
3. It provides support for entry-level CTTest validity by including essential tasks with written documentation provided in SOPs.
4. It can help evaluate both teamwork and individual task assignments.
5. Specific job tasks are evaluated based on assignment. A truck officer who would never be required to man a hoseline wouldn't be evaluated on a hoseline operation, but he may be evaluated on another task commensurate with his assigned duties.
6. Skill is evaluated in addition to physical work capacity.
7. All tasks required for each evolution are evaluated, not just the essential ones.
8. Task assignments match the job descriptions.
9. It meets the legal requirements of evaluating incumbents on essential tasks.
10. It provides an effective way to evaluate members (see Appendix E).

The disadvantages of this concept are as follows.

1. It requires the performance of multiple evolutions to encompass all of the essential tasks.
2. Some members may not perform all of the tasks. Although unnecessary, members could be rotated to ensure that most tasks are completed by everyone.
3. This method requires multiple evaluators to observe all of the company members performing simultaneously. This can be resolved by videotaping the participants, then evaluating them later.
4. It requires careful design, planning, and the development of company time standards.

A comparison between an entry-level CTTest and a company standards test is presented in Table 16-1.

TABLE 16-1

COMPARISON OF CTTEST AND COMPANY STANDARDS TEST

Entry-Level CTT	Company Standards Test for Incumbents
Nonskilled tasks.	Skill dependent.
Individual only.	Individual and team.
Essential tasks.	Essential tasks and marginal tasks.
Task sequence not job specific.	Tasks integrated into actual evolutions.
Tasks incorporated into one event.	Tasks incorporated into multiple evolutions.
Timed event.	Each evolution timed separately.
Scoring pass/fail or other.	Scoring pass/fail only.
Partial PPE worn.	Full PPE worn and used.
Simulated tools and props.	Actual tools used. May also use some props.

The company standards should reflect the actual job requirements, including staffing, the assignment of specific tasks, and the appropriate use of PPE. If a firefighter is routinely assigned to both engine and truck companies, he should be evaluated in both capacities. If a firefighter is assigned to one permanent position, the evaluation should be restricted to those parameters alone. Special assignments related to department operations, such as heavy rescue, haz mat, confined-space operations, SWAT teams, and the like, should incorporate essential tasks germane to those assignments that wouldn't ordinarily be required of new recruits and other members. If a given member is occasionally assigned alternate duties or is required to perform in a different capacity based on operational procedures, he should be evaluated in those alternative assignments.

The relationship between company standards and the CTTest. The development and validation of a CTTest should only be performed by professional test developers. During the job task analysis, department SOPs should be reviewed for identification of essential tasks. The procedure for determining these is illustrated in Figure 16-2.

In Figure 16-2, four sample company evolutions are presented, with five potential tasks identified for inclusion in the CTTest. The essential tasks can be pulled from any number of evolutions, and more than one task can be identified in any given evolution. In Company Standard Number 3, two tasks were extracted, and conceivably all of them could have been used. The second task in that evolution consists of carrying, raising, extending, and climbing a ladder. Since the ladder raise is a very skill-dependent task, it could not be selected for candidates. A portion of the task, however, could be used; i.e., the carry or the extension.

The CTTest is assembled from the essential tasks identified in the company standards. The tasks selected must meet the criteria for a CTTest battery.

1. Tasks that are critical and essential.
2. Tasks that require a minimal level of skill.
3. Tasks that a single firefighter can be expected to perform alone.
4. The most arduous tasks.
5. Tasks that match actual job performance requirements.

FIGURE 16-2

COMPANY STANDARDS AND CTTEST

Company Performance Standards	Physical Performance Standard
Engine Company, Standard No. 1: Hose Advance	**Criterion Task Test (CTTest)**
1. Engine lays in from hydrant.	
2. Water supply is established.	
3. Preconnect is advanced.	Task 1: Hose advance.
4. **Hoseline is charged for attack.**	
Engine Company, Standard No. 2: Standpipe	
1. Engine reverses from standpipe.	
2. **High-rise pack is carried to upper floor.**	Task 2: Stair climb.
3. Connections are made and water is supplied.	
4. Hose is charged and advanced on fire floor.	
Ladder Company, Standard No. 3: Rescue	
1. **Extension ladder is removed from truck.**	Task 3: Ladder carry.
2. Ladder is positioned and extended.	
3. Firefighter climbs ladder and enters structure.	
4. Floor is searched and victim is located.	
5. **Victim is carried or dragged to window.**	Task 4: Victim rescue.
6. Victim is rescued via ladder.	
Ladder Company, Standard No. 4: Ventilation	
1. Aerial ladder is spotted.	
2. Aerial is extended to roof of structure.	
3. Firefighter climbs aerial with full PPE and tools.	
4. **Firefighter ventilates the roof.**	Task 5: Simulated ventilation.

Since the CTTest is assembled from the company performance standards, the company standards can be used as a substitute for evaluating incumbents as long as all of the evolutions are performed. It is therefore unnecessary to administer the CTTest to incumbents on any given schedule, since the test elements are evaluated with company standards. The *Uniform Guidelines* do not specify any time frame for repeating a validation study, but they do provide a few comments that may indicate the need for a test and validation update.

1. Changes in equipment or tools, particularly with respect to weight.
2. Changes in staffing; i.e., apparatus assignment.
3. Changes in standard operating procedures that would alter the company standards.

4. Changes in member demographics that would alter the test validation sample.

If any of the above changes occur, the CTTest would have to be validated again, and members would be required to participate in the test sampling.

Evaluation. The most obvious question to ask when evaluating members using company standards in lieu of the CTTest is, "Can you effectively evaluate an individual?" The answer is yes. Using the subject matter expert, raters would employ a Likert-type scale to grade performance, as shown in Figure 16-3.

More specific language can be added to the scale for further clarification of point values.

The evaluation would require multiple graders. The resultant scores would then be averaged to arrive at a final grade. The team effort is graded on the basis of a time standard. Individual effort is evaluated by SME observation using the Likert-type scale shown above. I have employed this method of evaluation with recruit firefighters for years and have found it to be very effective and surprisingly accurate. The idea is quite simple. In any organization, if the members were assigned the task of ranking other members on job performance, three basic categories would evolve. We can all easily identify the top performers and the poor performers. For the majority of members who would fall between these two groups, there would be considerable variance. This variance for practical purposes is inconsequential. What's important are the groupings and, most importantly, the ability to identify the low performers.

I have employed this technique with recruit evaluations using officers who had neither personal experience with the recruits nor prior knowledge of their ability. The three groupings were easily identified, with the top and bottom performers appearing in almost perfect rank order. The key to the evaluation is using experienced raters, preferably training officers. However, any good officer with adequate experience will serve well.

Since one reason we're using the company standards test is to serve as a substitute for the CTTest, the low performers are the most important subjects to identify. Once identified, the proper action can be taken to modify performance.

A company is assigned the evolution and given the time standard for completion. The clock is started as soon as the evolution begins, and the raters observe both individual and group performance for proper procedures and the overall smoothness of the operation. Individual raters may be assigned one or two members to observe specifically. The individuals are then graded using a scale, and the performance time is recorded. A second evolution is assigned, and the raters are each given different members to observe and evaluate. The process continues for all subsequent evolutions until the predetermined battery has been completed. The use of video may expedite the process.

The performances are graded on a pass/fail system based on the time standard determined for each evolution. If the team doesn't meet the time standard, training is indicated. If unacceptable performance is attributed to one or more members, the individual problem should be identified. Note that an individual might score poorly even if the company meets the time standard. Individual scores should be averaged and the problem traced to one or more of the following.

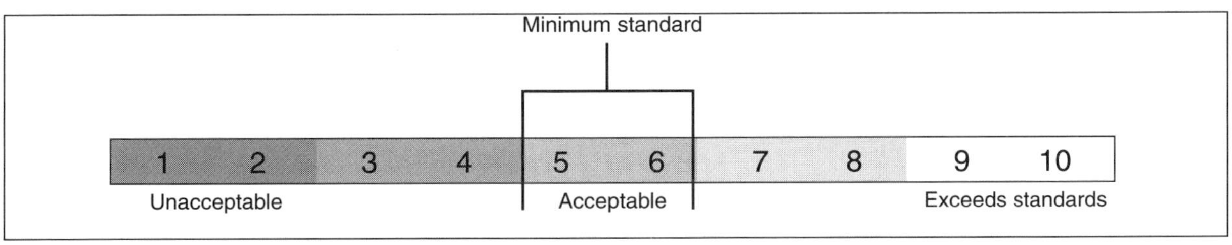

Fig. 16–3 Likert-Type Scale to Grade Performance

1. *A physical fitness or medical problem.* Offer medical referral when indicated. The physical component can be identified through comparison with the fitness profile, and exercise programming can be designed.
2. *Proficiency.* The remedy for a lack of skill could involve training, equipment modifications, or reassignment to another position. Individual training is the most common adjustment, since skills degrade rapidly in the context of less-busy companies and fewer training hours.
3. *Behavioral aspects.* When poor performance traces to behavioral causes, the employee assistance program may be the appropriate referral. However, disciplinary action may also be appropriate if the member simply refuses to perform.

Although this chapter has focused on fire department company standards, the basic concept applies to police departments as well. Law enforcement testing would involve tactical scenarios in lieu of company standards. Scenarios for incumbents could include emergency driving, arrest control, and weapons firing, for example, versus an entry-level test that relies on simulations. The tests should be tied to department SOPs and supported through actual incidents that officers have encountered in the line of duty.

RECOMMENDED READING

Fishkin, Gerald Loren (1988). *Police: Signs, Symptoms, and Solutions.* Los Angeles: Harcourt Brace Jovanovich, Legal and Professional Publications Inc.

Fishkin, Gerald (1989). *Firefighter and Paramedic BURNOUT.* Gardena, CA: Harcourt Brace Jovanovich, Legal and Professional Publications Inc.

Isernhagen, S.J. (1992) *Return-to-Work Testing: Functional Capacity and Work Capacity Evaluation.* Orthopaedic Physical Therapy Clinic, Industrial Physical Therapy. 1 (1).

Terra, Dale (1988). *Validation of Physical Ability Tests for Selection of Fire Fighters II and Fire Apparatus Engineers.* Criterion Validation Unit, California State Personnel Board. Sacramento, CA.

PART VI REFERENCES

AELE (1997). *Workshop on Discipline and Labor Problems.* Chicago, IL: Americans for Effective Law Enforcement Inc.; San Francisco Workshop, April 14–16.

American Heart Association (1998). *1998 Heart and Stroke Statistical Update.* Dallas, TX: American Heart Association 1997, p. 3.

Altug, Z.; Hoffman, J.L.; Slane, S.M.; Farabaugh, E.J.; Truschel, D.W.; and Bemesderfer, L.L. (1990) *Work-Circuit Training, Industrial Physical Therapy,* 10 (5): 41–48.

American Cancer Society. Guidelines on diet, nutrition, and cancer (1991). *Ca-A Cancer J Clin* 41 (6): 334–338.

Ardell, D. (1984) *The History and Future of Wellness.* Pleasant Hills, CA: Diablo Press.

Baltic, Scott (1996). All the learned and authentic fellows. *Fire Chief.* December, pp. 24–27.

Block, G.; Patterson B.; Subar, A. (1992) Fruit, vegetables, and cancer prevention: a review of epidemiological evidence. *Nutr Cancer* 18 (1): 29.

Bureau of Labor Statistics: *Reports on Employee Benefits in Medium and Large Private Establishments* (1995). Internet: http://stats.bls.gov/ebshome/htm.

Carafano, Peter (1990). Firefighter rehabilitation. *Fire Chief,* March.

Chaffin, D.B.; and Andersson, G.B.J. (1991) *Occupational Biomechanics, Second Edition.* John Wiley and Sons Inc.

Chaffin, D.B. (1992) *Occupational Health and Safety,* January, 61 (1).

Cooper, Kenneth (1982). *The Aerobics Program for Total Well-Being.* New York: Bantam, p. 11.

Cross, J.R.; and West, K.H., editors (1998). *The Source.* Sept/Oct. Manassas, VA: Infection Control/Engineering Concepts, p. 1.

Davis, P.O.; and LeCuyer, J.E. (1995) Fitness issues: performance and fitness are not the same. *Fire Chief.* August, pp. 20–23.

deVries, H. (1974) *Physiology of Exercise, Second Edition.* Dubuque, Iowa: Wm. C. Brown, pp. 228-229.

Dickinson, Arthur (1996). *Human Factors in Organizational Design and Management V: Combining Kinesiology, Aging Science and Research Design to Produce Valid and Defensible Job Analyses.* Proceedings of the Fifth International Symposium on Human Factors in Organizational Design and Management, held in Breckenridge, CO, July 31–August 3.

DOT (1993). *Selected Characteristics of Occupations Defined in the Revised DOT.* U.S. Department of Labor, Employment and Training Administration. Vol. 2, No. 1. Washington, DC: U.S. Government Printing Office.

Dreher, Henry (1988). *Your Defense Against Cancer: The Complete Guide to Cancer Prevention.* New York: Harper and Row Publishers.

Fairfax County Fire and Rescue Department (1993). *Wellness Survey.* Fairfax, VA. June 30.

Fisher, Thomas; and Reed, Deborah (1996). Lumbarsacral support belts: their role in back-injury prevention. *Professional Safety, American Society of Safety Engineers.* September, pp. 28–30.

Fishkin, Gerald Loren (1988). *Police Burnout: Signs, Symptoms, and Solutions.* Los Angeles: Harcourt Brace Jovanovich, Legal and Professional Publications Inc.

Fishkin, Gerald (1989). *Firefighter and Paramedic BURNOUT.* Gardena, CA: Harcourt Brace Jovanovich, Legal and Professional Publications Inc.

Fraser, T.M. (1992) *Fitness for Work.* Bristol, PA: Taylor and Francis Inc., pp. 39–44.

Gershoff, Stanley (1996). *The Tufts University Guide to Total Nutrition.* New York: Harper Perennial.

Healy, Thomas (1988). We're getting healthier. *Fire Command.* July, p. 11.

Heyward, Vivian (1984). *Designs for Fitness: A Guide to Physical Fitness Appraisal and Exercise Prescription.* New York: Macmillan Publishing Co., p. 75.

IAFF Department of Research, Health and Safety Division (1982). *Thermal Stress and the Firefighter.* ISBN 0-942920-04-X. Washington, DC: International Association of Fire Fighters.

IAFF (1999). *1998 Death and Injury Survey.* Washington DC: International Association of Fire Fighters.

Isernhagen, Dennis (1992). *The Continuum of Care.* Iglarsh and Richardson, editors, (p. 11). Orthopaedic Physical Therapy Clinic, Industrial Physical Therapy, 1 (1).

Isernhagen, S.J. (1992a) Functional capacity evaluation: rationale, procedure, utility of the kinesiophysical approach. *Journal of Occupational Rehabilitation,* 2 (3): 157–168.

Isernhagen, S. J. (1992b) *Return-to-Work Testing: Functional Capacity and Work Capacity Evaluation.* Iglarsh and Richardson, editors (p. 85). Orthopaedic Physical Therapy Clinic, Industrial Physical Therapy, 1 (1).

Isernhagen, S. J., guest editor (1992c). Iglarsh, Z.A.; and Richardson, J.K., consulting editors (pp.83–90). *Orthopaedic Physical Therapy Clinics of North America: Industrial Physical Therapy.* Philadelphia: W.B. Saunders Co., 1 (1).

Kipp, J.D.; and Loflin, M.E. (1996) *Emergency Incident Risk Management: A Safety and Health Perspective.* New York: Van Nostrand Reinhold.

Kopolow, Louis (1987). *Handling Stress.* National Institute of Mental Health, Plain-Talk Series. U.S. Department of Health and Human Services. Publication No. (ADM) 87–502. Rockville, MD.

Lafferty, Libby; and Dillard, Maureen (1985). Keep the fat out of the fire. *Fire Chief.* June, pp. 29–33.

Landy, Frank J. (1992) *Alternatives to Chronological Age in Determining Standards of Suitability for Public Safety Jobs, Volume I.* Technical report. Pennsylvania State University. January, p. 2–43.

LeCuyer, J.E. (1993) *A Bridge Between Industry and Rehabilitation.* Isernhagen Work Report. Fall, pp. 1and 4.

LeCuyer, J.E. (1994) Where tools and people meet. *Fire Chief.* February, pp. 34–39.

LeCuyer, J.E. (1997) The ins and outs of the return-to-work process. *Fire Chief.* February, Vol. 41, No. 2, pp. 64–68.

LeCuyer, J.E. (1999) Company standards: an alternative to physical performance testing. *Fire Engineering.* February, pp. 89–90, 92–94.

Levenstein, Mary K. (1992) *Effective Ways to Lower Your Odds of Getting Cancer: Everyday Cancer Risks and How to Avoid Them.* New York: Avery Publishing Group Inc.

McArdle, W.D.; Katch, F.I.; and Katch, V.L. (1986) *Exercise Physiology: Energy, Nutrition, and Human Performance, Second Edition. Philadelphia:* Lea and Febiger, p. 179.

Misner, J.E.; Plowman, S.A.; and Boileau, R. (1987) Performance differences between males and females on simulated fire fighter tasks. *Journal of Occupational Medicine.* 29: 801–805.

Morreale, S.J.; and Schwartz, N.E. (1995) Helping Americans eat right: developing practical and actionable public nutrition messages based on the ADA survey of American dietary habits. *J Am Diet Assoc.* 95: 305–308.

NFPA 1001, *Standard on Fire Fighter Professional Qualifications* (1997). Quincy, MA: National Fire Protection Association.

NFPA 1500, *Fire Department Occupational Safety and Health Program* (1997a). Quincy, MA: National Fire Protection Association, p. 22.

NFPA 1500, *Fire Department Occupational Safety and Health Program* (1997b). Quincy, MA: National Fire Protection Association. Appendix D-3.

NFPA 1561, *Standard on Fire Department Incident Management System* (1997). Quincy MA: National Fire Protection Association.

NFPA 1581, *Standard on Fire Department Infection Control Program* (1995). Quincy, MA: National Fire Protection Association.

NFPA 1582, *Medical Requirements for Fire Fighters* (1997). Quincy, MA: National Fire Protection Association.

NIOSH (1981). *A Work Practices Guide for Manual Lifting: Tech. Report No. 81–122.* U.S. Department of Health and Human Services. Cincinnati, OH: National Institute for Occupational Safety and Health.

NIOSH (1991). *Health Hazard Evaluation Report, HETA 90-395-2121.* Washington, DC: National Institute Occupational Safety and Health.

Ornish, D. (1982) *Stress, Diet, and Your Heart.* New York: Holt, Rinehart, and Winston.

Pagani, M.; Mazzuero, G.; and Ferrarri, A. (1991) Sympathovagal interaction during mental stress. *Circulation,* pp. 83: 43

Pritchard, R.E.; and Potter, G.C. (1990) *Fitness Inc.: A Guide to Corporate Health and Wellness Programs.* Homewood, IL: Dow-Jones Irwin.

Rasch, P.; and Burke, R. K. (1960) *Kinesiology and Applied Anatomy.* Philadelphia, PA: Lea and Febiger.

Rogers, Suzanne (1988). Job evaluation in worker fitness determination. *Occupational Medicine: State of the Art Reviews.* 3 (2): 219–237.

Ryden, L.; Molgoard, C.; and Bobbitt, S. (1988) Benefits of a back-care and light-duty health promotion program (in a) hospital setting. *Junior Community Health.* 13: 222–230.

Salvendy, Gauriel (1987). *Handbook of Human Factors.* New York: John Wiley and Sons.

Seaward, B. L. (1994) *Managing Stress.* Boston: Jones and Bartlett Publishers Inc., pp. 18–19.

Selye, H. (1956) *The Stress of Life.* New York: McGraw-Hill.

Shepard, R.J. (2000) Exercise and training in women, part I: influence of gender on exercise and training responses. *Canadian Journal of Applied Physiology.* 25 (1): 19–34.

Shrey, D.E.; and Mital, A. (1994) Disability management and the cardiac rehabilitation patient: job simulations and transitional work strategies. *Journal of Occupational Rehabilitation,* 4 (1): 39–52.

Sitterly, T.E. (1974) *Firefighter Skill Study.* Boeing Aerospace Company, Seattle, WA.

Skinner, James (1985). Coping with heat stress on the fireground: fighting the fire within. *Firehouse,* August.

Terra, Dale (1988). *Validation of Physical Ability Tests for Selection of Fire Fighters II and Fire Apparatus Engineers.* Criterion Validation Unit, California State Personnel Board. Sacramento, CA.

Title 29, Code of Federal Regulations, Part 1910.1030. *Occupational Exposure to Bloodborne Pathogens, Final Rule.* Federal Register, Vol. 56, No. 235, December 6, 1991.

United States Fire Administration (1991). *Stress Management: Model Program for Maintaining Firefighter Well-Being.* FA-100/February. Emmitsburg, MD: USFA.

U.S. Fire Administration (1992). *National Fire Academy: Guide to Developing and Managing an Emergency Service Infection Control Program.* Publication #FA-112. Emmitsburg MD: Publications Office, USFA.

U.S. Department of Health and Human Services, Public Health Service, Centers for Disease Control and Prevention(1989). *A Curriculum Guide for Public Safety and Emergency Response Workers, Prevention of Transmission of Human Immunodeficiency Virus and Hepatitis B Virus.* February. Washington DC: U.S. Government Printing Office.

U.S. Department of Labor, Bureau of Labor Statistics.(1997) *BLS Ready Facts.* Via fax-on-demand, (816) 426-3152.

Vincent, J. P.; and Knowles, L. (1979) Using performance standards to maintain manipulative skills. *Fire Command,* January, pp 14–15.

Wilkinson, W. J.; and Stelnicki, P.J. (1988) Substance abuse. *Chief Fire Executive.* August/September/October, pp. 34–37.

Williams, T.; and Evenson, S. (March, 1988) Physically fit for duty, part I: mandatory physical fitness programs. *Fire Chief,* pp. 43–46.

Yeung, A.; Vekshtein, V.; and Krantz, D. (1991) The effect of arteriosclerosis on the vasomotor response of coronary action in mental stress. *New England Journal of Medicine.* 325: 1551.

APPENDICES

Appendix A

A Basic Stretch Routine

The following illustrations depict a basic exercise routine using key stretches, progressing through lying, sitting, kneeling, and standing positions in a continuous fashion. It can be performed within about five minutes after a general-body warm-up and prior to and following a full workout. Additional stretching exercises can be added after the workout and performed for a longer duration. For an illustrated guide on the subject, consult Anderson, R. (1980) Stretching. Fullerton, CA: Shelter.

(0) **Start here**. Relax and stretch out in a supine position.

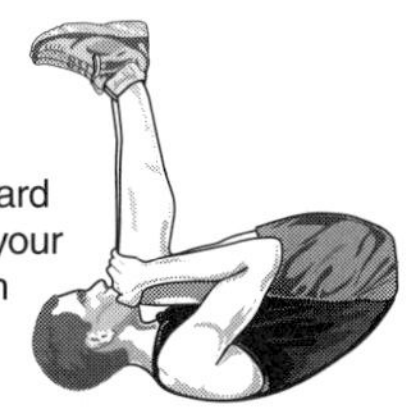

(1) **Back stretch**. Curl up, bringing both knees toward your chest. Rock. Keep your shoulders in contact with the floor.

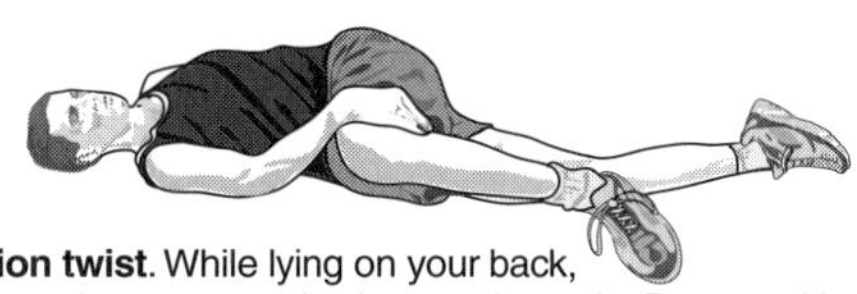

(3) **Rotation twist**. While lying on your back, rotate one leg over your body at a 90° angle. Repeat with the other leg. Keep your shoulders in contact with the floor.

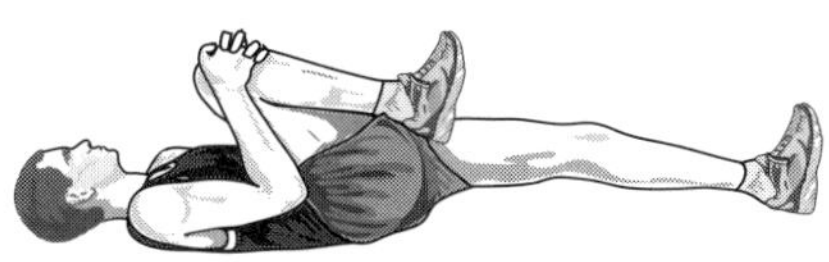

(2) **Lower-back stretch**. Pull one knee up to your chest and hold. Repeat with the other leg.

(4) **Groin stretch**. While in a seated position, pull both feet toward you and gently press down on your knees with your elbows.

(5) **Hamstring**. While seated, extend one leg and hold the other toward your groin. Reach for the extended leg and hold. Reverse leg positions and repeat.

(6) **Hip flexor**. Kneel as shown and push your hips forward while arching your back slightly. Reverse position and repeat with the other leg. You may perform this move from the calf-stretch position.

(7) **Calf stretch**. Stand as shown, keeping the heel of the extended leg in contact with the floor. Lean forward and bend your knee to stretch the calf and Achilles tendon. Repeat with the other leg.

(8) **Hamstring**. From a standing position, bend forward slowly and extend your arms to touch the floor. You may bend your knees slightly.

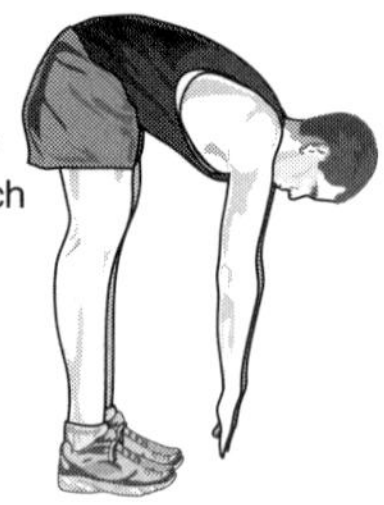

Rule: Stretch only what feels good for as long as it feels good.

You may perform additional stretches that work well for you personally, but always do the basic eight as described here.

On completion of the flexibility segment, proceed gradually into the aerobic phase of your workout. If you intend to run, walk or jog slowly before reaching your pace. If cycling, pedal at a low load for several minutes.

Appendix B

Health-Risk Appraisal and Preexercise Screen

HealthScreen is an easy-to-use software program that interprets ACSM guidelines and makes them easy to apply. It eliminates the need for other computer programs.

This program will tell you whether a person needs an exam before testing and whether a physician needs to be present. It will generate a report and even provide a letter for the attending physician. A sample printout appears below. The program is available through Human Kinetics Publishing. Reprinted with permission of the University of South Australia (c) 1999.

Preexercise Health Screening Report
John Dough

This is the report of your preexercise screening evaluation. The purpose of this screening is to assist you in safely undertaking an exercise program. The report summarizes your responses and leads to a recommendation about whether or not it is advisable for you to obtain medical clearance before proceeding with an exercise program. Your exercise specialist will help you interpret these recommendations and guide you in the next phase of your physical activity program.

Demographic Details
Subject name: John Dough
Gender: Male

Body mass (lbs.): 186.0
Height (in.): 70.0
Age: 31
Profession: Firefighter

Exercise
Exercise frequency: Four times per week
Exercise intensity: Moderate (e.g., walking, golf)
History of exercise pattern: 0–3 months
Type of exercise: Walk, job
Exercise duration (min.): 40
Intended intensity: Moderate

The recent U.S. Surgeon General's report on physical activity recommends that people try to accumulate at least thirty minutes of physical activity per day on most days of the week. Health benefits can be gained from a variety of activities, including brisk walking.

Medical Information
Do you have established, long-term diabetes?
Have you had a stroke?
Do you have heart disease?
Do you have bad asthma?
Do you believe you are pregnant?
Do you have any other diseases or conditions?

List any medications that you're taking.

The preceding questions are included to determine whether there are any known medical conditions that might cause problems when you begin or upgrade an exercise program, or when you undergo an exercise test. If you answered yes to any of these questions, it is recommended that, as part of the preexercise evaluation, you seek advice from your physician.

Signs and Symptoms
Do you experience chest pains?
Have you had episodes of shortness of breath?
Have you had episodes of severe dizziness?
Do you experience dyspnea (breathing difficulty)?
Do you experience ankle edema (swelling in and around your ankles)?
Have you ever had heart palpitations (unusual heart racing or pounding)?
Do you regularly get claudication (sharp pains in your legs)?
Has your doctor told you that you have a heart murmur?
Do you experience undue fatigue?

The preceding questions are included to determine whether there are any signs or symptoms of underlying cardiopulmonary disease, even if it hasn't been diagnosed. If you answered yes to any of these questions, it is recommend that, as part of the preexercise evaluation, you seek advice from your physician.

Cardiovascular Risk Factors	**Your Values**	**Risk Factor Limits**
Family member with cardiovascular disease?	No	
Smoking habit?	Yes, 40 cigarettes per day	
Systolic blood pressure (mmHg)*	130	140 or greater
Diastolic blood pressure (mmHg)*	80	90 or greater
Total cholesterol (mmol/L)*	190.00 (C)	Greater than 5.2
HDL cholesterol (mmol/L)*	33.00	Less than 0.9
Triglycerides (mmol/L)*	160.00	
Diabetes as a risk factor?	No	
Sedentary lifestyle?	No	
Premature menopause?	No	

* Default values based on population norms may have been substituted for missing data.

The preceding questions, together with the measurements taken, have been included to determine whether there are risk factors that might mean that you are predisposed to cardiovascular disease. Reducing these risk factors—for example, by incorporating regular physical activity and a healthy diet into your lifestyle—will help to reduce your risk of developing cardiovascular disease. In the case of HDL cholesterol, the higher the value, the better.

The risk-factor limits shown indicate the recommended cutoffs where the risk of developing cardiovascular disease significantly increases. Having risk factors may mean that you need to seek medical clearance. This depends on your age and exercise intentions. Advice about how to reduce risk factors is available from your health specialist or physician.

According to the ACSM guidelines, you are in the "apparently healthy" category.

Recommendations are as follows:
1. You do not need a medical check-up and exercise ECG before testing or exercise.
2. An exercise professional may conduct a maximal test.
3. An exercise professional may conduct submaximal testing.

Signature: ______________________________ Date: ______________

Witness: ______________________________ Date: ______________

PAR-Q & YOU
PHYSICAL ACTIVITY READINESS QUESTIONNAIRE

PAR-Q is designed to help you help yourself. Many health benefits are associated with regular exercise. Completing PAR-Q is a sensible first step to take if you are planning to increase the amount of physical activity in your life. For most people, physical activity should not pose any problem or hazard. PAR-Q has been designed to identify the small number of adults for whom physical activity might be inappropriate or those who should have medical advice concerning the type of activity most suitable for them. Common sense is your best guide in answering these few questions. Please read them carefully and circle **YES** or **NO** to each question.

YES **NO** 1. Has your physician ever said you have heart trouble?

YES **NO** 2. Do you frequently have pains in your heart and chest?

YES **NO** 3. Do you often feel faint or have spells of severe dizziness?

YES **NO** 4. Has a physician ever said your blood pressure was too high?

YES **NO** 5. Has your physician ever told you that you have a bone or joint problem such as arthritis that has been aggravated by exercise or might be made worse by exercise?

YES **NO** 6. Is there a good physical reason not mentioned here why you should not follow an Activity program even if you wanted to?

YES **NO** 7. Are you over age 65 and not accustomed to vigorous exercise?

IF YOU ANSWERED YES TO ONE OR MORE QUESTIONS:
If you have not recently done so, consult with your physician by telephone or in person **BEFORE** increasing your physical activity or taking a fitness test. Tell him or her what questions you answered YES on PAR-Q, or show your copy.

Program
After medical evaluation, seek advice from your physician as to your suitability for:
Unrestricted physical activity, probably on a gradually increasing basis.
Restricted or supervised activity to meet your specific needs, at least on an annual basis.
Check in your community for special programs or services.

IF YOU ANSWERED NO TO ALL QUESTIONS:
If you answered PAR-Q accurately, you have reasonable assurance that you are suited for:

A GRADUATED EXERCISE PROGRAM
A gradual increase in proper exercise promotes good fitness development while minimizing or eliminating discomfort.

AN EXERCISE TEST
Simple tests of fitness or more complex types may be undertaken if you so desire.

Postpone
If you have a temporary minor illness, such as a common cold, vigorous exercise or exercise testing should be postponed.

Physical Activity Readiness Medical Examination (revised 1995)

PARmed-X PHYSICAL ACTIVITY READINESS MEDICAL EXAMINATION

Following is a checklist of medical conditions for which a degree of precaution and/or special advice should be considered for those who answered "YES" to one or more questions on the PAR-Q, and people over the age of 69. Conditions are grouped by system. Three categories of precautions are provided. Comments under Advice are general, since details and alternatives require clinical judgement in each individual instance.

	Absolute Contraindications	Relative Contraindications	Special Prescriptive Conditions	ADVICE
	Permanent restriction or temporary restriction until condition is treated, stable, and/or past acute phase.	Highly variable. Value of exercise testing and/or program may exceed risk. Activity may be restricted. Desirable to maximize control of condition. Direct or indirect medical supervision of exercise program may be desirable.	Individualized prescriptive advice generally appropriate: • limitations imposed; and/or • special exercises prescribed. May require medical monitoring and/or initial supervision in exercise program.	
Cardiovascular	❑ aortic aneurysm (dissecting) ❑ aortic stenosis (severe) ❑ congestive heart failure ❑ crescendo angina ❑ myocardial infarction (acute) ❑ myocarditis (active or recent) ❑ pulmonary or systemic embolism—acute ❑ thrombophlebitis ❑ ventricular tachycardia and other dangerous dysrhythmias (e.g., multi-focal ventricular activity)	❑ aortic stenosis (moderate) ❑ subaortic stenosis (severe) ❑ marked cardiac enlargement ❑ supraventricular dysrhythmias (uncontrolled or high rate) ❑ ventricular ectopic activity (repetitive or frequent) ❑ ventricular aneurysm ❑ hypertension—untreated or uncontrolled severe (systemic or pulmonary) ❑ hypertrophic cardiomyopathy ❑ compensated congestive heart failure	❑ aortic (or pulmonary) stenosis—mild angina pectoris and other manifestations of coronary insufficiency (e.g., post-acute infarct) ❑ cyanotic heart disease ❑ shunts (intermittent or fixed) ❑ conduction disturbances • complete AV block • left BBB • Wolff-Parkinson-White syndrome ❑ dysrhythmias—controlled ❑ fixed rate pacemakers	• clinical exercise test may be warranted in selected cases, for specific determination of functional capacity and limitations and precautions (if any). • slow progression of exercise to levels based on test performance and individual tolerance. • consider individual need for initial conditioning program under medical supervision (indirect or direct).
			❑ intermittent claudication	progressive exercise to tolerance
			❑ hypertension: systolic 160-180; diastolic 105+	progressive exercise; care with medications (serum electrolytes; post-exercise syncope; etc.)
Infections	❑ acute infectious disease (regardless of etiology)	❑ subacute/chronic/recurrent infectious diseases (e.g., malaria, others)	❑ chronic infections ❑ HIV	variable as to condition
Metabolic		❑ uncontrolled metabolic disorders (diabetes mellitus, thyrotoxicosis, myxedema)	❑ renal, hepatic & other metabolic insufficiency	variable as to status
			❑ obesity ❑ single kidney	dietary moderation, and initial light exercises with slow progression (walking, swimming, cycling)
Pregnancy		❑ complicated pregnancy (e.g., toxemia, hemorrhage, incompetent cervix, etc.)	❑ advanced pregnancy (late 3rd trimester)	refer to the "PARmed-X for PREGNANCY"

References:

Arraix, G.A., Wigle, D.T., Mao, Y. (1992). Risk Assessment of Physical Activity and Physical Fitness in the Canada Health Survey Follow-Up Study. **J. Clin. Epidemiol.** 45:4 419-428.

Mottola, M., Wolfe, L.A. (1994). Active Living and Pregnancy, In: A. Quinney, L. Gauvin, T. Wall (eds.), **Toward Active Living: Proceedings of the International Conference on Physical Activity, Fitness and Health**. Champaign, IL: Human Kinetics.

PAR-Q Validation Report, British Columbia Ministry of Health, 1978.

Thomas, S., Reading, J., Shephard, R.J. (1992). Revision of the Physical Activity Readiness Questionnaire (PAR-Q). **Can. J. Spt. Sci.** 17:4 338-345.

The PAR-Q and PARmed-X were developed by the British Columbia Ministry of Health. They have been revised by an Expert Advisory Committee assembled by the Canadian Society for Exercise Physiology and the Fitness Program, Health Canada (1995).

You are encouraged to copy the PARmed-X, but only if you use the entire form

Disponible en français sous le titre *«Évaluation médicale de l'aptitude à l'activité physique (X-AAP)»*.

Physical Activity Readiness
Medical Examination
(revised 1995)

	Special Prescriptive Conditions	ADVICE
Lung	❑ chronic pulmonary disorders	special relaxation and breathing exercises
	❑ obstructive lung disease ❑ asthma	breath control during endurance exercises to tolerance; avoid polluted air
	❑ exercise-induced bronchospasm	avoid hyperventilation during exercise; avoid extremely cold conditions; warm up adequately; utilize appropriate medication.
Musculoskeletal	❑ low back conditions (pathological, functional)	avoid or minimize exercise that precipitates or exasperates e.g., forced extreme flexion, extension, and violent twisting; correct posture, proper back exercises
	❑ arthritis—acute (infective, rheumatoid; gout)	treatment, plus judicious blend of rest, splinting and gentle movement
	❑ arthritis—subacute	progressive increase of active exercise therapy
	❑ arthritis—chronic (osteoarthritis and above conditions)	maintenance of mobility and strength; non-weightbearing exercises to minimize joint trauma (e.g., cycling, aquatic activity, etc.)
	❑ orthopaedic	highly variable and individualized
	❑ hernia	minimize straining and isometrics; stregthen abdominal muscles
CNS	❑ convulsive disorder not completely controlled by medication	minimize or avoid exercise in hazardous environments and/or exercising alone (e.g., swimming, mountainclimbing, etc.)
	❑ recent concussion	thorough examination if history of two concussions; review for discontinuation of contact sport if three concussions, depending on duration of unconsciousness, retrograde amnesia, persistent headaches, and other objective evidence of cerebral damage
Blood	❑ anemia—severe (< 10 Gm/dl) ❑ electrolyte disturbances	control preferred; exercise as tolerated
Medications	❑ antianginal ❑ antiarrhythmic ❑ antihypertensive ❑ anticonvulsant ❑ beta-blockers ❑ digitalis preparations ❑ diuretics ❑ ganglionic blockers ❑ others	NOTE: consider underlying condition. Potential for: exertional syncope, electrolyte imbalance, bradycardia, dysrhythmias, impaired coordination and reaction time, heat intolerance. May alter resting and exercise ECG's and exercise test performance.
Other	❑ post-exercise syncope	moderate program
	❑ heat intolerance	prolong cool-down with light activities; avoid exercise in extreme heat
	❑ temporary minor illness	postpone until recovered
	❑ cancer	if potential metastases, test by cycle ergometry, consider non-weight bearing exercises; exercise at lower end of prescriptive range (40-65% of heart rate reserve), depending on condition and recent treatment (radiation, chemotherapy); monitor hemoglobin and lymphocyte counts; add dynamic lifting exercise to strengthen muscles, using machines rather than weights.

**Refer to special publications for elaboration as required*

The following companion forms are available by contacting the Canadian Society for Exercise Physiology (address below):

The **Physical Activity Readiness Questionnaire (PAR-Q)** - a questionnaire for people aged 15-69 to complete before becoming much more physically active.

The **Physical Activity Readiness Medical Examination for Pregnancy (PARmed-X for PREGNANCY)** - to be used by physicians with pregnant patients who wish to become more physically active.

To order multiple printed copies of the PARmed-X and/or any of the companion forms (for a nominal charge), please contact the:

Canadian Society for Exercise Physiology
1600 James Naismith Dr., Suite 311
Gloucester, Ontario CANADA K1B 5N4
Tel. (613) 748-5768 FAX: (613) 748-5763

Note to physical activity professionals...

It is a prudent practice to retain the completed Physical Activity Readiness Conveyance/Referral Form in the participant's file.

Source: PARmed-X, 1995. Reprinted with permission of the Canadian Society for Exercise Physiology.

Physical Activity Readiness Conveyance/Referral Form

Based upon a current review of the health status of ______________________, I recommend:

- ❑ No physical activity
- ❑ Only a medically-supervised exercise program until further medical clearance
- ❑ Progressive physical activity
  - ❑ with avoidance of: ______________________
  - ❑ with inclusion of: ______________________
  - ❑ with Physical Therapy: ______________________
- ❑ Unrestricted physical activity — start slowly and build up gradually

Further Information:
- ❑ Attached
- ❑ To be forwarded
- ❑ Available on request

Physician/clinic stamp:

______________________ M.D.

______________________ 19____

(date)

RISKO Michigan Heart Association
With modifications by ARA Human Factors, reproduced with permission
From Health Metrics

Scoring:
6 - 11 - Risk well below average
12 - 17 - Risk below average
18 - 24 - Risk generally average
25 - 31 - Risk moderate
32 - 40 - Risk greater than average--dangerous.
41 - 62 - Danger! Urgent! See your physician immediately.

AGE	10 to 20 1	21 to 30 2	31 to 40 3	41 to 50 4	51 to 60 6	61 and over 8
HEREDITY	No known history of heart disease 1	1 relative with cardiovascular disease over 60 2	2 relatives with cardiovascular disease, over 60 3	1 relative with cardiovascular disease under 60 4	2 relatives with cardiovascular disease, under 60 6	3 relatives with cardiovascular disease, under 60 7
WEIGHT (% Fat)	Males: < 16% Females: <20% 0	Males: <20% Females: <24% 1	Males: < 24% Females: <28% 2	Males: < 28% Females: <32% 3	Males: < 32% Females: <36% 5	Males: ≥ 32% Females:≥36% 7
TOBACCO SMOKING	Non-user 0	Cigar and/or pipe 1	10 cigarettes or less a day 2	20 cigarettes a day 4	30 cigarettes a day 6	40 cigarettes a day or more 10
EXERCISE	Intensive occupational and recreational exertion 1	Moderate occupational and recreational exertion 2	Sedentary work and intense recreational exertion 3	Sedentary work and moderate recreational exertion 5	Sedentary work and light recreational exertion 6	Complete lack of all exertion 8
CHOL-ESTEROL AND/OR % FAT IN DIET	Cholesterol below 180 mg%. Di ɪt contains no animal or solid fats. 1	Cholesterol 181-205 mg%. Diet contains 10% animal or solid fats. 2	Cholesterol 206-230 mg%. Diet contains 20% animal or solid fats. 3	Cholesterol 231-255 mg%. Diet contains 30% animal or solid fats. 4	Cholesterol 256-280 mg%. Diet contains 40% animal or solid fats. 5	Cholesterol 281-300 mg%. Diet contains 50% animal or solid fats. 7
BLOOD PRESSURE	100 upper reading 1	120 upper reading 2	140 upper reading 3	160 upper reading 4	180 upper reading 6	200 or over upper reading 8
SEX	Female under 40 1	Female 40 - 50 2	Female over 50 3	Male 5		

Source: ARA Human Factors–Fitness Coordinator Manual (12th ed.). Reprinted with permission of Health Metrics.

APPENDIX C

STANDARD TABLES

A sample of tests and standards using normative data is provided on the following pages. The norm tables can be used as standards for those selecting the basic-level approach (LOP-1) to standards, as described in Chapter 11. Norms for other tests can be found in *Advanced Fitness Assessment and Exercise Prescription*, Third Edition (Heyward, V.) and *Y's Way to Physical Fitness*, Third Edition (Golding et al.). The criterion-referenced standards in Tables C-5 through C-8 can be adopted as is or used as a guide for developing your own standards for those selecting LOP-2 or LOP-3 levels of standard development.

TABLE C-1
NORMS AND HEALTH-BENEFIT ZONES BY AGE GROUPS AND GENDER

Age 20-29

Test >	Grip Strength* (kg)		Push-Ups (#)		Trunk Fwd Flexion (cm)		Partial Curl-Up (#)		Vertical Jump (cm)	
Gender	**M**	**F**	**M**	**F**	**M**	**F**	**M**	**F**	**M**	**F**
Zone										
Excellent	≥124	≥71	≥36	≥30	≥40	≥41	≥25	≥25	≥56	≥40
Very Good	113–123	65–70	29–35	21–29	34–39	37–40	23–24	23–24	39–55	28–39
Good	106–112	61–64	22–28	15–20	30–33	33–36	21–22	19–22	30–38	20–27
Fair	97-105	55–60	17–21	10–14	25–29	28–32	13–20	13–18	21–29	15–19
Needs Improvement	≤96	≤54	≤16	≤9	≤24	≤27	≤12	≤12	≤20	≤14

*Combined right and left hand.

Age 30–39

Test >	Grip Strength* (kg)		Push-Ups (#)		Trunk Fwd Flexion (cm)		Partial Curl-Up (#)		Vertical Jump (cm)	
Gender	**M**	**F**	**M**	**F**	**M**	**F**	**M**	**F**	**M**	**F**
Zone										
Excellent	≥123	≥73	≥30	≥27	≥38	≥41	≥25	≥25	≥51	≥32
Very Good	113–122	66–72	22–29	20–26	33–37	36–40	23–24	22–24	37–50	24–31
Good	105–112	61–65	17–21	13–19	28–32	32–35	21–22	16–21	27–36	16–23
Fair	97–104	56–60	12–16	8–12	23–27	27–31	13–20	10–15	18–26	11–15
Needs Improvement	≤96	≤55	≤11	≤7	≤22	≤26	≤12	≤10	≤17	≤10

*Combined right and left hand.

Age Group (yrs) 40–49

Test >	Grip Strength* (kg)		Push-Ups (#)		Trunk Fwd Flexion (cm)		Partial Curl-Up (#)		Vertical Jump (cm)	
Gender	**M**	**F**	**M**	**F**	**M**	**F**	**M**	**F**	**M**	**F**
Zone										
Excellent	≥119	≥73	≥22	≥24	≥35	≥38	≥25	≥25	≥43	≥28
Very Good	110–118	65–72	17–21	15–23	29–34	34–37	22–24	21–24	31–42	19–27
Good	102–109	59–64	13–16	11–14	24–28	30–33	16–21	13–20	23–30	15–18
Fair	94–101	55–58	10–12	5–10	18–23	25–29	10–15	6–12	16–22	10–14
Needs Improvement	≤93	≤54	≤9	≤4	≤17	≤24	≤10	≤5	≤15	≤9

*Combined right and left hand.

Age Group (yrs) 50–59

Test >	Grip Strength* (kg)		Push-Ups (#)		Trunk Fwd Flexion (cm)		Partial Curl-Up (#)		Vertical Jump (cm)	
Gender	**M**	**F**	**M**	**F**	**M**	**F**	**M**	**F**	**M**	**F**
Zone										
Excellent	≥110	≥65	≥21	≥21	≥35	≥39	≥25	≥25	≥38	≥25
Very Good	102–109	59–64	13–20	11–20	28–34	33–38	20–24	16–24	29–37	17–23
Good	96–101	55–58	10–12	7–10	24–27	30–32	14–20	9–15	20–28	13–16
Fair	87–95	51–54	7–9	2–6	16–23	25–29	9–13	4–8	14–19	9–12
Needs Improvement	≤86	≤50	≤6	≤1	≤15	≤24	≤8	≤3	≤13	≤8

*Combined right and left hand.

Source: The Canadian Physical Activity, Fitness, and Lifestyle Appraisal: CSEP's Plan for Healthy Active Living, 1996. Reproduced by permission from the Canadian Society for Exercise Physiology. Note: Norms for age groups 15–19 and 60–69 are available but not shown, since they would not be applicable for public safety officers.

TABLE C-2

NORMATIVE DATA FOR SELECTED TESTS BY AGE GROUP AND GENDER

1. THE 1-RM BENCH PRESS, RELATIVE SCORE*

Age/Male Category	20–29	30–39	40–49	50–59
Superior	≥1.32	≥1.12	≥1.00	≥0.90
Excellent	1.15–1.31	0.99–1.11	0.89–0.99	0.80–0.89
Good	1.00–1.14	0.89–0.98	0.81–0.88	0.72–0.79
Fair	0.89–0.99	0.79–0.88	0.73–0.80	0.64–0.71
Poor	≤0.88	≤0.78	≤0.72	≤0.63

Age/Female Category	20–29	30–39	40–49	50–59
Superior	≥ 0.81	≥ 0.71	≥ 0.63	≥ 0.56
Excellent	0.71–0.80	0.61–0.70	0.55–0.62	0.49–0.55
Good	0.60–0.70	0.54–0.60	0.51–0.54	0.44–0.48
Fair	0.52–0.59	0.48–0.53	0.44–0.50	0.40–0.43
Poor	≤ 0.51	≤ 0.47	≤ 0.43	≤ 0.39

2. THE 1–RM LEG–PRESS, RELATIVE SCORE*

Age/Male Category	20–29	30–39	40–49	50–59
Superior	≥ 2.13	≥ 1.93	≥ 1.82	≥1.71
Excellent	1.98–2.12	1.78–1.92	1.69–1.81	1.59–1.70
Good	1.84–1.97	1.66–1.77	1.58–1.68	1.47–1.58
Fair	1.64–1.83	1.53–1.65	1.45–1.57	1.33–1.46
Poor	≤ 1.63	≤ 1.52	≤ 1.44	≤ 1.32

Age/Female Category	20–29	30–39	40–49	50–59
Superior	≥ 1.68	≥ 1.47	≥ 1.37	≥ 1.25
Excellent	1.51–1.67	1.34–1.46	1.24–1.36	1.11–1.24
Good	1.38–1.50	1.22–1.33	1.14–1.23	1.00–1.10
Fair	1.23–1.37	1.10–1.21	1.03–1.13	0.89–.099
Poor	≤ 1.22	≤ 1.09	≤ 1.02	≤ 0.88

TABLE C-2, CONTINUED

3. AEROBIC SCORES: VO_2 (ML • KG^{-1} • MIN^{-1})

Age/Male Category	20–29	30–39	40–49	50–59
Superior	≥ 49	≥ 48	≥ 45	≥ 42
Excellent	45–48	43–47	41–44	38–41
Good	42–44	40–42	38–40	35–37
Fair	38–41	36–39	34–37	31–34
Poor	≤ 37	≤ 35	≤ 33	≤ 30

Age/Female Category	20–29	30–39	40–49	50–59
Superior	≥ 42	≥ 40	≥ 37	≥ 33
Excellent	38–41	36–39	33–36	30–32
Good	35–37	33–35	31–32	28–29
Fair	32–34	30–32	28–30	25–27
Poor	≤ 31	≤ 29	≤ 27	≤ 24

4. MUSCULAR ENDURANCE SCORES: PUSH–UPS

Age/Male Category	20–29	30–39	40–49	50–59
Superior	≥ 48	≥ 40	≥ 31	≥ 26
Excellent	38–47	31–39	25–30	20–25
Good	30–37	25–30	19–24	14–19
Fair	23–29	18–24	12–18	10–13
Poor	≤ 22	≤ 17	≤ 11	≤ 9

Age/Female Category	20–29	30–39	40–49	50–59
Superior	≥ 37	≥ 32	≥ 25	≥ 21
Excellent	31–36	25–31	19–24	18–20
Good	24–30	20–24	14–18	13–17
Fair	18–23	12–19	7–13	7–12
Poor	≤ 17	≤ 11	≤ 6	≤ 6

*Relative Score: Strength-to-body weight ratio

Source: The Physical Fitness Specialist Certification Manual, The Cooper Institute for Aerobics Research, 1997.

TABLE C-3

NORMATIVE DATA FOR SELECTED TESTS FROM THREE FIREFIGHTER TEST GROUPS

Test	Group 1, N=225 Mean	Group 1, N=225 Std.Dev.	Group 2, N=67 Mean	Group 2, N=67 Std.Dev.	Group-3, N=198 Mean	Group-3, N=198 Std.Dev.	Total N
1-RM Bench Press (VRM)*	1.32	.31	1.46	.34	0	0	292
1-RM Leg Press (VRM)*	2.27	.39	2.68	.56	0	0	292
Vo_2max	49	6.11	49	4.90	47.8	3.91	490
WD-Sit & Reach (inches)	20	2.82	19	3	20	4.20	490
Push-Ups (reps)	47	17	52	18	52	14	490
Sit-Ups (reps)	61	18	67	11	57	17	490

* VRM= Variable-resistance machine.

Source: Data gathered from the testing of 490 structure firefighters (1988-89).

TABLE C-4

NORMATIVE DATA ON SELECTED TESTS FOR POLICE OFFICERS

1. 1-RM Bench Press/Relative Score

Age/Male	20–29	30–39	40–50+
Level			
1	> 1.35	> 1.30	> 1.25
2	1.25–1.34	1.20–1.29	1.15–1.24
3	1.15–1.24	1.10–1.19	1.05–1.14
4	1.05–1.14	0.90–1.09	0.95–1.04
5	0.95–1.04	0.80–0.64	0.85–0.94

Age/Female	20–29	30–39	40–50+
Level			
1	> 0.88	> 0.85	> 0.81
2	0.81–0.87	0.78–0.84	0.75–0.80
3	0.74–0.80	0.72–0.77	0.68–0.74
4	0.68–0.73	0.65–0.71	0.62–0.67
5	0.61–0.67	0.58–0.64	0.55–0.61

TABLE C-4, CONTINUED

2. Sit-and-Reach/Flexibility (15" Zero Axis)

Age/Male	20–29	30–39	40–50+
Level			
1	>25	>24	>23
2	24	23	22
3	22–23	21–22	20–21
4	19–21	19–20	18–19
5	15–19	14–18	13–17

Age/Female	20–29	30–39	40–50+
Level			
1	>26	>25	>24
2	25	24	23
3	24	23	22
4	21–23	20–21	19–20
5	16–20	15–19	14–18

3. Aerobic: 1.5-Mile Run Time*

Age/Male	20–29	30–39	40–50+
Level			
1	< 9:00	< 10:00	< 11:00
2	09:55–09:01	10:55–10:01	11:55–11:01
3	10:50–09:56	11:50–10:56	12:50–11:56
4	11:45–10:51	12:45–11:51	13:45–12:51
5	12:40–11:46	13:40–12:46	14:40–13:46

Age/Female	20–29	30–39	40–50+
Level			
1	< 10:48	< 12:00	< 13:12
2	11:53–10:49	13:05–12:01	14:17–13:13
3	12:58–11:54	14:10–13:06	15:22–14:18
4	14:03–12:59	15:15–14:11	16:27–15:23
5	15:08–14:04	16:20–15:16	17:32–16:28

*Refer to Appendix Table J-1 for conversion of 1.5-mile run time to aerobic score (mL • kg^{-1} • min^{-1}).

Table C-4, Continued

4. Body Composition/Percent Body Fat

Age/Male	20–29	30–39	40–50+
Level			
1	< 7.0	< 9.0	< 11.0
2	09.2–07.1	11.2–09.1	13.2–11.1
3	11.4–09.3	13.4–11.3	15.4–13.3
4	13.6–11.5	15.6–13.5	17.6–15.5
5	15.8–13.7	17.8–15.7	19.8–17.7
6	18.0–15.9	20.0–17.9	22.0–19.9

Age/Female	20–29	30–39	40–50+
Level			
1	< 14.0	< 16.0	< 18.0
2	16.2–14.1	18.2–16.1	20.2–18.1
3	18.4–16.3	20.4–18.3	22.4–20.3
4	20.6–18.5	22.6–20.5	24.6–22.5
5	22.8–20.7	24.8–22.7	26.8–24.7
6	25.0–22.9	27.0–24.9	29.0–26.9

Source: *Physical Efficiency Battery* (manual), Federal Law Enforcement Training Center. Glynco, GA. 1988.

Using the data from Table C-3, Group 1 (n = 225), a sample Pompier scale as described in the text is constructed below. The standard deviation for aerobic (6.11) was rounded to 6, and the standard deviation for the sit and reach (2.82) was rounded up to 3 for practical purposes and used in the table. The strength ratios weren't rounded, since the standard deviations are smaller increments and easy to use.

Table C-5

Recommended Criterion Standards Using the Pompier Scale

Fitness Level	Aerobic	Bench Press*	Leg Press*	Flex	Sit-Ups	Push-Ups	Category
1	>56	>1.64	>2.80	>22	>67	>60	Superior
2	49–55	1.32–1.63	2.40–2.79	19–21	48–66	42–59	Excellent
3	42–48	1.00–1.31	2.00–2.39	15–18	30–47	25–41	Good (standard)
4	37–41	0.69–0.99	1.61–1.99	12–14	12–29	8–24	Fair
5	31–36	0.31–0.68	1.22–1.60	11–13	N/A	N/A	Poor

*Variable resistance machine.

TABLE C-6
SAMPLE POMPIER TABLE WITH SLIDING SCALE (CRITERION)

Age Group 40+ Fitness Level	Flex.	Vo_2	Bench press	Leg press	Sit-ups	Push-ups
1	>21	54	1.72	2.84	60	61
2	18	48	1.36	2.42	45	43
3	**15**	**42**	**1.00**	**2.00**	**30**	**25** (Criterion std)
4	12	36	0.64	1.58	15	7
5	9	30	0.28	1.16	0	0
	(<11)	(<29)	(<0.63)	(<1.57)	(<14)	(<6)**

Age Group 30–39 Fitness Level	Flex.	Vo_2	Bench press	Leg press	Sit-ups	Push-ups
1	>22	56	1.84	2.98	65	67
2	19	50	1.48	2.56	50	49
3	**16**	**44**	**1.12**	**2.14**	**35**	**31** (Criterion +1/3 std. dev.)
4	13	38	0.76	1.72	20	13
5	10	32	0.40	1.30	5	0
	(<12)	(<37)	(<0.75)	(<1.71)	(<19)	(<12)**

Age Group 20–29 Fitness Level	Flex.	Vo_2	Bench press	Leg press	Sit-ups	Push-ups
1	>23	58	1.96	2.92	70	73
2	20	52	1.60	2.50	55	55
3	**17**	**46**	**1.24**	**2.28**	**40**	**37** (Criterion +2/3 std. dev.)
4	14	40	0.88	2.04	25	19
5	11	34	0.52	1.80	10	0
	(<13)	(<39)	(<0.87)	(<1.79)	(<24)	(<18)**

* The scale shown was constructed using the method described in Chapter 11. The basic table places the criterion at Fitness Level 3 for the oldest age group, then the basic scale was "slipped upward" one-third of the standard deviation for each of the remaining age groups. The remaining tables were completed in the same manner as the basic table—by adding and subtracting from the criterion.

**The numbers in parentheses can be substituted for the Level-5 scores, since any score less than the Level-4 number would be classified as a Level 5.

The following test notation and the standard deviation for each test was used to develop the table. The data was derived from the testing of 245 firefighters.

Test	Std. dev.	Test	Std. dev.
Flex. = Flexibility, or sit-and-reach test.	3	Leg press = Leg press ratio.	0.42
Vo_2 = Aerobic score in mL • kg^{-1} • min^{-1}.	6	Sit-ups = Number of sit-ups.	15
Bench press = Bench press ratio.	0.36	Push-ups = Number of push-ups.	18

TABLE C-7

SUGGESTED CRITERION STANDARDS FOR SELECTED TEST MODES

Strength and Endurance Test Modes

Program: Calisthenic-No equipment

Test: Body weight (calisthenic type)

Standard	Time	Test
25 reps	60 sec.	Push-ups
30 reps	60 sec.	Sit-ups
45 reps	120 sec.	Sit-ups
20 reps	60 sec.	Chin-up (modified)
6 reps	No time	Chin-up (standard)
20 reps*	60 sec.	One-leg squat
45 reps	120 sec.	Squat thrusts
20 reps	60 sec.	Chair dips

*Combined score (10 reps each leg), 30 sec. Each 60-sec. total.

Test: Body weight (Isometric)

Standard/Time	Test	Body Part
30 seconds	45 degree sit-up and hold	Midsection
30 seconds	Flexed-arm hang (chin-up)	Upper body
60 seconds	Wall-sit, 90 degree	Lower body

Program: Calisthenic-test equipment

Maximum reps with % of body weight (weights)*

Standard	Time	% of Body Weight	Test
30 reps	60 sec.	50%	Bench press
45 reps	120 sec.	100%	Leg press
25 reps	60 sec.	25%	Bar curl
25 reps	60 sec.	50%	Lat pulldown
25 reps	60 sec.	40%	Seated military press

*Max-rep weighted tests using a percentage of one's body weight can be used in calisthenic-based programs where weight equipment is available and used for test purposes only.

Program: Weight-training equipment available for training and testing purposes

Test Weight Equipment (variable-resistance machine or free weights)

Test	Standard: 1RM	VRM	Free weights	VRM	Free weights
	Estimate 1 RM from reps	Relative score		Absolute score	
Bench press*	3–5 reps	1.00	0.70	145 lbs.	100 lbs.
Leg press*	5–10 reps	2.00	1.40	290 lbs.	200 lbs.
Lat pulldown	3–5 reps	0.95	N/A	95 lbs.	N/A
Bar curl	5–10 reps	0.45	0.45	65 lbs.	65 lbs.
Seated military	5–10 reps	0.50	0.40	75 lbs.	65 lbs.
Upright row	5–10 reps	0.40	0.40	60 lbs.	60 lbs.

*See Table C-8 for a standard-table example.

TABLE C-8

SAMPLE STANDARD CRITERION TABLE FOR BENCH PRESS AND LEG PRESS AT VARIOUS BODY WEIGHTS

Free Weights	Relative (ratio)		Absolute (lbs.)	
Test>	**Bench press**	**Leg press**	**Bench press**	**Leg press**
Body weight	0.70a	1.40	100	200
≤145d	100*	200*	100	200
160	112	224	100	200
190	133	266	100	200
>225	158*	315	100	200

VRMachine	Relative (ratio)		Absolute (lbs.)	
Test>	**Bench press**	**Leg press**	**Bench press**	**Leg press**
Body-Weight*	1.00b	2.00c	145	290
≤145d	145	290	145	290
160	160	320	145	290
190	190	380	145	290
225	225	450	145	290

Note: Table selection is based on the type of equipment used: free weights vs. variable-resistance machine.
As shown, relative scores vary with individual body size, and the absolute scores remain the same for all. If desired, both tables can be used with an *either/or* option, by which an individual must meet standards at the relative or absolute level. In this manner, the discriminatory influence of body weight is reduced without compromising the minimum requirement.

*Weight rounded off.

A conversion factor is applied to the variable-resistance machine to provide the equivalent free-weight lift on a VRM. The factor varies with equipment, and the table should be adjusted accordingly. In this example, a 145-lb. bench press on a VRM (.70 factor) equates to a 100-lb. free-weight lift.

The criterion standard of 1.00 [VRM], 100 lbs.[FW] for the bench press is based on the DOT classification for very heavy work: a minimum of ≥100 lbs. absolute and the expectation that a public safety officer should be able to handle his own body weight. (The relative factor is 1.0, since it is unlikely that a public safety officer would weigh less than 100 lbs.)

The 2.0 leg-press ratio (VRM); 200 lbs. (FW) is the expected leg press comparative to the 1.0 bench press (Heyward, 1984, p. 77).

Given that public safety officers wear protective equipment in their normal duties, an additional weight is added to the 100-lb. requirement to establish the minimum body weight on the table. Firefighters wear approximately 45 lbs. (Police officers carry about 25 lbs.) Thus, the 145-lb. starting weight sets the minimum standard for anyone weighing 145 lbs. or less.

APPENDIX D

EXERCISE RECOMMENDATION FORMS

The three basic forms needed for exercise recommendation (EXRX) at the very basic level are provided in this appendix. The first, the Standard Score Card, is applicable for those who opt to use criterion standards. This form is appropriate for use with either the T scale or the Pompier method. Instructions for completing it are provided in Chapter 11. The EXRX card is standard and complete. The last card, entitled Fitness Profile and Exercise Recommendation, is basic and can be adapted by the user. Together, these three cards complete the EXRX process at Level 1. The procedure is simple: (1) Use your completed table to identify a participant's score, (2) refer to the EXRX card for the corresponding column and recommendation, and (3) transfer the information to the fitness profile for use by the participant.

Standard Score Card

Fitness level	Test 1 (Aerobic)	Test 2 (Strength 1)	Test 3 (Strength 2)	Test 4 (Strength 3)	Test 5 (Endurance)	Test 6 (Flexion)	Test 7 (Body comp)	Points	Total
1									
2									
3									
4									
5									

Total points	
Average	
Fitness level	

Fig. D–1 Standard Score Card The standard score card can be completed using methods such as the Pompier scale.

Fitness Level	Aerobic	Strength and Endurance		Flexibility
		Weights	*Body weight*	
1	85-95% MPHR 30-45 minutes 3-5 days per week Mode: run, walk, etc.	80-100% 1RM Pyramid 8/6/4/2 4- to 6-day split Mode: free weight	80-100% reps 4-5 sets 3-4 days per week Mode: calisthenics	Below pain threshold One set, 10-12 stretches 10-45 seconds, 3-5 days Mode: static
2	75-85% MPHR 20-30 minutes 3-5 days per week	75-85% 1RM Pyramid 8/6/4/2 4-day split	75-85% reps 3-4 sets 3 days per week Mode: calisthenics	Below pain threshold One set, 10-12 stretches 10-45 seconds, 3-5 days Mode: static
3	60-70% MPHR 30-45 minutes 4-5 days per week	65-75% 1RM 3 sets, 8-12 reps 3 days per week Total body routine	65-75% reps 3 sets 3 days per week Mode: calisthenics	Below pain threshold One set, 10-12 stretches 30-60 seconds, 5-6 days Mode: static
4	55-65% MPHR 45-60 minutes 5-6 days per week	60-70% 1RM 2 sets, 10-15 reps 3 days per week	60-70% reps 2 sets 3 days per week Mode: calisthenics	Below pain threshold One set, 10-12 stretches 30-60 seconds, 5-6 days Mode: static
5	50-60% MPHR 45-60 minutes 5-6 days per week	50-60% 1RM 2 sets, 12-20 reps 3 days per week	50-60% reps 2 sets 3 days per week Mode: calisthenics	Below pain threshold One set, 10-12 stretches 40-60 seconds, 7 days Mode: static

Consider all percentages to be starting figures. Make adjustments based on individual response.

Fig. D–2 Exercise Recommendation

Fitness Profile and Exercise Recommendation

Profile	**Fitness level**	
Name: ______________________	**Test**	**Raw score**
Age:	Aerobic:	1.
Weight:	Strength:	1.
Training heart rate:		2.
Blood pressure		3.
RISKO	Flexibility:	1.
Medical exam: Y N	Percent body fat:	1.

Aerobic Recommendation
Frequency:
Intensity (heart-rate range):
Duration:
Mode:

Strength Recommendation
Frequency:
Intensity:
Duration:
Mode:

Flexibility Recommendation
Frequency:
Intensity:
Duration:
Mode:

Attach exercise log form.

Fitness coordinator ______________________
Date: ______________________
Retest date: ______________________

Fig. D–3 Fitness Profile and Exercise Recommendation

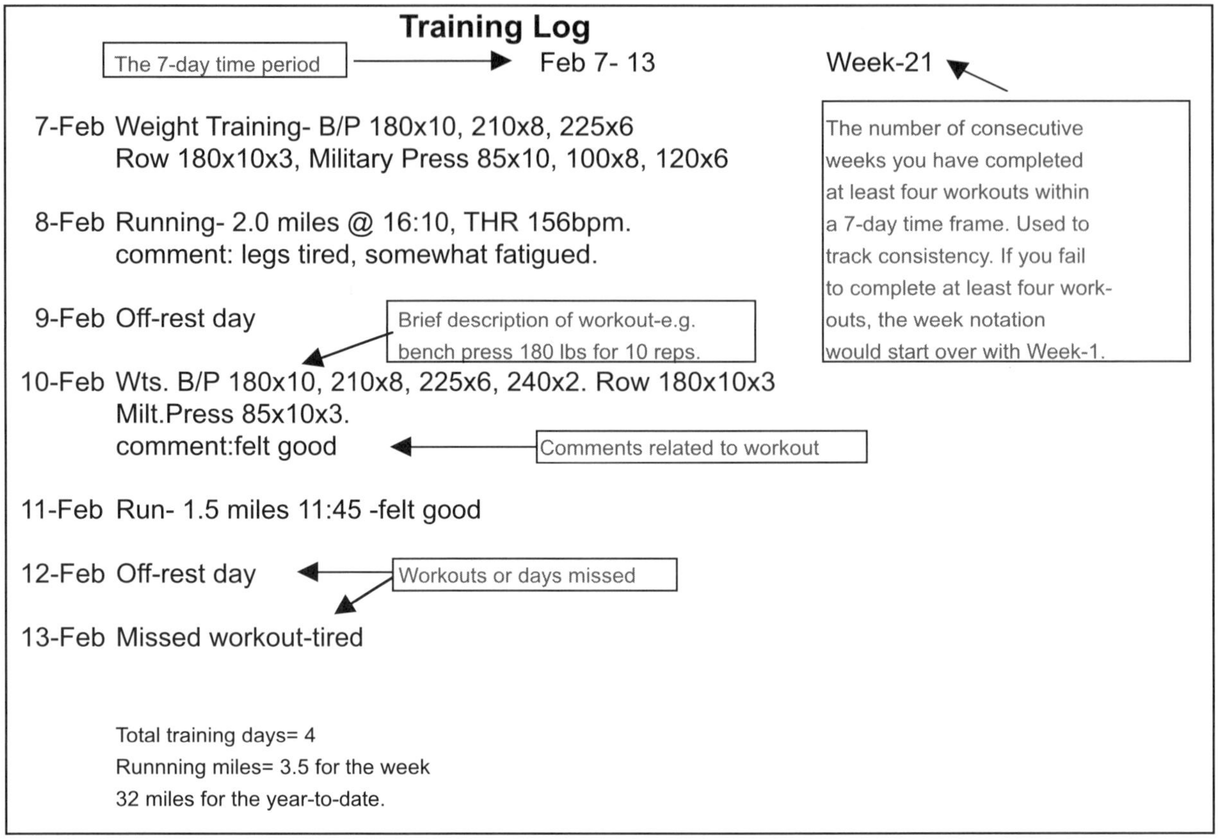

Fig. D–4 **A Simple One-Page Training Log** example that covers a 7-day time period. Enables one to track and compare workout weights, run, etc. at a glance from week to week, while monitoring program adherence by checking the number of consecutive weeks one has completed four workouts in a seven day time-frame.

APPENDIX E

FITNESS, PERFORMANCE, AND TESTING

Internal Report
Company Standards: Supplement to 1998 Training Report
Prepared by the author

Introduction

In 1988, company standards were introduced, and all companies were required to perform each evolution. Following the administration of that test process, the top-scoring companies (10 percent) were compared with the lowest performers (10 percent), using performance time as the criterion.

This informal analysis revealed that the companies housed at the busiest stations outperformed those assigned to the stations with the fewest alarms, and the individual team members' training reports from the top performers logged twice the total training hours as compared with those assigned to the slowest stations. The total training hours and the number of alarms appeared directly related to the performance time on the company standard evolutions.

In an effort to verify the relationship between training hours, alarms, and performance times, a similar analysis was conducted following the administration of company standards for 1998. The key differences between the '88 and '98 study are as follows.

1. A sample was selected in 1998 instead of the entire department. The primary reason was logistical. Additional items were evaluated that required inspection of approximately 160 hard-copy files. Investigating the entire department would have been impractical.
2. The determination as to what constitutes a busy company was based on the number of alarms, taken from monthly reports, as opposed to station assignments, as used in 1988.
3. Only fire alarms were counted in 1998, versus all alarms in 1988. Since the company standards are firefighting evolutions, it appeared logical to exclude EMS calls.
4. Only one evolution was selected in 1998, versus all evolutions in the previous analysis, and only ten companies (20 percent) again for logistical reasons related to the collection of data.

5. Dimensions that were not evaluated in 1988 but were added in 1998 are as follows.
    a. Physical fitness scores.
    b. Physical performance scores.
    c. Physical fitness hours.
    d. The total number of alarms.

Procedure

The evolutions selected initially were No. 9 (reverse lay from an elevated platform to a hydrant) and No. 4 (reverse lay from a gated wye to a hydrant). The No. 9 evolution reported a 50 percent failure rate and the No. 4 evolution a 48 percent failure rate. These were the two highest rates of failure. Since both evolutions are similar and only eight companies performed the No. 9, whereas twenty performed the No. 4, the latter was selected for analysis.

The twenty companies that performed No. 4 were then rank ordered, based on performance times. The top five and bottom five were selected for analysis. Individual team members (n = 40) were identified from the score sheets, and individual files were reviewed. Training hours were taken from the pretest month of July 1998. The data collected were as follows.

1. Total training hours.
2. Physical fitness hours.
3. Performance test scores (most current).
4. Fitness test scores (most current).
5. Fire alarms by company (July report).

The top-performing companies (n = 5) were designated Group A, and the lower-performing companies (n = 5) were designated Group B.

Results

Group A responded to more fire alarms and logged more training hours than Group B. The total training hours for Group A was 325, and the total for Group B was 251. Fire alarms for Group A were 214 versus 131 for Group B (Figure E-1).

The total fitness hours for both groups was almost identical. Group A had 251 hours, and Group B had 256 hours. Both training and fitness hours are presented in Figure E-2.

Although the fitness hours are comparable, the fitness scores are significantly different. Group A averaged a fitness score of 8.2 (Fitness Level 2), and Group B's average score was 7.0 (Fitness Level 3), the minimum standard for the department. The individual team members' performance scores were also inspected, and the results were similar to the fitness test. Group A averaged 4:33 on the performance test, with Group B a minute slower and below the department average of 5:36. The slowest performer in Group A was below average yet still outperformed 50 percent of the members in Group B. Both fitness and performance scores appear in Figure E-3. (Lower point = higher score.)

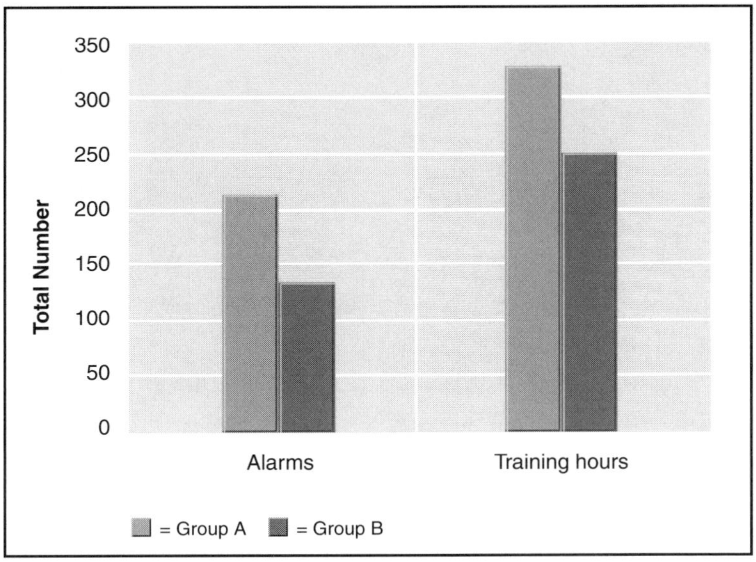

Fig. E–1 Training Hours and Fire Alarms

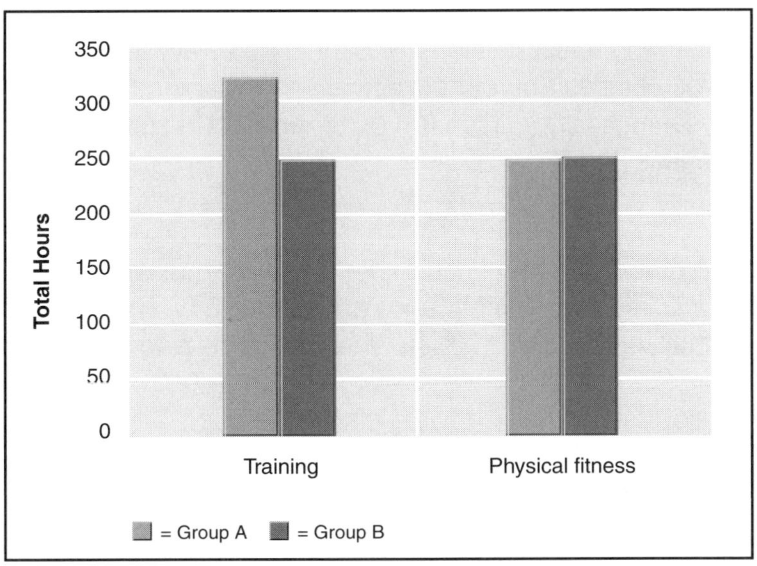

Fig. E–2 Training and Fitness Hours

Discussion

When comparing training hours and alarm responses, the results were the same as those observed in 1988: The busy companies train more than the slow companies. The obvious conclusion is that the more you run and train, the better you perform.

The most interesting observation not evaluated in 1998 was the relationship between fitness and performance. Although high-fit individuals appeared in both groups, the most fit (Fitness Level 1) were in Group A *only*. In fact, six of the twenty firefighters in Group A were at Fitness Level 1.

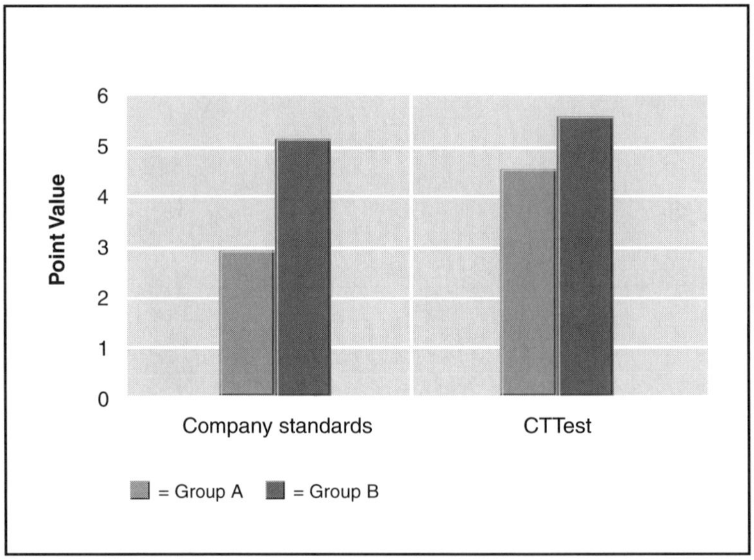

Fig. E–3 Comparison of Group A and Group B on Standards and Performance Tests

Although a cause-and-effect relationship cannot be established, the data does imply a connection between fitness and performance. An earlier study with one recruit academy produced a meaningful correlation (r = 0.78), enough to support the hypothesis that fitness is related to task performance time.

Conclusion

It can be stated that fitness is related to performance and, when combined with skills training, it can result in improved performance and productivity. The data from this analysis appear to support that conclusion. Therefore, I will make the following recommendations.

1. Identify the slowest company (fewest alarms) through the monthly run report, and schedule additional training to ensure that an acceptable skill level is maintained.
2. Recognize the fitness/performance relationship, and ensure that the members adhere to the department's mandatory fitness policy. Although the total hours logged were similar for the two groups, some members recorded zero fitness hours, offset only by a few high-level performers. Logged hours aren't necessarily evidence of meaningful participation, as seen with the group comparison.
3. Include the fitness test score in each individual's annual performance appraisal. Remember, the lower-fit group failed the company standard evolution. It recorded below-average scores in both the performance and the fitness tests. In spite of the group meeting the department standards, when all factors are considered and combined (fewer alarms and training hours with lower fitness scores), the result can lead to less than desirable performance, as demonstrated. Including the fitness score in the annual appraisal will help reinforce the importance of maintaining an acceptable level of fitness, and it will hopefully result in improved performance.

Internal Report
An Injury/Fitness Comparison
Prepared by the Author

The relationship between back injuries and fitness components, specifically flexibility, has been the subject of investigation for years. Researchers have reported conflicting opinions, and no definitive connection has been established. Several studies are currently in progress, but the methodology doesn't appear to be much different from that of previous inquiries. Therefore, the results aren't expected to shed any new light on the problem. A brief and informal investigation was performed for the purpose of identifying one potential approach that could hopefully lead to a more effective method for researchers.

Purpose

To determine whether a meaningful relationship exists between significant back injuries and individual fitness dimensions.

Method

Review injury reports and fitness profiles of the firefighters who have suffered back injuries. Record the data and make an informal observation to determine whether there is potential for further investigation using specific injuries and fitness data.

Procedure

Injuries that require surgery or that are classified as serious (defined as a cost in excess of $4,000) are identified and selected for review. The individual's fitness profile is then retrieved and reviewed.

Injury Group Data: The fitness profiles of the back-injured members were retrieved, and the scores of each component were recorded. A comparison of the selected member profiles (n = 10) revealed similarities with two fitness components: upper-body strength (bench-press score) and flexibility (sit-and-reach score). The results were as follows.

1. Bench press: Mean = 1.02 (minimum standard).
2. Flexibility: Mean = 16.3 (within the minimum-standard bracket).

Department Profile: Using the department's fitness score statistics as a baseline for investigation, the following data on the two components were evaluated.

1. Bench press: Mean = 1.32 (relative score), standard deviation = 0.30 (n = 225).
2. Flexibility: Mean = 20″, standard deviation = 2.8″ (n = 238).

Observation

The two components from the injured group were compared to department data.

1. Bench press: 90 percent of the scores were below the department average, and 30 percent were one standard deviation below the mean. Only one score (1.44) was above the mean of 1.32.
2. Flexibility: 100 percent of the scores were below the mean; 30 percent were one standard deviation below the mean, and another 30 percent were two standard deviations below.

Discussion

Since a relatively small number (n = 10) of members were included in the analysis, the relationship between back injury and the two fitness dimensions may be somewhat arbitrary or just coincidental. Since the injuries reviewed were incurred prior to implementation of the current record-analysis system, accurate records for workers' compensation costs and specific reports weren't available for all injuries, and the cause of injury was undetermined in some cases. Therefore, some members may have unintentionally been excluded. Other back injuries—those not meeting the criteria of surgery or costing more than $4,000—might have produced different results had they been included, since some were incurred by higher-fit individuals.

In spite of the limitations, there does appear to be some connection between flexibility and back injury, based on the cases reviewed. There also seems to be some relationship between upper-body strength and back injury—an unexpected pattern. Unfortunately, causal factors haven't been identified, and it is therefore difficult to extrapolate, particularly from such a small number of cases. However, when viewing some standard firefighter tasks, the potential for back injury as a result of low upper-body strength can be postulated. For example, a ladder raise that requires above-average upper-body strength, performed in a pressing motion similar to that of a bench-press exercise, could produce an arch in the lower back, and thus greater stress, if the subject's level of strength were inadequate. Similarly, lifting, dragging, or carrying a heavy object with a rounded back and extended arms could increase back stress if the firefighter had inadequate upper-body strength to pull the weight higher and hold the load in a more desirable position. In both cases, the back is compensating for a deficiency in upper-body strength, resulting in the use of poor body mechanics.

Conclusion

The observation is interesting and may warrant further investigation. If so, more accurate information with a detailed analysis of each injury will be necessary to derive some conclusive evidence. Since injury records seem to lump all types of back injuries into one general category, significant relationships may be difficult to discern. Not all back injuries are the same, and the data used in the analysis should distinguish between cases that are related to trauma, versus those related to overexertion, versus chronic problems reported as a result of an unrelated event. The preinjury fitness status of the injured person must also be available for comparison. A larger number of cases is also desirable, although uncontrollable. From a human perspective, they are undesirable. In subsequent investigations, a long time frame or access to good data will be necessary. Still, I believe that the evidence in this report is adequate to support the use of the following approach.

1. Investigate, record, and review all back injuries thoroughly.
2. Sort and categorize the types of back injuries, and include the causal factors.
3. Maintain up-to-date fitness data on all the members.
4. Keep accurate records on all injuries, including costs, lost workdays, and so on.
5. Select, review, and compare the data. If possible, use current available data.

Sample Test Validation Procedure
Letter to the Author From Arthur Dickinson, Ph.D., FACSM.

The following is an excerpt from a letter to the author, dated October 1998, from Dr. Arthur Dickinson, University of Colorado Human Performance and Applied Physiology Laboratory. The subject is the development of physical performance tests—an alternative method to that used by human factors specialists and ergonomists who present backgrounds in either mechanical engineering or psychometry.

... In both disciplines, it was vexing to observe that, in performing a job-task analysis, the human body—the essential reason for analysis and the most important variable—was considered as little more than a box displayed on the computer terminal screen.

The process of objectively determining job tasks by including the kinesiologist (human-movement scientist) in the analysis incorporates for the first time information relative to the ranges of physical abilities that employees possess. Whether the objective of the job analysis is to reduce work-related injury, establish minimum physical ability necessary to perform a job satisfactorily, or to change equipment or standard operating procedures, the extent to which the physical potential and limitations of the employees presently performing the work is known, the more precise and accurate the job analysis. The more physically demanding the work task—requiring work output closer to the employee's maximum ability—the more critical is an accurate assessment of work demands and the worker's maximum physiological capability.

Presently, electronic equipment is available to establish, for the employee of any job unit, both worker maximum capability and the job task's demands. The former measure—worker capability—has been evaluated in laboratory surroundings for some time. Aerobic capacity is found in treadmill or bicycle ergometry, where maximum oxygen uptake is measured by indirect calorimetry. The preferred measures of muscular strength/endurance is through four resistance exercises: bench press, bicep (preacher) curl, and latissimus pulldown for upper-body measures, and leg and hip extensions for the lower body, which correlates well with total lower-body strength.

Until recently, work demand has been difficult to accurately assess. Employee interviews, psychophysiological scales (perceived exertion), heart rate measures, and video film analysis were the principal means to secure useful information about the job's physical demands. However, there was no common denominator in measurement between worker capability and work demands. The development and validation of metabolic energy analyzers weighing about one kilogram and compact in size (5″ X 3″ X 8″) that use either telemetry or stored physiological data from work bouts ranging from 15 minutes to one hour in duration have been used in the CU lab for the past three years.

Employed in actual work or work-simulated conditions, energy costs for individuals to perform one or a series of work tasks can be accurately measured. Pace (rate of work), rest periods, differences in the equipment used or clothing worn can be nicely assessed. Differences in the work cost of any absolute task relative to body size, skill or experience level, age (though rarely a variable), gender (closely related to size), and individual strength and aerobic condition can be studied.

The jobs or professions where more physically demanding tasks are unalterable are served best by the analysis techniques described here. Firefighting and other emergency response work is an ideal example. Some of the observations made during the testing of firefighters from three metro Denver departments appear below.

1. Demonstrated the lack of importance of a high aerobic capacity to perform on a criterion-task test. Expressed in $mL \cdot kg^{-1} \cdot min^{-1}$, correlations of $r = 0.46$, $r = 0.35$, and $r=0.36$ were found.

2. The relationship between elapsed time to complete a criterion task test and annual job performance reports was $r = 0.95$ for the single department for which this measure was calculated.

3. The relationship between muscular strength and criterion task test performance time was $r = -0.85$, $r = -0.68$, and $r = -0.67$. The highest value was reported using one test measure: absolute bench-press score.

4. The correlation between criterion task test time and age was poor ($r = 0.25$), demonstrating that little need or concern exists for an aging member's ability to meet the standard.

Controlling the work environment is difficult to nearly impossible. The first ten minutes on the fireground or on the emergency scene often require a work effort at near-maximal level to accomplish the mission. Virtually all of the various tasks are critical in nature and, for most firefighters, the effort level is close to their maximum ability to produce work. Risks for musculoskeletal injury and cardiovascular insufficiency increase as the margin of safety narrows between the level of exertion and the ability to perform work. Employment of these techniques have resulted in procedures, protocols, test batteries, and job-specific fitness testing that is fair to employees, valid in practice, and legally defensible.

APPENDIX F

JOB DESCRIPTION FOR HEALTH AND SAFETY OFFICER

Position: Health and Safety Officer
Job Code: DOT-373.364-010 (fire); DOT-375.263-014 (police)
Department: Training

Authorization

Assigned and authorized by the fire chief/police chief to manage the department's health and safety program.

Primary Duties and Responsibilities

Serves as the department's health and safety officer; manages the health and fitness program in cooperation with the department's designated physician; develops, maintains, and continually revises physical conditioning and rehabilitation programs; works in liaison with physical therapists, occupational therapists, physicians, and safety specialists to provide a complete medical and physical management program; maintains a health and safety data file that includes occupational injuries, the results of fitness and performance tests, and the results of medical and physical exams; supervises certified fitness coordinators in the delivery of fitness programs; investigates occupational injuries and accidents in cooperation with the risk management office and medical providers; serves on the department health and safety committee; applies ergonomic concepts as an injury intervention tool; makes recommendations to reduce risk factors associated with department operations; identifies needs, criteria, standards, guidelines, and requirements for research matters relating to member health, safety, and emergency operations; writes standards, operating procedures, review, summaries, and reports related to health and safety; interacts with all department members through in-service training and personal counseling, which includes physical, medical, and psychological components of wellness and exercise programming.

Working Conditions

Physical demands: Light office work to heavy physical work. Includes essential tasks of the occupation as specified in department Physical Performance Assessment and department Physical Fitness Program Standards. DOT physical require code 6V-OOOOO-FFOO-FFN-FFF-OFF-ONFF-4NF-NO-FNFFF (fire); 6M-OOOOO-OOON-FFN-FFF-FFF-ONNN-3NN-NN-NNNNO (police).

Work environment: Includes office, training facilities, and emergency scenes; exposure to extreme weather conditions, hazardous environments (chemical and physical); degrees of visual impairment, auditory extremes, and psychological stress associated with traumatic events; secondary environment imposed by personal protective equipment (forty-five pounds for fire; twenty-five pounds for police).

Equipment used: Computers and ancillary equipment; medical and physical diagnostic equipment; exercise equipment; training aids and devices; personal protective equipment; departmental tools, vehicles, and apparatus.

Minimum Qualifications**

Education: Bachelor's degree in health-related field. Graduate level preferred. Graduation from department basic training program.

Experience: Five years' of service with the department, and five years' experience with a health and fitness program.

Knowledge: All phases of the specific principles and practices associated with management of an exercise program, exercise training, and recommendation of department operations; relevant laws, rules, and regulations; standards and standard operating procedures and guidelines, as defined by professional organizations; familiarity with the general principles of occupational and physical therapy; test development and validation; rehabilitation and wellness concepts; computer applications.

Abilities: Ability to establish and maintain an effective working relationship with members and professional affiliates; effective oral communication; strong writing skills; ability to work independently as well as within a team; self-motivated; organized; above-average leadership skills.

Licenses and Certifications Required

State-certified fire or police officer; emergency medical technician (EMT B); certified fitness coordinator; instructor.

Review and Approval
Department Director ______________________________
Date ____________
Internal Services Director ______________________________
Date ____________

This description includes the major duties and responsibilities of the job and is not to be construed as being all-inclusive of the occupational demands.

Appendix G

Fitness Coordinator Training Courses

The following seven certification courses can provide the proper training for a department fitness coordinator. For more specific course descriptions, requirements, schedules, and offerings, contact the individual organization directly. For more options, a directory of professional fitness certifications is available through AAHPERD at 1-800/321-0789.

Oklahoma State University Wellness Center
1415 West Hall of Fame
Stillwater, OK 74078-8041
Phone: 405/744-5727; 1-800/304-5727
Fax: 405/744-7377

Course: Fitness Instructor Training and Certification. This course is designed to provide participants with the knowledge and skills needed to become a fitness/health leader within their fire department, city, and community. Successful completion will qualify the participant to be certified by the Oklahoma State University Wellness Center. The program covers five areas of academic instruction: fitness, exercise physiology, biomechanics, nutrition, and stress management.

Audience: Firefighters who desire exercise leadership training and certification.

Prerequisite: Instructional techniques or an instructional methodology equivalent.

Certification: Successful completion requires attendance at all classes and functions. A score of 70 percent is required to pass all written and practical exams.

Cost: The five-day session is #395 for OSU students and $475 for out-of-state participants.

The Cooper Institute for Aerobics Research
12330 Preston Road
Dallas, TX 75230
Phone: 1-800/635-7050
Fax: 972/341-3224
Internet: http://www.cooperinst.org

Course: Physical Fitness Specialist. This course provides training in fitness leadership and the technical skills necessary for implementing an individualized fitness program that is safe, effective, and motivational.

The topics covered include screening options, fitness assessment, setting goals, exercise prescription, nutrition programs, motivation techniques, feedback methods, and safety programming.

Audience: Personal trainers, as well as health and fitness professionals in wellness centers and government agencies, including fire and police departments. Participants must be eighteen years of age or older.

Certification: Provides lifetime certification with a minimum of eight hours' annual continuing education.

Cost: This five-day course costs $595 and includes a resource manual.

Other programs: The Cooper Institute offers a variety of course and certifications in the areas of nutrition, cycling, aquatics, and well-being, and an advanced physical fitness specialist certification for public safety organizations. The public safety course is offered on-site or on a contract basis at your facility for a cost of $575.

The National Strength and Conditioning Association
P.O. Box 38909
Colorado Springs, CO 80937
Phone: 719/632-6722
Fax: 719/632-6367

The NSCA is considered the worldwide authority on strength and conditioning for improved physical performance. Although the association offers no basic training courses, its Professional Education Series, held quarterly around the U.S., is a hands-on workshop available to NSCA members.

Audience: Sport coaches, strength coaches, personal trainers, athletic trainers, and rehabilitation specialists. The prerequisites are an accredited college B.A. or B.S., as well as CPR certification.

Certification: The Certified Strength and Conditioning Specialist (CSCS) credential recognizes those professionals who possess the knowledge to design and implement safe, effective strength training and conditioning programs primarily for athletics. The NSCA Certification Commission is located in Lincoln, Nebraska. The commission offers two of the finest credentials available among fitness-related certifications: CSCS and NSCA-Certified Personal Trainer.

NSCA Certification Commission
920 O Street, Fourth Floor
P.O. Box 83469
Lincoln, NE 68501
Phone: 1-888/746-2378

Course: The NSCA offers study packages and practice exams. The association provides its members with two publications, Strength and Conditioning and Journal of Strength and Conditioning Research.

Cost: CSCS exam fee of $200 for members and $275 for nonmembers.

Health Metrics Inc.
3602 Equestrian Drive
West Richland, WA 99353-7710
Phone: 509/628-0215

Course: Certified Fitness Coordinator (CFC). The basic course provides the fundamentals of exercise physiology, exercise testing, and prescription. It features the On-target fitness program and FitScan software for exercise testing and evaluation. Two nutritional software programs are also available. Classes are offered around the country, and arrangements can be made to host them at your facility.

Audience: Firefighters and other public safety officers. There are no prerequisites.

Certification: The fitness coordinator certification requires attendance at all classes and the successful completion of a written exam.

Cost: The five-day workshop is $550 for the first person from a department and $500 for each additional member. The cost includes a manual. Recertification is recommended every three years at a cost of $275.

The American College of Sports Medicine
P.O. Box 1440
Indianapolis, IN 46206-1440
Phone: 317/637-9200
Fax: 317/634-7817
Internet: http://www.acsm.org/sportsmed

Course: Workshops that provide comprehensive preparation for certification exams.

Audience: Health and fitness personnel working in a fitness setting where the participants are apparently healthy, or in a clinical setting where the participants are engaged in cardiac or pulmonary rehabilitation or have chronic disease. The only prerequisite is current CPR verification, though ACSM strongly recommends a level of education or experience for each certification.

Certification: The ACSM offers three levels of certification within two specific tracks. The offerings in the Health and Fitness Track include Exercise Leader, Health/Fitness Instructor, and Health/Fitness Director. The offerings in the Clinical Track include Exercise Test Technologist, Exercise Specialist, and Program Director. All certified ACSM individuals are also eligible for a Certificate of Enhanced Qualification (CEQ), offered in a one-day workshop on select topics, such as nutrition and exercise.

Cost: The cost range for the workshop (n = 10) is $75 to $399, and certification exams range from $99 to $375 for members and $99 to $425 for nonmembers.

Other information: The ACSM offers five types of new memberships: professional, professional in training, graduate student, undergraduate student, and associate. Member cost and benefits vary, depending on type. The college publishes a monthly scientific journal, Medicine and Science in Sports and Exercise, as well as other bulletins and reviews.

The YMCA
101 North Wacker Drive
Chicago, IL 60606
Phone: 1-800/USA-YMCA (872-9622)

Course: Fitness Specialist. This course prepares participants to work with members in physical fitness assessments and individualized exercise plans. The course includes twenty-two hours of lecture, fifteen hours of practical lab experience, and four hours of testing. Participants will be certified to conduct the Y's Way to Physical Fitness Assessment protocol. Prerequisites are current CPR certification, some practical experience with YMCA health and fitness principles, and successful completion of a pretest. An exercise degree or related experience is desired.

Audience: YMCA staff members. Nonmembers should contact their local YMCA for eligibility and course options.

Certification: Successful completion of certification requires passing a written and practical examination, plus the submission of ten fitness evaluations after returning home.

Cost: $325, which includes the certification fee, plus all testing equipment, materials, and forms.

Other information: The YMCA offers a wide variety of courses, materials, and publications. A program store is available to members at 1-800/747-0089.

The Canadian Society for Exercise Physiology
185 Somerset Street West, Suite 202
Ottawa, Ontario, Canada K2P 0J2

Course: National Fitness Appraisal Certification and Accreditation (FACA) program. This voluntary national program provides standardized physical activity, fitness, and lifestyle appraisal guidelines across Canada. The program recognizes two levels: the Certified Fitness Consultant (CFC) and the Professional Fitness and Lifestyle Consultant (PFLC).

Audience: The CFC program is designed for individuals who have little formal preparation in fitness appraisal and exercise counseling. The PFLC program is designed for the advanced preparation of fitness consultants with formal university training in exercise science and lifestyle counseling.

Certification: A resource kit titled the *Canadian Physical Activity, Fitness, and Lifestyle Appraisal (CPAFLA)*, which contains background material for certification as a CFC, is available. The Professional Fitness and Lifestyle Consultant Resource Manual outlines appraisal protocols for which PFLCs are responsible during PFLC workshops and apprenticeships. Members receive a certificate of recognition. Additional, ACSM Exercise Specialist and Program Director certifications are offered in Canada through a joint agreement with the CSEP and the ACSM.

Cost: The PFLC workshop and apprenticeship are available in Canada only, and costs vary among the provinces, ranging from $350 to $400. The length of courses also varies from thirty-six to forty hours.

Other information: CSEP members receive the *Canadian Journal of Applied Physiology: Physical Activity, Health, and Fitness,* as well as newsletters, a member directory, and information on the annual scientific conference.

APPENDIX H

WORK SHEETS

Sample work sheets are provided to help you estimate costs for various levels of participation, as described in Chapters 6 and 7 of the text.

FIGURE H-1

MEDICAL EXAM WORK SHEET

LOP Options	Cost and No. of Members	Total Cost of Options	Notes
Level 1, Health Risk Appraisal High-risk RISKO score >31 at 60% Vo_2. All others use standard test procedure. All asymptomatic members are tested.	No pretest exam	Level 1 cost = $0	If at any time during, prior to, or following the test a member experiences dizziness, short ness of breath, or chest pain, terminate the test and refer him to a physician.
Level 1A, PAR-Q and RISKO Administer PAR-Q and refer all positives for exam. Estimate 10% of members if unknown. Basic exam (resting EKG) ____ Test RISK scores >31% at 60% Vo_2.	Basic exam M1 = A = $	Level 1A cost = $ M1 X A = $	Administer PAR-Q to all members prior to pro gram start to determine the exact number needing referral in place of the 10% estimate recommended.
Level 2, PAR-Q/RISKO or HealthScreen PAR-Q and HS. Refer all positives for exam. Estimate 10% of members. Basic exam plus graded exercise test.	Basic exam + GXT M1 = B = $	Level 2 cost = $ M1 X B = $	

FIGURE H-1, CONTINUED

Level 2A, PAR-Q or HS PAR-Q or HS and refer all positives for exam.	Basic exam + GXT and >40 (50)	Level 2A Cost = $	
Estimate 10% of members.	M1 =	M1 X A = $	
Refer all men over 40 and women over 50.			
Determine number of members.	M2 =	M2 X B = $	Cost-reduction alternative: Ifinsurance reimbursement is available, reduce cost accordingly.
Basic exam resting EKG.	A = $		
Basic exam plus graded exercise test.	B = $	Total M1 + M2 = $	
Level 3, Exam for All Members	Basic exam + GXT for All Members	Level 3 Cost = $	
5. Basic exam (resting EKG)	A = $	A X M = $	If using age-related exam, schedule as stated in NFPA 1582. Adjust annual cost to reflect schedule.
6. Basic exam plus graded exercise test.	B = $		
Total number of department members.	M =		
Level 3A, Exam + GXT for All Members		Level 3A Cost = $	If using a once-every-three-year schedule, the actual cost will be 33% of the estimates determined on this work sheet.
Total number of department members	M =	B X M = $	

Figure H-2
Staff Work Sheet

LOP Options	Cost of Staff	Notes	Cost
Level 1: Physician on a referral basis	No staffing cost at the basic level .Referral fee is included on the Medical Work Sheet.	Since the physician is not employed, the cost at Level 1 is the same as for a Level 1 medical exam option.	None
Level 2: Physician on a contract basis	Annual fee of $10,000 to $20,000	Cost of the physician is for a fixed amount, as determined by contract.	
Level 3: Full-time physician	Approx. $150,000	$150,000 represents a starting figure.	
Level 1: Program director (collateral duty)	20% of salary	A physician may assume the role of program director if he is a full-time employee.	
Level 2: Full-time manager (uniformed)	Salaried full-time employee	Salary based on rank.	
Level 3: Civilian professional	Salaried full-time employee	Salary would vary with experience and education, but comparable to that of a uniformed member at Level 2.	
Level 1: Certified fitness coordinator (department member, certified collateral duty)	Est. $0 to $500	A small department may have one CFC who performs all duties. If the CFC has a degree, no training costs required.	
Level 2: Fitness coordinator (uniformed or civilian with a degree; collateral or part-time duty; additional CFCs on staff.	Est. $500 and percent of salary. Est. $500 each for training costs.	An in-house survey can help identify qualified members who can perform this collateral duty.	
Level 3: Fitness coordinator (graduate-level civilian; additional CFCs on staff)	Salaried full-time employee. Est. $500 each for training costs.	The full-time CFC may assume the role of program director alone or with assistance from noncompensated coordinators.	

FIGURE H-3

EQUIPMENT/FACILITY WORK SHEET

LOP Options	Equipment	Facility	Cost
Level 1	None	In-house	None
Level 2	Basic complement of free weights, times the number of units needed.	In-house, in a central facility, or by contract with an outside facility.	
Level 3	Full complement of free weights, variable-resistance machines, etc.	Multiple facilities, whether contracted, rented, or constructed	

Appendix I

Test Protocols

Following are procedural descriptions, with illustrations, of nine test protocols. Additional illustrations for seven more tests referenced in the text are also provided, with a brief description for each.

TEST: ONE-MILE WALK, ROCKPORT VERSION

Purpose: A field test to estimate aerobic power and capacity.

Equipment: A measured track or route and a stopwatch. A heart-rate monitor is optional to assess pulse rate at the completion of each walk.

Procedure: Wear good walking shoes and loose-fitting clothes. Walk one lap (1/4 mile) slowly to warm up and stretch for five to ten minutes both before and after the test.

At the start of the test, walk one mile as fast as you can, maintaining a steady pace. Record your time to the nearest second. Most people walk between 3.0 and 6.0 miles per hour, so it should take ten to twenty minutes to walk the mile. A target time for a good effort would be twelve to thirteen minutes. Be sure to keep an even pace, and avoid speeding up or slowing down anytime during the walk, particularly toward the end of the fourth lap.

Record your heart rate immediately at the end of the mile (it will begin to slow almost as soon as you stop). Count your pulse for fifteen seconds and multiply by four, then record this number. This is your heart rate per minute after your walk test. You may choose to wear a heart-rate monitor for a quick and accurate assessment.

Score: Besides the heart rate and walk time, the participant's age and weight are logged and entered into the chosen formula for calculating VO2max.

Relative measure (r = 0.88):

(0.0769 x body weight) + (0.3877 x age) + (3.2649 x time) + (0.1565 x heart rate) – (132.853 + 6.315, male only) = Vo_2max $mL \cdot kg^{-1} \cdot min^{-1}$.

Absolute measure (r = 0.93):

6.9652 + (0.0091 x body weight) – (0.0257 x age) – (0.5955, male only) – (0.2240 x time) – (0.0115 x heart rate) = Vo_2max L/min.

Body weight = Body weight in pounds.
Heart rate = Beats per minute.
Time = One-mile walk time converted to decimals (12:15 = 12.25; 13:30 = 13.5, etc.).
Age = Age in years.

Standard: $Vo_2max = 42\ mL \cdot kg^{-1} \cdot min^{-1}$, or 3.0 L/min.

References: Kline et. al. (1987) Estimation of Vo_2max from a one-mile track walk, gender, age, and body weight. Medicine and Science in Sport and Exercise. 19 (3): 253–259.

LeCuyer, J.E. (1997) Assessing firefighter aerobic capacity: the Rockport field test. *Fire Engineering*. February, pp. 49–52, 54, and 56.

TEST: SIT AND REACH, STANDARD VERSION (WELLS-DILLON)

Purpose: To assess the flexibility or range of motion of the lower back and hamstring muscle group. This is important because reduced flexibility is associated with an increased risk of low-back injury.

Equipment: A yardstick and a box. The box should be approximately 12 inches high and 18 inches wide, with sufficient length to provide seating for the evaluator, if desired.

Secure the yardstick along the centerline of the top of the box, with the 1-inch mark pointing toward the subject and the 15-inch mark aligned at the edge where the subject's feet are positioned.

Procedure: Have the subject perform stretching exercises for the lower back and hamstring group prior to the assessment. Seat the subject on the floor, feet flat on the floor and against the box at right angles, approximately eight inches apart and without shoes. Instruct the subject to place one hand on top of the other and, without bouncing, to lean forward slowly and touch the yardstick with his fingers. The subject should hold that position for one or two seconds, then retract his body *slightly* to allow the hamstrings to relax, then slowly lean forward again. Repeat for three trials.

Score: Record the farthest distance reached within one-half inch.

Standard: The criterion standard is the ability to touch one's toes, indicated by a score of 15 inches.

Reference: Wells, K.F.; and Dillon, E.K. (1952) Sit and reach: a test of back and leg flexibility. Research Quarterly. 23: 115–118.

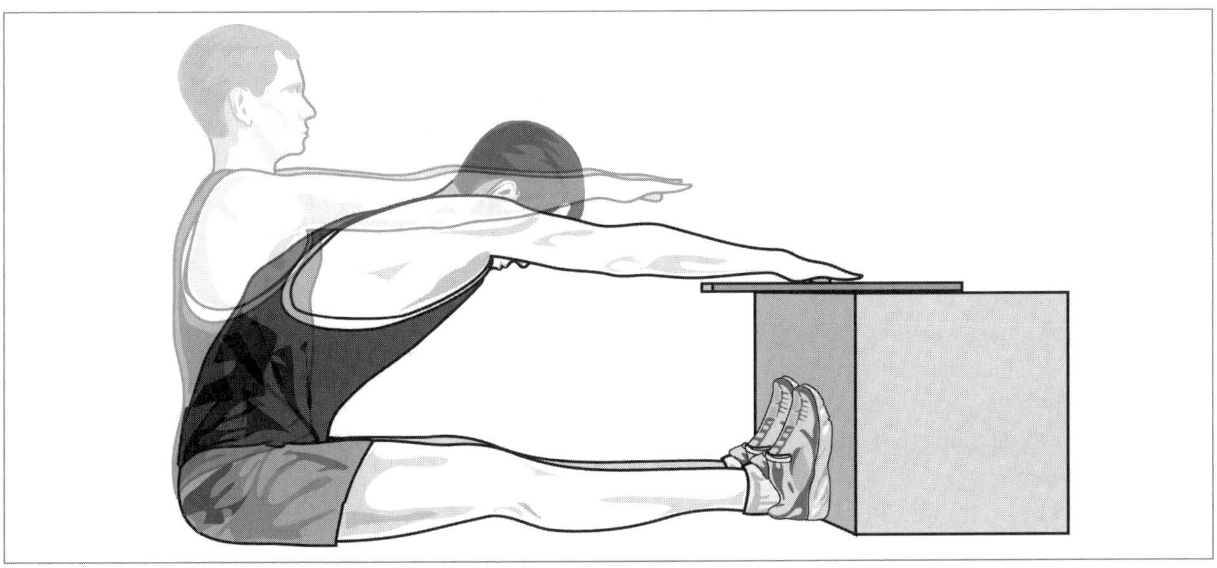

Fig. I–1 Standard Sit-and-Reach Test

Fig. I–2 Bench Press

TEST: BENCH PRESS, VRM VERSION

Purpose: To measure upper-body strength (chest, shoulders, and triceps) using an estimated one-repetition maximum effort.

Procedure: Adjust the bar on the variable-resistance machine or free-weight bench stands to suit the individual. Demonstrate the proper technique. The subject's back should be flat against the bench, feet flat on the floor. Proper technique is to grasp the bar at shoulder width, with the thumbs around the bar, particularly for free weights. Inform the subject not to move his feet or arch his back during the lift. The subject should avoid holding his breath. He should exhale on extension.

Have the subject warm up by performing ten to fifteen repetitions with a light weight that he can handle easily. Then, increase the weight 10 to 20 percent, depending on individual response. The subject then performs eight to ten repetitions. Continue increasing the weight in this manner until the subject can perform no more than three to five repetitions. Three to five reps is the target range. Allow a rest period of about two minutes between sets.

Use a spotter with both free weights and a variable-resistance machine. A liftoff is recommended when the subject is approaching his three-to-five-rep range. When using a VRM, the bar should be lifted (spotter's job) to the up position (arms extended) to start.

The bar should be lowered to the point of contact or in line with the chest, then raise it until the arms are once again fully extended. This constitutes one repetition.

Score: Record the maximum weight in pounds. The estimated one-rep max is determined using the rep factor formula shown below. Multiply the weight times the factor for the corresponding number of repetitions. For example, 180 lbs. for five reps equals 180 x 1.18 = estimated one-rep max of 212 lbs.

Reps	Factor
2	1.07
3	1.12
4	1.15
5	1.18
6	1.21
7	1.24
8	1.27
9	1.30
10	1.33

Absolute score: The one-rep max weight is the final score.

Relative score: The ratio of the one-rep max divided by the individual's body weight is used as the final score.

Standard: Relative score of 1.00 (VRM) or 0.70 (free weights), or an absolute score of 200 lbs. (VRM) or 145 lbs. (free weights).

References: Altug, Z., et al. (1987) A test selection guide for assessing and evaluating athletes. *NSCA Journal,* 9 (3): 6–66.

Jackson, A.; Watkins, M.; and Patton, R. (1980) A factor analysis of twelve selected maximal isotonic strength performances on the Universal Gym. *Medicine and Science in Sports and Exercise,* 12: 274.

TEST: LEG PRESS, PLATE-LOADED HIP SLED VERSION

Purpose: To measure lower-body strength (hip, thighs, gluteals) using an estimated one-repetition maximum effort.

Equipment: Plate-loaded hip sled, variable-resistance machine, or free weights and power rack. A body-weight scale. An adequate assortment of weight plates and three sizes of weight-lifting belts, if applicable.

Procedure: Instruct and demonstrate the proper form to be used for the test. Adjust the starting position of the sled to suit the subject, and instruct him on the proper use of the safety lock device. Perform several practice reps with minimal or no weight to ensure proper use of the safety device and to identify the correct start position.

Begin with the legs extended, feet flat against the platform at shoulder width. Slowly lower the weight until the knees form a 90-degree angle. Push the weight back to the start position, extending the legs fully without locking the knees. This constitutes one repetition.

Begin with a warm-up set of ten to fifteen reps. Add 10 to 15 percent more weight, and perform three to five reps for each set until the subject feels that he has approached a maximum weight. Attempt to perform as many reps as possible, up to ten, at this weight. Rest approximately two minutes between sets.

Score: Record the amount of weight lifted and the number of correct repetitions. The estimated one-rep max is determined by multiplying the weight times the factor for the corresponding number of reps. For example, 290 lbs. x eight reps = 290 x 1.27 = estimated one-rep max of 368 lbs.

Reps	Factor
2	1.07
3	1.12
4	1.15
5	1.18
6	1.21
7	1.24
8	1.27
9	1.30
10	1.33

Fig. I–3 Leg Press Sled

Absolute score: The one-rep max weight is the final score.

Relative score: The ratio of the one-rep-max weight divided by the individual's body weight is the final score. For example, a one-rep max of 368 divided by the subject's body weight of 185 lbs. = 1.98.

Standard: Relative score of 2.00 (VRM) or 1.40 (free weights), or an absolute score of 290 lbs. (VRM) or 200 lbs. (free weights).

References: Altug, Z., et al. (1987) A test selection guide for assessing and evaluating athletes. NSCA Journal, 9 (3): 6–66.

Jackson, A.; Watkins, M.; and Patton, R. (1980) A factor analysis of twelve selected maximal isotonic strength performances on the Universal Gym. *Medicine and Science in Sports and Exercise*, 12: 274.

TEST: SIT-UPS, 60-SECOND CRUNCH-STYLE VERSION

Purpose: To assess the muscular endurance (strength) of the abdominal muscles.

Equipment: A stopwatch, chair, and an exercise mat (optional).

Procedure: Instruct the subject to lie on his back and place his feet on the chair, with his legs bent at right angles. The subject should place his hands on the side of his head for the start position, but he should not interlock his fingers behind his head.

On the go command, the participant curls upward until the scapula (shoulder blade) is off the floor and the elbows touch the legs. Count one repetition when the subject returns to the start position. The subject should continue performing repetitions until instructed or until he can no longer continue. Correct the participant's form during the test, if necessary, and count only the repetitions that are performed properly. It's best to provide instruction before the start of the test and to allow the participant to demonstrate the proper form.

Score: Count the number of correct repetitions performed in a one-minute interval.

Standard: A minimum of thirty sit-ups.

References: Juker, D.; McGill, S.; Kripf, P.; and Steffen, T. (1998) Quantitative intramuscular myoelectric activity of lumbar portions of psoas and the abdominal wall during a wide variety of tasks. Medicine and Science in Sports and Exercise, 30 (2): 301–310.

Fig. I–4 Crunch-Style Sit-Up

TEST: BODY COMPOSITION, ANTHROPOMETRIC MEASURE

Purpose: To determine the percentage of body fat relative to total body weight.

Equipment: Measuring tape, preferably a cloth tape with a spring attachment to ensure proper tension. Body composition reference tables for males and females.

Procedure: Take the measurements directly against the skin, not over clothing. The tape measure should be flat, not twisted, and you should take at least two readings at each site. If you don't obtain a consistent measurement, take a third reading and average the three.

For males, measure the waist over the umbilicus, and measure the neck just below the larynx. For females, measure the natural waistline, or the narrowest point between the rib cage and the hips. Measure the neck just below the larynx, and measure the hips at the widest part below the waist.

Score: Record the tape readings to the nearest half-inch. For males, the circumference value equals the abdomen measure minus the neck measure. For females, the circumference value equals the waist measure plus the hip, minus the neck. Cross-reference the subject's circumference value with his or her height in inches to identify the percentage of body fat.

Standard: Although there are no standard requirements, the desirable percentages are 15 to 18 percent for males, 20 to 25 percent for females.

References: ARA Human Factors (1994). *Certified Fitness Coordinators Manual, Eleventh Edition.* Burtonsville, MD: Author.

Hodgdon, J. (1992) *Body Composition in the Military Services: Standards and Methods in Body Composition and Physical Performance, Applications for the Military Services.* B. Marriot and J. Grumstrop-Scott, editors (p. 65). Washington, DC: National Academy Press.

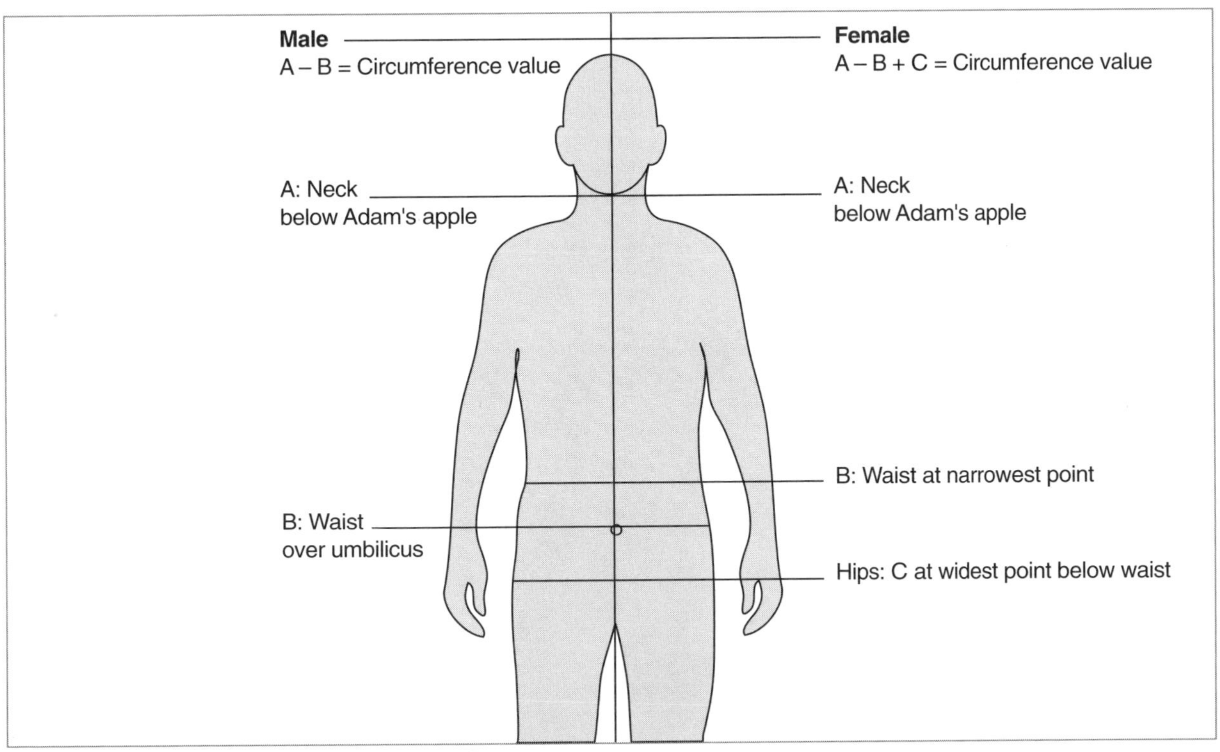

Fig. I–5 Measurements for Estimating Body Composition

TEST: AEROBIC CAPACITY, 1.5-MILE RUN VERSION

Purpose: A field test to estimate aerobic power and capacity.

Equipment: A measured track and a stopwatch. Aerobic reference tables.

Procedure: Allow the participant time to stretch and warm up prior to testing. A slow jog is recommended for the warm-up. Then, instruct the subjects to run as fast as possible for the requisite distance.

Score: Aerobic capacity is determined by consulting a table. Select the Vo_2 score nearest to the recorded time of completion. Round off down to the nearest score. An altitude adjustment to the time is subtracted from the ending time if appropriate. Refer to Appendix Table J-1.

Standard: The minimum standard is a Vo_2max of 42 $mL \cdot kg^{-1} \cdot min^{-1}$. Refer to conversion chart for time (Table J-1).

Reference: Cooper, K. (1982) The Aerobics Program for Total Well-Being. New York: Bantam Books.

TEST: TREADMILL

TEST: SUBMAXIMAL TREADMILL, YOUNG PROTOCOL VERSION

Stage	Duration	Speed mph	Grade	METS	Vo_2 mL • kg^{-1} • min^{-1}
1	1-Minute	4.5	0%	6.4	22.4
2	↑	4.5	2%	7.3	25.6
3		5	2%	8.9	31.1
4		5	4%	9.6	33.6
5		5.5	4%	10.9	38.2
6		5.5	6%	11.7	40.9
7		6	6%	12.6	44.1
8		6	8%	13.4	46.9
9		6.5	8%	14.5	50.7
10		6.5	10%	15.4	53.9
11		7	10%	16.5	57.7
12		7	12%	17.5	61.2
13		7.5	12%	18.6	65.0
14		7.5	14%	19.6	68.6
15		8	14%	20.9	73.1
16		8	16%	22.0	77.0
17		8.5	16%	23.3	81.3
18		8.5	18%	24.5	85.7
19	↓	9	18%	25.8	90.0
20	Each Stage	9	20%	27.0	94.5

Source: Phoenix Fire Dept.

TEST: SUBMAXIMAL TREADMILL, BRUCE PROTOCOL VERSION

Stage	Duration	Speed mph	Grade	METS	Vo_2 mL • kg^{-1} • min^{-1}
1	3:00	1.7	10%	4.6	16.1
2	6:00	2.5	12%	7.0	24.5
3	9:00	3.4	14%	10.2	35.7
4	12:00	4.2	16%	12.1	42.4
5	15:00	5.0	18%	14.9	52.2
6	18:00	5.5	20%	17.0	59.5
7	21:00	6.0	22%	19.3	67.6

TEST: AEROBIC CAPACITY, PAPER/PENCIL VERSION

The paper/pencil test to determine aerobic capacity requires either an assessment of the percentage of body fat or a determination of the body-mass index, plus the completion of an activity inquiry. The data from the two assessments is then plugged into the appropriate formula.

CALCULATING VO_2MAX FROM BODY-FAT PERCENTAGE

Male: Vo_2max = 56.376 + 1.589 (physical activity rating) – 0.289 (age) – 0.552 (% fat).
Female: Vo_2max = 50.513 + 1.589 (physical activity rating) – 0.289 (age) – 0.552 (% fat).

CALCULATING VO_2MAX FROM BODY-MASS INDEX

Male: Vo_2max = 67.350 + 1.921 (physical activity rating) – 0.381 (age) – 0.754 (body-mass index).
Female: Vo_2max = 56.363 + 1.921 (physical activity rating) – 0.381 (age) – 0.754 (body-mass index).

PHYSICAL ACTIVITY RATING

Select the number (0 to 7) that best describes your activity in the previous month.

Do not participate in regularly programmed recreation, sport, or heavy physical activity.

0 Avoid walking or exertion; always use the elevator; drive whenever possible.

1 Walk for pleasure; routinely use stairs; occasionally exercise sufficiently to cause heavy breathing or perspiration.

Participate regularly in recreation or work requiring modest physical activity, such as golf, horseback riding, calisthenics, gymnastics, table tennis, bowling, weightlifting, yard work.

2 10 to 60 minutes per week.

3 More than one hour per week.

Participate regularly in heavy physical exercise, such as running, jogging, swimming, cycling, rowing, skipping rope, running in place, or vigorous aerobic activity, such as tennis, basketball, or handball.

4 Run less than one mile per week or spend less than 30 minutes per week in a comparable activity.

5 Run one to five miles per week or spend 30 to 60 minutes per week in a comparable activity.

6 Run five to ten miles per week or spend one to three hours per week in a comparable activity.

7 Run over ten miles per week or spend over three hours per week in a comparable activity.

Source: Jackson, A.S., et al. (1990) Prediction of functional aerobic capacity without exercise testing. *Medicine and Science in Sports and Exercise*. 22 (6): 863–870.

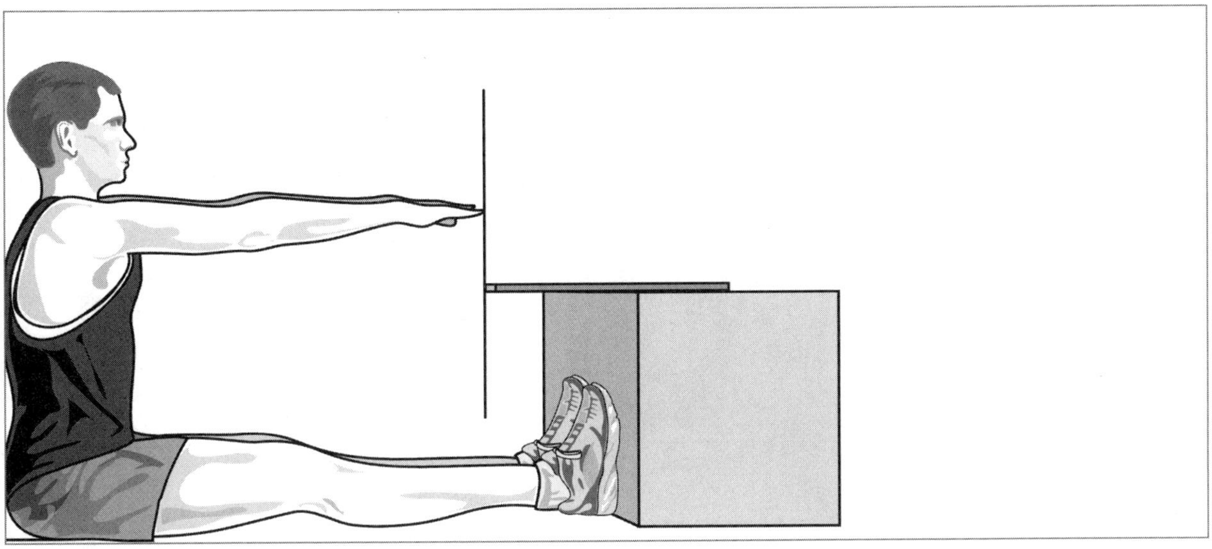

Fig. I–6 Modified Sit and Reach

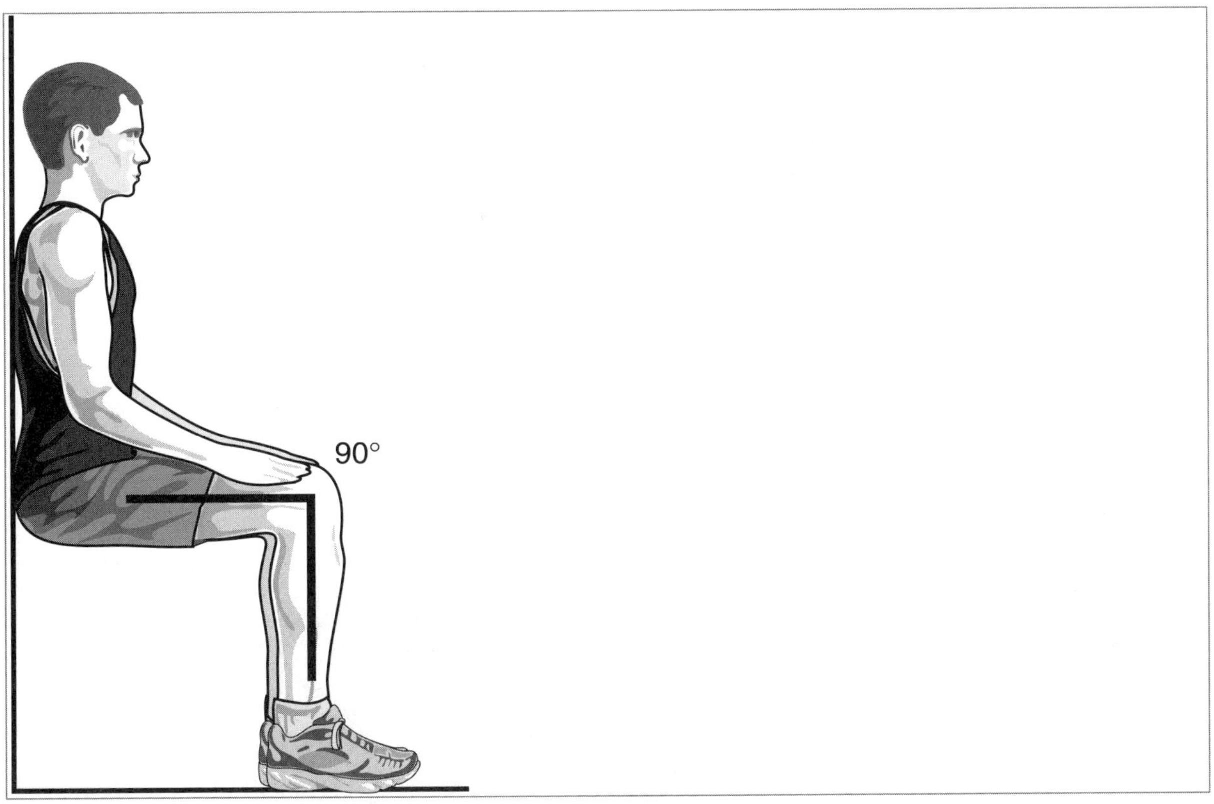

Fig. I–7 Wall Sit (Isometric)

Fig. I–8 Flex Hang

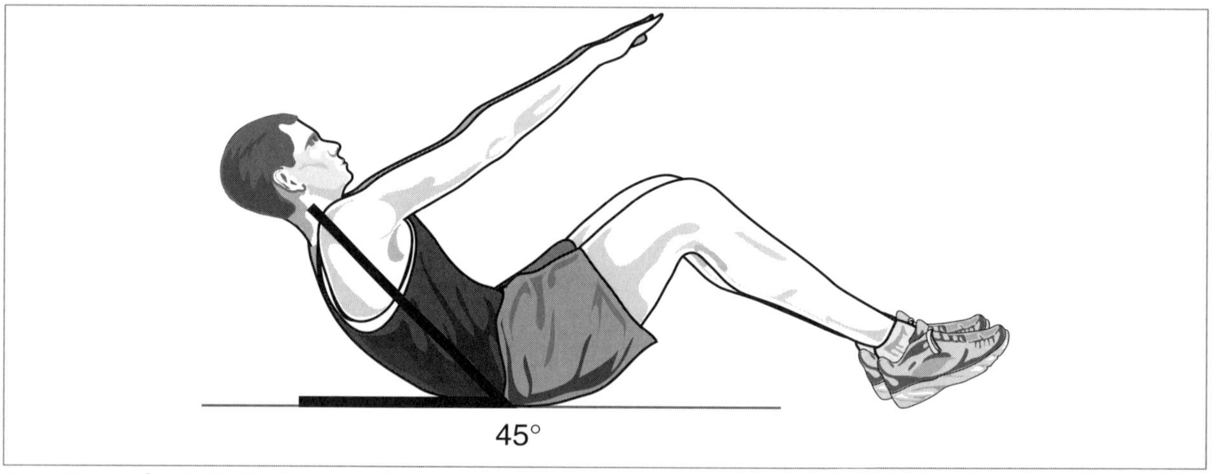

Fig. I–9 The 45 Sit-Up (Isometric Sit-Up)

Fig. I–10 Burpee (Squat Thrust)

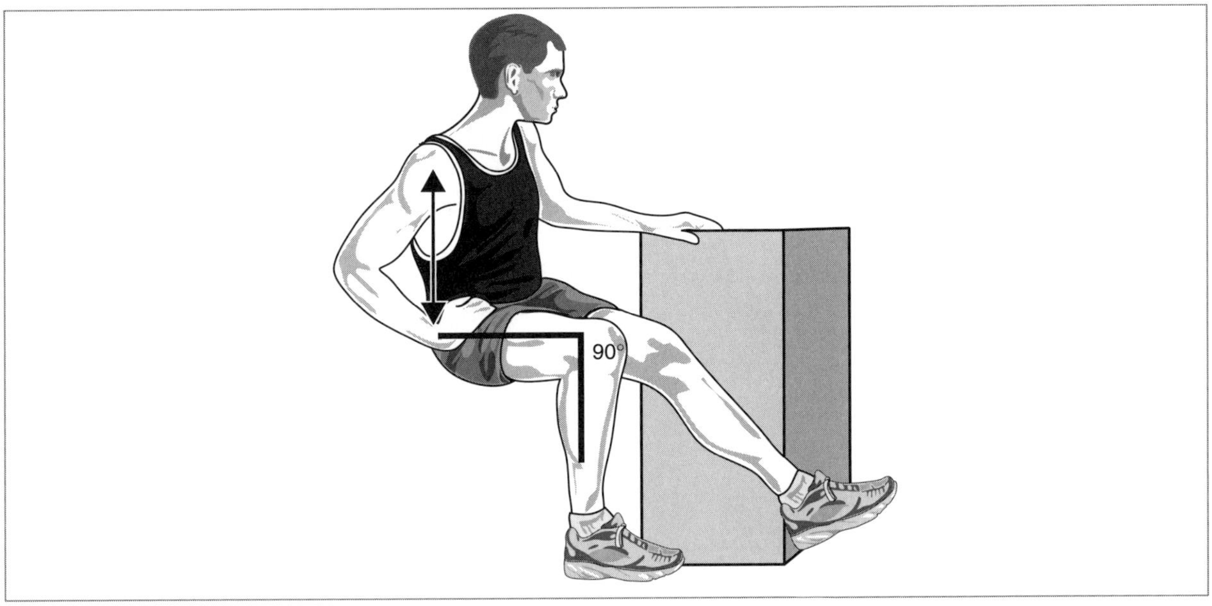

Fig. I–11 One-Leg Squat

Fig. I–12 Modified Chin-Up

Appendix J

Table J-1

Maximal Met Level Based on 1.5-Mile Run Time

Minutes/seconds	0	10	20	30	40	50
7:00	22.44	21.95	21.47	21.00	20.54	20.10
8:00	19.67	19.24	18.83	18.43	18.04	17.67
9:00	17.30	16.94	16.59	16.25	15.92	15.60
10:00	15.28	14.98	14.69	14.40	14.12	13.85
11:00	13.59	13.33	13.08	12.84	12.61	12.38
12:00	12.16	11.95	11.74	11.54	11.34	11.15
13:00	10.96	10.78	10.61	10.44	10.27	10.11
14:00	9.95	9.80	9.65	9.50	9.36	9.22
15:00	9.08	8.95	8.82	8.69	8.56	8.44
16:00	8.31	8.19	8.07	7.95	7.83	7.72
17:00	7.60	7.48	7.37	7.25	7.13	7.02

The above table is useful for estimating metabolic levels corresponding to times for the 1.5-mile run. The table may also be used to estimate Vo_2max levels by multiplying the recorded metabolic value by 3.5. For example, suppose an individual ran 1.5 miles in 10:40. The estimated met equivalent for that time is 14.12, and the corresponding Vo_2max value is $14.12 \times 3.5 = 49.42\ mL \cdot kg^{-1} \cdot min^{-1}$.

Time Adjustment for Run Tests Administered at High Altitudes

The completion time of the 1.5-mile run should be adjusted for altitude according to the following criteria.

Under 5,000 feet	no adjustment
5,000 feet	subtract 30 sec.
6,000 feet	subtract 40 sec.
7,000 feet	subtract 50 sec.
8,000 feet	subtract 60 sec.

Source: Reprinted from Federal Emergency Management Agency, United States Fire Administration. *Physical Fitness Coordinator's Manual for Fire Departments.* Emmitsburg, MD. FA/September 1990, pp. 4-11.

TABLE J-2

CALCULATION OF VO_2MAX $mL \bullet kg^{-1} \bullet min^{-1}$ FROM/TO LITERS PER MIN ($L \bullet Min^{-1}$)

Kg.	Lbs.	2	2.1	2.2	2.3	2.4	2.5	2.6	2.7	2.8	2.9	3	3.1	3.2	3.3	3.4	3.5	3.6	3.7	3.8	3.9	4	4.1	4.2	4.3	4.4	4.5	4.6	4.7
50	100	40	42	44	46	48	50	52	54	56	58	60	62	64	66	68	70												
52	115	38	40	42	44	46	48	50	52	54	56	58	60	62	63	65	67	69											
54	119	37	39	41	43	44	46	48	50	52	54	56	57	59	61	63	65	67	69										
56	123	36	38	40	41	43	45	46	48	50	52	54	56	57	59	61	63	65	67	69									
58	128	34	36	38	40	41	43	45	46	48	50	52	53	57	59	60	62	64	66	67	69								
60	132	33	35	37	38	40	42	43	45	47	48	50	52	53	55	57	58	60	62	65	67	68							
62	167	32	34	35	37	39	40	42	44	45	48	50	52	53	55	58	58	60	61	63	65	66	68						
64	141	31	33	34	36	38	39	41	42	44	45	47	48	50	52	53	55	56	58	59	61	63	64	66					
66	146	30	32	33	35	36	38	39	41	42	44	45	47	48	50	52	53	55	56	58	59	61	62	64	65				
68	150		31	32	34	35	37	38	40	41	43	44	46	47	49	50	51	53	54	56	57	59	60	62	63	65			
70	154		30	31	33	34	36	37	39	40	41	43	44	46	47	49	50	51	53	54	56	57	59	60	61	63	64		
72	159			31	32	33	35	36	38	40	42	44	44	46	47	49	50	51	53	54	56	57	58	60	61	63	65	67	
74	163			30	31	32	34	35	36	38	39	41	42	43	45	47	49	50	51	53	54	55	57	58	59	61	62	64	65
76	158				30	32	33	34	36	37	38	39	41	42	43	45	46	47	49	50	51	53	54	55	57	58	59	61	62
78	172					31	32	33	35	36	37	38	40	41	42	44	45	46	47	49	50	51	53	54	55	56	58	59	60
80	176					30	31	33	34	35	36	38	39	40	41	43	44	45	46	48	49	50	51	53	54	55	56	58	59
82	181						30	32	33	34	35	37	38	39	40	41	43	44	45	46	48	49	50	51	52	54	56	57	59
84	185						30	31	32	33	35	36	37	38	39	40	42	43	44	45	46	48	49	50	51	52	54	55	56
86	190							30	31	33	34	35	36	37	38	40	41	42	43	44	45	47	48	49	50	51	52	53	55
88	194							30	31	32	33	34	35	36	38	39	40	41	42	43	44	45	47	48	49	50	51	52	53
90	198								30	31	32	33	34	36	37	38	39	40	41	42	43	44	46	47	48	49	50	51	52
92	203									30	32	33	34	35	36	37	38	39	40	41	42	43	45	46	47	48	49	50	51
94	207									30	31	32	33	34	35	36	37	38	39	40	41	43	44	45	46	47	48	49	50
96	212										30	31	32	33	34	35	36	38	39	40	41	42	43	44	45	46	47	48	49
98	216										30	31	32	33	34	35	36	37	38	39	40	41	42	43	44	45	46	47	48
100	220											30	31	32	33	34	35	36	37	38	39	40	41	42	43	44	45	46	47

TABLE J-3

BODY COMPOSITION TABLES—MALES

Height

Circumference		60.0	60.5	61.0	61.5	62.0	62.5	63.0	63.5	64.0	64.5	65.0	65.5	66.0	66.
	11.0	3	2	2	2	2	1	1	1	1	1	0	0	-	-
	11.5	4	4	4	3	3	3	3	2	2	2	2	2	1	1
	12.0	6	5	5	5	5	4	4	4	4	3	3	3	3	3
	12.5	7	7	6	6	6	6	6	5	5	5	5	4	4	4
	13.0	8	8	8	8	7	7	7	7	6	6	6	6	6	5
	13.5	10	9	9	9	9	8	8	8	8	8	7	7	7	7
	14.0	11	11	10	10	10	10	10	9	9	9	9	8	8	8
	14.5	12	12	12	11	11	11	11	11	10	10	10	10	9	9
	15.0	13	13	13	13	12	12	12	12	11	11	11	11	11	10
	15.5	15	14	14	14	14	13	13	13	12	12	12	12	12	12
	16.0	16	15	15	15	15	15	14	14	13	14	13	13	13	13
	16.5	17	17	16	16	16	16	15	15	14	14	14	14	14	14
	17.0	18	18	17	17	17	17	16	16	15	16	16	15	15	15
	17.5	19	19	19	18	18	18	18	17	16	17	17	16	16	16
	18.0	20	20	20	19	19	19	19	18	17	18	18	17	17	17
	18.5	21	21	21	20	20	20	20	19	18	19	19	18	18	18
	19.0	22	22	22	21	21	21	21	20	19	18	20	19	19	19
	19.5	23	23	23	22	22	22	22	21	20	21	21	20	20	20
	20.0	24	24	23	23	23	23	22	22	21	22	22	21	21	21
	20.5	25	25	24	24	24	24	23	23	22	23	22	22	22	22
	21.0	26	26	25	25	25	25	24	24	23	24	23	23	23	23
	21.5	27	26	26	26	26	25	25	25	24	24	24	24	24	24
	22.0	28	27	27	27	27	26	26	26	25	25	25	25	25	24
	22.5	28	28	28	28	27	27	27	27	26	26	26	26	25	25
	23.0	29	29	29	29	28	28	28	28	26	27	27	27	26	26
	23.5	30	30	30	29	29	29	29	28	27	28	28	27	27	27
	24.0	31	31	30	30	30	30	29	29	28	29	28	28	28	28
	24.5	32	31	31	31	31	30	30	30	29	29	29	29	29	29
	25.0	33	32	32	32	31	31	31	31	30	30	30	30	29	29
	25.5	33	33	33	33	32	32	32	31	31	31	31	31	30	30
	26.0	34	34	34	33	33	33	32	32	32	32	32	31	31	31
	26.5	35	35	34	34	34	33	33	33	33	32	32	32	32	32
	27.0	36	35	35	35	34	34	34	34	33	33	33	33	32	32
	27.5	36	36	36	35	35	35	35	34	34	34	34	33	33	33
	28.0	37	37	36	36	36	36	35	35	35	35	34	34	34	34
	28.5	38	37	37	37	37	36	36	36	36	35	35	35	34	34
	29.0	38	38	38	38	37	37	37	37	36	36	36	36	35	35
	29.5	39	39	39	38	38	38	37	37	37	37	36	36	36	36
	30.0	40	39	39	39	39	38	38	38	38	37	37	37	37	36
	30.5	-	-	40	40	39	39	39	39	38	38	38	38	37	37
	31.0	-	-	-	-	40	40	39	39	39	39	38	38	38	38
	31.5	-	-	-	-	-	-	-	40	40	39	39	39	39	38
	32.0	-	-	-	-	-	-	-	-	-	40	40	39	39	39
	32.5	-	-	-	-	-	-	-	-	-	-	-	-	40	40

Note: Circumference value = abdomen II circumference - neck circumference (inches).

Circumference	Height 67.0	67.5	68.0	68.5	69.0	69.5	70.0	70.5	71.0	71.5	72.0	72.5	73.0	73.5
11.5	1	1	1	0	0	-	-	-	-	-	-	-	-	-
12.0	2	2	2	2	2	1	1	1	1	1	0	0	0	-
12.5	4	4	3	3	3	3	3	2	2	2	2	2	1	1
13.0	5	5	5	5	4	4	4	4	4	3	3	3	3	3
13.5	6	6	6	6	6	5	5	5	5	5	4	4	4	4
14.0	8	8	7	7	7	7	7	6	6	6	6	6	5	5
14.5	9	9	9	8	8	8	8	8	7	7	7	7	7	6
15.0	10	10	10	10	9	9	9	9	9	8	8	8	8	8
15.5	11	11	11	11	11	10	10	10	10	9	9	9	9	9
16.0	12	12	12	12	12	11	11	11	11	11	10	10	10	10
16.5	14	13	13	13	13	13	12	12	12	12	12	11	11	11
17.0	15	14	14	14	14	14	13	13	13	13	13	12	12	12
17.5	16	16	15	15	15	15	14	14	14	14	14	13	13	13
18.0	17	17	16	16	16	16	15	15	15	15	15	14	14	14
18.5	18	18	17	17	17	17	16	16	16	16	16	15	15	15
19.0	19	19	18	18	18	18	17	17	17	17	17	16	16	16
19.5	20	19	19	19	19	19	18	18	18	18	18	17	17	17
20.0	21	20	20	20	20	20	19	19	19	19	18	18	18	18
20.5	22	21	21	21	21	20	20	20	20	20	20	19	19	19
21.0	22	22	22	22	22	21	21	21	21	20	20	20	20	20
21.5	23	23	23	23	22	22	22	22	22	21	21	21	21	21
22.0	24	24	24	24	23	23	23	23	22	22	22	22	22	21
22.5	25	25	25	24	24	24	24	23	23	23	23	23	22	22
23.0	26	26	25	25	25	25	25	24	24	24	24	23	23	23
23.5	27	26	26	26	26	26	25	25	25	25	24	24	24	24
24.0	27	27	27	27	27	26	26	26	26	25	25	25	25	25
24.5	28	28	28	28	27	27	27	27	26	26	26	26	26	25
25.0	29	29	29	28	28	28	28	27	27	27	27	27	26	26
25.5	30	30	29	29	29	29	28	28	28	28	28	27	27	27
26.0	31	30	30	30	30	29	29	29	29	29	28	28	28	28
26.5	31	31	31	31	30	30	30	30	29	29	29	29	29	28
27.0	32	32	32	31	31	31	31	30	30	30	30	30	29	29
27.5	33	33	32	32	32	32	31	31	31	31	30	30	30	30
28.0	33	33	33	33	33	32	32	32	32	31	31	31	31	31
28.5	34	34	34	33	33	33	33	33	32	32	32	32	31	31
29.0	35	35	34	34	34	34	33	33	33	33	33	32	32	32
29.5	35	35	35	35	35	34	34	34	34	33	33	33	33	33
30.0	36	36	36	35	35	35	35	35	34	34	34	34	33	33
30.5	37	37	36	36	36	36	35	35	35	35	35	34	34	34
31.0	37	37	37	37	37	36	36	36	36	35	35	35	35	34
31.5	38	38	38	37	37	37	37	36	36	36	36	36	35	35
32.0	39	38	38	38	38	38	37	37	37	37	36	36	36	36
32.5	39	39	39	39	38	38	38	38	37	37	37	37	37	36
33.0	40	40	39	39	39	39	39	38	38	38	38	37	37	37
33.5	-	-	-	40	40	39	39	39	39	38	38	38	38	38
34.0	-	-	-	-	-	40	40	39	39	39	39	39	38	38
34.5	-	-	-	-	-	-	-	-	40	40	39	39	39	39
35.0	-	-	-	-	-	-	-	-	-	-	40	40	40	39

Height

Circumference	74.0	74.5	75.0	75.5	76.0	76.5	77.0	77.5	78.0	78.5	79.0	79.5
13.5	4	4	3	3	3	3	3	2	2	2	2	2
14.0	5	5	5	4	4	4	4	4	3	3	3	3
14.5	6	6	6	6	5	5	5	5	5	5	4	4
15.0	7	7	7	7	7	6	6	6	6	6	6	5
15.5	9	8	8	8	8	8	7	7	7	7	7	6
16.0	10	9	9	9	9	9	8	8	8	8	8	8
16.5	11	11	10	10	10	10	10	9	9	9	9	9
17.0	12	12	11	11	11	11	11	10	10	10	10	10
17.5	13	13	12	12	12	12	12	11	11	11	11	11
18.0	14	14	13	13	13	13	13	12	12	12	12	12
18.5	15	15	14	14	14	14	14	13	13	13	13	13
19.0	16	16	15	15	15	15	15	14	14	14	14	14
19.5	17	17	16	16	16	16	16	115	15	15	15	15
20.0	18	17	17	17	17	17	16	16	16	16	16	16
20.5	19	18	18	18	18	18	17	17	17	17	17	16
21.0	19	19	19	19	19	18	18	18	18	18	18	17
21.5	20	20	20	20	20	19	19	19	19	19	18	18
22.0	21	21	21	21	20	20	20	20	20	19	19	19
22.5	22	22	22	21	21	21	21	21	20	20	20	20
23.0	23	23	22	22	22	22	22	21	21	21	21	21
23.5	24	23	23	23	23	23	22	22	22	22	22	21
24.0	24	24	24	24	24	23	23	23	23	23	22	22
24.5	25	25	25	25	24	24	24	24	24	23	23	23
25.0	26	26	26	25	25	25	25	25	24	24	24	24
25.5	27	27	26	26	26	26	26	25	25	25	25	25
26.0	27	27	27	27	27	26	26	26	26	26	25	25
26.5	28	28	28	28	27	27	27	27	27	26	26	26
27.0	29	29	28	28	28	28	28	27	27	27	27	27
27.5	30	29	29	29	29	29	28	28	28	28	28	27
28.0	30	30	30	30	29	29	29	29	29	28	28	28
28.5	31	31	31	30	30	30	30	30	29	29	29	28
29.0	32	31	31	31	31	31	30	30	30	30	30	29
29.5	32	32	32	32	31	31	31	31	31	30	30	30
30.0	33	33	33	32	32	32	32	32	31	31	31	31
30.5	34	33	33	33	33	33	32	32	32	32	32	31
31.0	34	34	34	34	33	33	33	33	33	32	32	32
31.5	35	35	34	34	34	34	34	33	33	33	33	33
32.0	36	35	35	35	35	34	34	34	34	34	33	33
32.5	36	36	36	35	35	35	35	35	34	34	34	34
33.0	37	37	36	36	36	36	35	35	35	35	35	34
33.5	37	37	37	37	36	36	36	36	36	35	35	35
34.0	38	38	37	37	37	37	37	36	36	36	36	36
34.5	39	38	38	38	38	37	37	37	37	37	36	36
35.0	39	39	39	38	38	38	38	38	37	37	37	37
35.5	-	-	39	39	39	39	38	38	38	38	38	37
36.0	-	-	40	40	39	39	39	39	39	38	38	38
36.5	-	-	-	-	40	40	39	39	39	39	39	38
37.0	-	-	-	-	-	-	-	40	40	39	39	39
37.5	-	-	-	-	-	-	-	-	-	40	40	40

BODY COMPOSITION TABLES—FEMALES

Circumference	Height									
	58.0	58.5	59.0	59.5	60.0	60.5	61.0	61.5	62.0	62.5
34.5	1	-	-	-	-	-	-	-	-	-
35.0	2	1	1	1	-	-	-	-	-	-
35.5	3	2	2	2	1	1	-	-	-	-
36.0	4	3	3	3	2	2	1	1	1	-
36.5	5	4	4	4	3	3	2	2	2	1
37.0	6	5	5	4	4	4	3	3	3	2
37.5	7	6	6	5	5	5	4	4	4	3
38.0	7	7	7	6	6	6	5	5	5	4
38.5	8	8	8	7	7	7	6	6	5	5
39.0	9	9	9	8	8	7	7	7	6	6
39.5	10	10	9	9	9	8	8	8	7	7
40.0	11	11	10	10	10	9	9	8	8	8
40.5	12	12	11	11	10	10	10	9	9	9
41.0	13	12	12	12	11	11	11	10	10	10
41.5	14	13	13	13	12	12	11	11	11	10
42.0	14	14	14	13	13	13	12	12	12	11
42.5	15	15	15	14	14	13	13	13	12	12
43.0	16	16	15	15	15	14	14	14	13	13
43.5	17	17	16	16	15	15	15	14	14	14
44.0	18	17	17	17	16	16	16	15	15	14
44.5	19	18	18	17	17	17	16	16	16	15
45.0	19	19	19	18	18	17	17	17	16	16
45.5	20	20	19	19	19	18	18	18	17	17
46.0	21	20	20	20	19	19	19	18	18	18
46.5	22	21	21	20	20	20	19	19	19	18
47.0	22	22	22	21	21	20	20	20	19	19
47.5	23	23	22	22	22	21	21	21	20	20
48.0	24	23	23	23	22	22	22	21	21	21
48.5	25	24	24	23	23	23	22	22	22	21
49.0	25	25	25	24	24	23	23	23	22	22
49.5	26	26	25	25	24	24	24	23	23	23
50.0	27	26	26	26	25	25	24	24	24	23
50.5	27	27	27	26	26	26	25	25	24	24
51.0	28	28	27	27	27	26	26	25	25	25
51.5	29	28	28	28	27	27	27	26	26	25
52.0	29	29	29	28	28	28	27	27	27	26
52.5	30	30	29	29	29	28	28	28	27	27
53.0	31	30	30	30	29	29	29	28	28	27
53.5	31	31	31	30	30	30	29	29	28	28
54.0	32	32	31	31	31	30	30	30	29	29
54.5	33	32	32	32	31	31	31	30	30	29
55.0	33	33	33	32	32	32	31	31	30	30
55.5	34	34	33	33	33	32	32	31	31	31
56.0	35	34	34	33	33	33	32	32	32	31
56.5	35	35	34	34	34	33	33	33	32	32
57.0	36	35	35	35	34	34	34	33	33	33
57.5	36	36	36	35	35	35	34	34	34	33
58.0	37	37	36	36	36	35	35	35	34	34
58.5	38	37	37	37	36	36	35	35	35	34

Note: Circumference value = waist + hip - neck circumference (inches).

Height

Circumference	63.0	63.5	64.0	64.5	65.0	65.5	66.0	66.5	67.0	67.5	68.0	68.5	69.0	69.5	70.0
37.0	2	2	1	1	1	-	-	-	-	-	-	-	-	-	-
37.5	3	3	2	2	2	1	1	1	-	-	-	-	-	-	-
38.0	4	3	3	3	2	2	2	1	1	1	-	-	-	-	-
38.5	5	4	4	4	3	3	3	2	2	2	1	1	1	1	-
39.0	6	5	5	5	4	4	4	3	3	3	2	2	2	2	0
39.5	7	6	6	6	5	5	5	4	4	4	3	3	3	3	1
40.0	7	7	7	6	6	6	5	5	5	4	4	4	3	4	2
40.5	8	8	8	7	7	7	6	6	6	5	5	5	4	5	3
41.0	9	9	8	8	8	7	7	7	6	6	6	5	5	5	4
41.5	10	10	9	9	9	8	8	8	7	7	7	6	6	6	5
42.0	11	10	10	10	9	9	9	8	8	8	8	7	7	7	6
42.5	12	11	11	11	10	10	10	9	9	9	8	8	8	8	7
43.0	12	12	12	11	11	11	10	10	10	9	9	9	9	9	8
43.5	13	13	13	12	12	12	11	11	11	10	10	10	9	9	9
44.0	14	14	13	13	13	12	12	12	11	11	11	10	10	10	9
44.5	15	15	14	14	14	13	13	13	12	12	12	11	11	11	9
45.0	16	15	15	15	14	14	14	13	13	13	12	12	12	12	10
45.5	16	16	16	15	15	15	14	14	14	13	13	13	12	13	11
46.0	17	17	17	16	16	16	15	15	15	14	14	14	13	13	12
46.5	18	18	17	17	17	16	16	16	15	15	15	14	14	14	13
47.0	19	18	18	18	17	17	17	16	16	16	15	15	15	15	13
47.5	19	19	19	18	18	18	17	17	17	16	16	16	15	16	14
48.0	20	20	20	19	19	18	18	18	18	17	17	17	16	16	15
48.5	21	21	20	20	20	19	19	19	18	18	18	17	17	17	16
49.0	22	21	21	21	20	20	20	19	19	19	18	18	18	18	16
49.5	22	22	22	21	21	21	20	20	20	19	19	19	18	18	17
50.0	23	23	22	22	22	21	21	21	20	20	20	19	19	19	18
50.5	24	23	23	23	22	22	22	21	21	21	20	20	20	20	18
51.0	24	24	24	23	23	23	22	22	22	21	21	21	20	21	19
51.5	25	25	24	24	24	23	23	23	22	22	22	21	21	21	20
52.0	26	25	25	25	24	24	24	23	23	23	22	22	22	22	21
52.5	26	26	26	25	25	25	24	24	24	23	23	23	22	23	21
53.0	27	27	26	26	26	25	25	25	24	24	24	23	23	23	22
53.5	28	27	27	27	26	26	26	25	25	25	24	24	24	24	23
54.0	28	28	28	27	27	27	26	26	26	25	25	25	24	24	23
54.5	29	29	28	28	28	27	27	27	26	26	26	25	25	25	24
55.0	30	29	29	29	28	28	28	27	27	27	26	26	26	26	24
55.5	30	30	30	29	29	29	28	28	28	27	27	27	26	26	25
56.0	31	31	30	30	30	29	29	29	28	28	28	27	27	27	26
56.5	32	31	31	31	30	30	30	29	29	29	28	28	28	28	26
57.0	32	32	32	31	31	31	30	30	30	29	29	29	28	28	27
57.5	33	32	32	32	31	31	31	30	30	30	30	29	29	29	28
58.0	33	33	33	32	32	32	31	31	31	30	30	30	29	29	28
58.5	34	34	33	33	33	32	32	32	31	31	31	30	30	30	29

Height

Circumference	70.5	71.0	71.5	72.0	72.5	73.0	73.5	74.0	74.5	75.0	75.5	76.0	76.5	77.0	77.5
40.0	3	2	2	2	1	1	1	-	-	-	-	-	-	-	-
40.5	3	3	3	2	2	2	2	1	1	1	-	-	-	-	-
41.0	4	4	4	3	3	3	2	2	2	2	1	1	1	-	-
41.5	5	5	4	4	4	4	3	3	3	2	2	2	2	1	1
42.0	6	6	5	5	5	4	4	4	4	3	3	3	2	2	2
42.5	7	6	6	6	6	5	5	5	4	4	4	3	3	3	3
43.0	8	7	7	7	6	6	6	5	5	5	5	4	4	4	3
43.5	8	8	8	7	7	7	7	6	6	6	5	5	5	5	4
44.0	9	9	9	8	8	8	8	7	7	6	6	6	6	5	5
44.5	10	10	9	9	9	8	8	8	8	7	7	7	6	6	6
45.0	11	10	10	10	10	9	9	9	8	8	8	7	7	7	7
45.5	12	11	11	11	10	10	10	9	9	9	9	8	8	8	7
46.0	12	12	12	11	11	11	10	10	10	10	9	9	9	8	8
46.5	13	13	12	12	12	12	11	11	11	10	10	10	9	9	9
47.0	14	13	13	13	13	12	12	12	11	11	11	11	10	10	10
47.5	15	14	14	14	13	13	13	12	12	12	12	11	11	11	10
48.0	15	15	15	14	14	14	13	13	13	13	12	12	12	11	11
48.5	16	16	15	15	15	14	14	14	14	13	13	13	12	12	12
49.0	17	16	16	16	15	15	15	15	14	14	14	13	13	13	13
49.5	17	17	17	17	16	16	16	15	15	15	14	14	14	14	13
50.0	18	18	18	17	17	17	16	16	16	15	15	15	15	14	14
50.5	19	19	18	18	18	17	17	17	16	16	16	16	15	15	15
51.0	20	19	19	19	18	18	18	17	17	17	17	16	16	16	15
51.5	20	20	20	19	19	19	18	18	18	17	17	17	17	16	16
52.0	21	21	20	20	20	19	19	19	18	18	18	18	17	17	17
52.5	22	21	21	21	20	20	20	19	19	19	19	18	18	18	17
53.0	22	22	22	21	21	21	20	20	20	20	19	19	19	18	18
53.5	23	23	22	22	22	21	21	21	20	20	20	20	19	19	19
54.0	24	23	23	23	22	22	22	21	21	21	21	20	20	20	19
54.5	24	24	24	23	23	23	22	22	22	21	21	21	21	20	20
55.0	25	24	24	24	24	23	23	23	22	22	22	22	21	21	21
55.5	25	24	24	24	24	23	23	23	22	22	22	22	22	22	21
56.0	26	26	25	25	25	25	24	24	24	23	23	23	22	22	22
56.5	27	26	26	26	25	25	25	25	24	24	24	23	23	23	23
57.0	27	27	27	26	26	26	25	25	25	25	24	24	24	23	23
57.5	28	28	27	27	27	26	26	26	26	25	25	25	24	24	24
58.0	29	28	28	28	27	27	27	26	26	26	26	25	25	25	24
58.5	29	29	29	28	28	28	27	27	27	26	26	26	26	25	25

TABLE J-4
HEAT STRESS AND WIND CHILL INDEX

HEAT STRESS INDEX

Relative Humidity

Temperature (F)	10%	20%	30%	40%	50%	60%	70%	80%	90%
104	98	104	110	120	132				
102	97	101	108	117	125				
100	95	99	105	110	120	132			
98	93	97	101	106	110	125			
96	91	95	98	104	108	120	128		
94	89	93	95	100	105	111	122		
92	87	90	92	96	100	106	115	122	
90	85	88	90	92	96	100	106	114	122
88	82	86	87	89	93	95	100	106	115
86	80	84	85	87	90	92	96	100	109
84	78	81	83	85	88	89	91	95	99
82	77	79	80	81	84	86	89	91	95
80	75	77	78	79	81	83	85	86	89
78	72	75	77	78	79	80	81	83	85
76	70	72	75	76	77	77	77	78	79
74	68	70	73	74	75	75	75	76	77

Note: Add 10 degrees when wearing protective clothing

Humiture	Danger Category	Threat
Below 60°F	None	Little or no danger under normal circumstances.
60–80°F		
80–90°F	Caution	Fatigue possible with prolonged exposure or physical activity.
90–105°F	Extreme caution	Heat cramps and heat exhaustion possible if exposure is prolonged and there is physical activity.
105–130°F	Danger	Heat cramps or exhaustion likely. Heat stroke possible.
Above 130°F	Extreme danger	Heat stroke imminent.

WIND CHILL INDEX

Wind speed (mph)	Temperature (F) 50	40	30	20	10	0	−10	−20	−30	−40	−50
Calm	50	40	30	20	10	0	−10	−20	−30	−40	−50
5	48	37	27	16	6	−5	−15	−26	−36	−47	−57
10	40	28	16	4	−9	−21	−33	−46	−58	−70	−83
15	36	22	9	−5	−18	−38	−45	−58	−72	−85	−99
20	32	18	4	−10	−25	−39	−53	−67	−82	−96	−110
25	30	16	0	−15	−29	−44	−59	−74	−88	−104	−118
30	28	13	−2	−18	−33	−48	−63	−79	−94	−109	−125
35	27	11	−4	−20	−35	−49	−67	−82	−98	−113	−129
40	26	10	−6	−21	−37	−53	−69	−85	−100	−116	−132

A B C

A: Little danger if properly clothed.
B: Increasing danger; flesh may freeze.
C: Great danger; flesh may freeze in thirty seconds.

Note: Wind speeds of greater than 40 mph may have little additional effect.

TABLE J-5

ESTIMATING THE ONE-REP MAX FROM REPETITIONS

Weight (lbs.)	Repetitions 2	3	4	5	6	7	8	9	10
100	107	112	115	118	121	124	127	130	133
110	118	123	127	130	133	136	140	143	146
120	128	134	138	142	145	149	152	156	160
130	139	146	150	153	157	161	165	169	173
140	150	157	161	165	169	174	178	182	186
150	161	168	173	177	182	186	191	195	200
160	171	179	184	189	194	198	203	208	213
170	182	190	196	201	206	211	216	221	226
180	193	202	207	212	218	223	229	234	239
190	203	213	219	224	230	236	241	247	253
200	214	224	230	236	242	248	254	260	266
210	225	235	230	236	254	260	267	273	279
220	235	246	253	260	266	273	279	286	293
230	246	258	265	271	278	285	292	299	306
240	257	269	276	283	290	297	305	312	319
250	268	280	286	295	303	310	318	325	333
260	278	291	299	309	315	322	330	338	346
270	289	302	310	319	327	335	343	351	359
280	300	314	322	330	339	347	356	364	372
290	310	325	336	342	351	360	368	377	386
300	321	336	345	354	363	372	381	390	399
Factor:	1.07	1.12	1.15	1.18	1.21	1.24	1.27	1.30	1.33

Appendix K

National Organizations

International Association of Fire Fighters
Department of Occupational Health and Safety
1750 New York Avenue N.W.
Washington, DC 20006
(202) 737-8484

International Association of Fire Chiefs
1329 18th ST. N.W.
Washington, DC 20036
(202) 833-3420

International Association of Chief of Police
515 N. Washington Street
Alexandria, VA. 22314-2357

National Fire Protection Association
1 Batterymarch Park
Quincy, Massachusetts 02269-9904
(800) 344-3555 (to order documents)

U.S. Department of Labor
Occupational Health and Safety Administration
200 Constitution Avenue N.W.
Washington, DC 20210

United States Fire Administration
Office of Firefighter Health and Safety
16825 South Seton Avenue
Emmitsburg, MD 21727

The National Fire Academy and
Federal Emergency Management Agency
(same as USFA, above)

American College of Sports Medicine
P.O. Box 1440
Indianapolis, IN 46206-1440
(317) 637-9200

Canadian Society for Exercise Physiology
185 Somerset St. West
Suite 202
Ottawa, Ontario CANADA K2P 0J2
(613) 234-3755

American Heart Association
National Center
7272 Greenville Avenue
Dallas, TX 75231-4596
1-800-AHA-USA1

American Alliance for Health, Physical Education,
Recreation and Dance (AAHPERD)
1900 Association Drive
Reston, VA 20191
1-800-321-0789

INDEX

A

B

C

D

E

F

G

H

I

J

K

L

O

P

Q

R

S

T

U

V

W

Y